Thomas Kreis

Handbook of Holographic Interferometry
Optical and Digital Methods

Thomas Kreis

Handbook of Holographic Interferometry

Optical and Digital Methods

WILEY-VCH Verlag GmbH & Co. KGaA

Author
Dr. Thomas Kreis
BIAS – Bremer Institut für angewandte Strahltechnik
Klagenfurter Str. 2, 28359 Bremen, Germany
e-mail: kreis@bias.de

Cover Picture
Holographic interference pattern of a deformed satellite tank. Deformation caused by variation of internal pressure. Frequency doubled Nd:YAG-laser of 532 nm wavelength used.

All books published by Wiley-VCH are carefully produced. Nevertheless, authors, editors, and publisher do not warrant the information contained in these books, including this book, to be free of errors. Readers are advised to keep in mind that statements, data, illustrations, procedural details or other items may inadvertently be inaccurate.

Library of Congress Card No.: applied for

British Library Cataloging-in-Publication Data:
A catalogue record for this book is available from the British Library

Bibliographic information published by Die Deutsche Bibliothek
Die Deutsche Bibliothek lists this publication in the Deutsche Nationalbibliografie; detailed bibliographic data is available in the Internet at <http://dnb.ddb.de>.

© 2005 WILEY-VCH GmbH & Co. KGaA, Weinheim

All rights reserved (including those of translation into other languages). No part of this book may be reproduced in any form – nor transmitted or translated into machine language without written permission from the publishers. Registered names, trademarks, etc. used in this book, even when not specifically marked as such, are not to be considered unprotected by law.

Printed in the Federal Republic of Germany
Printed on acid-free paper

Printing Strauss Offsetdruck GmbH, Mörlenbach
Bookbinding Litges & Dopf Buchbinderei GmbH, Heppenheim

ISBN 3-527-40546-1

Contents

Preface		**XI**
1	**Introduction**	**1**
	1.1 Scope of the Book	1
	1.2 Historical Developments	3
	1.3 Holographic Interferometry as a Measurement Tool	6
2	**Optical Foundations of Holography**	**9**
	2.1 Light Waves	9
	2.1.1 Solutions of the Wave Equation	9
	2.1.2 Intensity	12
	2.2 Interference of Light	13
	2.2.1 Interference of Two Waves with Equal Frequency	13
	2.2.2 Interference of Two Waves with Different Frequencies	14
	2.2.3 Interference of Two Waves with Different Amplitudes	15
	2.3 Coherence	16
	2.3.1 Temporal Coherence	17
	2.3.2 Spatial Coherence	19
	2.4 Scalar Diffraction Theory	21
	2.4.1 Fresnel-Kirchhoff Diffraction Formula	21
	2.4.2 Fresnel Approximation	23
	2.4.3 Fraunhofer Approximation	25
	2.4.4 Thin Lens	26
	2.4.5 Propagation of Light Waves as a Linear System	29
	2.5 Speckles	30
	2.5.1 Statistics of Speckle Intensity and Phase	30
	2.5.2 Speckle Size	34
	2.6 Holographic Recording and Optical Reconstruction	36
	2.6.1 Hologram Recording	36
	2.6.2 Optical Reconstruction of a Wave Field	40
	2.6.3 Holographic Imaging Equations	44
	2.6.4 Types of Holograms	47
	2.7 Elements of the Holographic Setup	53
	2.7.1 Laser	53
	2.7.2 Recording Media	58

		2.7.3	Optical Components	61
		2.7.4	Beam Modulating Components	62
	2.8	CCD- and CMOS-Arrays	65
		2.8.1	CCD Concept	66
		2.8.2	CCD Array Performance Parameters	70
		2.8.3	CMOS Image Sensors	73
		2.8.4	Spatial Sampling with CCD-Arrays	74
		2.8.5	Color Still Cameras	76

3 Digital Recording and Numerical Reconstruction of Wave Fields 81

	3.1	Digital Recording of Holograms	81
		3.1.1	CCD Recording and Sampling	81
		3.1.2	Reduction of the Imaging Angle	84
		3.1.3	Reference Waves	89
	3.2	Numerical Reconstruction by the Fresnel Transform	93
		3.2.1	Wave Field Reconstruction by the Finite Discrete Fresnel Transform	93
		3.2.2	Real and Virtual Image	97
		3.2.3	Digital Fourier Transform Holography	100
		3.2.4	The D.C.-Term of the Fresnel Transform	102
		3.2.5	Suppression of the D.C.-Term	105
		3.2.6	Suppression of the Twin Image	107
		3.2.7	Variation of the Reference Wave	108
		3.2.8	Anamorphic Correction	114
	3.3	Numerical Reconstruction by the Convolution Approach	115
		3.3.1	The Diffraction Integral as a Convolution	115
		3.3.2	Size of the Image Field	117
		3.3.3	Shifting of the Image Field	118
		3.3.4	Scaling of the Image Field	120
	3.4	Further Numerical Reconstruction Methods	124
		3.4.1	Phase-Shifting Digital Holography	124
		3.4.2	Local Amplitude and Phase Retrieval	129
		3.4.3	Wavelet Approach to Numerical Reconstruction	132
		3.4.4	Comparison of Reconstruction Methods	134
		3.4.5	Hologram Recording Using Consumer Cameras	139
	3.5	Wave-Optics Analysis of Digital Holography	140
		3.5.1	Frequency Analysis of Digital Holography with Reconstruction by Fresnel Transform	141
		3.5.2	Frequency Analysis of Digital Holography with Reconstruction by Convolution	148
		3.5.3	The Transfer Function as a Filter	151
	3.6	Non-Interferometric Applications of Digital Holography	159
		3.6.1	Particle Analysis by Digital Holography	160
		3.6.2	Microscopy by Digital Holography	169
		3.6.3	Data Encryption with Digital Holography	180

4 Holographic Interferometry — 185

- 4.1 Generation of Holographic Interference Patterns — 186
  - 4.1.1 Recording and Reconstruction of a Double Exposure Holographic Interferogram — 186
  - 4.1.2 Recording and Reconstruction of a Real-Time Holographic Interferogram — 188
  - 4.1.3 Time Average Holography — 190
  - 4.1.4 Interference Phase Variation Due to Deformation — 191
  - 4.1.5 Interference Phase Variation Due to Refractive Index Variation — 194
  - 4.1.6 Computer Simulation of Holographic Interference Patterns — 196
- 4.2 Variations of the Sensitivity Vectors — 198
  - 4.2.1 Optimization of the Holographic Arrangement — 198
  - 4.2.2 Two Reference Beam Holographic Interferometry — 201
- 4.3 Fringe Localization — 203
  - 4.3.1 Fringe Formation with Diffusely Scattering Surfaces — 203
  - 4.3.2 Fringe Localization with Collimated Illumination — 206
  - 4.3.3 Fringe Localization with Spherical Wave Illumination — 211
  - 4.3.4 Fringe Localization with Phase Objects — 211
  - 4.3.5 Observer Projection Theorem — 214
- 4.4 Holographic Interferometric Measurements — 215
  - 4.4.1 Qualitative Evaluation of Holographic Interferograms — 215
  - 4.4.2 Holographically Measurable Physical Quantities — 216
  - 4.4.3 Loading of the Objects — 218

5 Quantitative Determination of the Interference Phase — 221

- 5.1 Role of Interference Phase — 221
  - 5.1.1 Sign Ambiguity — 222
  - 5.1.2 Absolute Phase Problem — 224
- 5.2 Disturbances of Holographic Interferograms — 225
  - 5.2.1 Varying Background Illumination — 226
  - 5.2.2 Electronic Noise — 226
  - 5.2.3 Speckle Decorrelation — 227
  - 5.2.4 Digitization and Quantization — 227
  - 5.2.5 Environmental Distortions — 228
- 5.3 Fringe Skeletonizing — 229
  - 5.3.1 Pattern Preprocessing — 229
  - 5.3.2 Fringe Skeletonizing by Segmentation — 231
  - 5.3.3 Skeletonizing by Fringe Tracking — 233
  - 5.3.4 Other Fringe Skeletonizing Methods — 233
  - 5.3.5 Fringe Numbering and Integration — 234
- 5.4 Temporal Heterodyning — 235
  - 5.4.1 Principle of Temporal Heterodyning — 235
  - 5.4.2 Technical Realization of Temporal Heterodyning — 237
  - 5.4.3 Errors of Temporal Heterodyning — 238
  - 5.4.4 Experimental Application of Temporal Heterodyning — 240

- 5.5 Phase Sampling Evaluation 242
  - 5.5.1 Phase Shifting and Phase Stepping 243
  - 5.5.2 Solution of the Phase Sampling Equations with Known Phase Shifts 245
  - 5.5.3 Solution of the Phase Sampling Equations with Unknown Phase Shifts 248
  - 5.5.4 Application of Phase Shift Evaluation Methods 251
  - 5.5.5 Discussion of Phase Shift Evaluation Methods 255
- 5.6 Fourier Transform Evaluation 256
  - 5.6.1 Principle of the Fourier Transform Evaluation Method .. 256
  - 5.6.2 Noise Reduction by Spatial Filtering 258
  - 5.6.3 Spatial Filtering and Sign Ambiguity 260
  - 5.6.4 Fourier Transform Evaluation of Phase Shifted Interferograms 261
  - 5.6.5 Spatial Heterodyning 263
  - 5.6.6 Spatial Synchronous Detection 265
- 5.7 Dynamic Evaluation 266
  - 5.7.1 Principles of Dynamic Evaluation 266
  - 5.7.2 Dynamic Evaluation by a Scanning Reference Beam 268
- 5.8 Digital Holographic Interferometry 269
  - 5.8.1 Digital Phase Subtraction 269
  - 5.8.2 Enhancement of Interference Phase Images by Digital Filtering 273
  - 5.8.3 Evaluation of Series of Holograms 275
  - 5.8.4 Compensation of Motion Components 278
  - 5.8.5 Multiplexed Holograms Discriminated in Depth 280
  - 5.8.6 Multiplexed Holograms with Discrimination by Partial Spectra 282
- 5.9 Interference Phase Demodulation 287
  - 5.9.1 Prerequisites for Interference Phase Demodulation 287
  - 5.9.2 Path-Dependent Interference Phase Demodulation 288
  - 5.9.3 Path-Independent Interference Phase Demodulation 289
  - 5.9.4 Interference Phase Demodulation by Cellular Automata .. 292
  - 5.9.5 Further Approaches to Interference Phase Demodulation . 294

6 Processing of the Interference Phase 297
- 6.1 Displacement Determination 297
  - 6.1.1 Displacement Determination with Known Reference Displacement 298
  - 6.1.2 Displacement Determination with Unknown Reference Displacement 299
  - 6.1.3 Elimination of Overall Displacement 301
  - 6.1.4 Non-Vibration Isolated Objects 303
- 6.2 The Sensitivity Matrix 306
  - 6.2.1 Determination of the Sensitivity Vectors 306
  - 6.2.2 Correction of Perspective Distortion 307
  - 6.2.3 Condition of the Sensitivity Matrix 310
- 6.3 Holographic Strain and Stress Analysis 311
  - 6.3.1 Definition of Elastomechanical Parameters 311
  - 6.3.2 Beams and Plates 314
  - 6.3.3 Numerical Differentiation 317
  - 6.3.4 Fringe Vector Theory 318

1.2 Historical Developments

To make the work with this book more comfortable, the references are given in the sequence of occurrence at the end of the book. This is accompanied by an alphabetically ordered author/coauthor index. A subject index lists a number of terms; these are printed in italics in the text to make their identification easier.

1.2 Historical Developments

Holography got its name from the Greek words 'holos' meaning whole or entire and 'graphein' meaning to write. It is a means for recording and reconstructing the whole information contained in an optical wavefront, namely amplitude and phase, and not just intensity as ordinary photography does. Holography essentially is a clever combination of interference and diffraction, two phenomena based on the wave nature of light.

Diffraction was first noted by F. M. Grimaldi (1618 – 1663) as the deviation from rectilinear propagation, and the interference generated by thin films was observed and described by R. Hooke (1635 – 1703). I. Newton (1642 – 1727) discovered the composition of white light from independent colors. The mathematical basis for the wave theory describing these effects was founded by Chr. Huygens (1629 – 1695), who further discovered the polarization of light. The interference principle introduced by Th. Young (1773 – 1829) and the Huygens principle were used by A. J. Fresnel (1788 – 1827) to calculate the diffraction patterns of different objects. Since about 1850 the view of light as a transversal wave won against the corpuscular theory. The relations between light, electricity, and magnetism were recognized by M. Faraday (1791 – 1867). These phenomena were summarized by J. C. Maxwell (1831 – 1879) in his well known equations. A medium supporting the waves was postulated as the all pervading ether. The experiments of A. A. Michelson (1852 – 1931), published in 1881, and the work of A. Einstein (1879 – 1955) were convincing evidence that there is no ether.

In 1948 D. Gabor (1900 – 1979) presented holography as a lensless process for image formation by reconstructed wavefronts [1–3]. His goal was to improve electron microscopy, using this new approach to avoid the previous aberrations. However, a successful application of the technique to electron microscopy has not materialized so far because of several practical problems. The validity of Gabor's ideas in the optical field was recognized and confirmed by, for example, G. L. Rogers [4], H. M. A. El-Sum and P. Kirkpatrick [5], and A. Lohmann [6]. But the interest in holography declined after a few years, mainly because of the poor quality of the holographic images obtained in those days. The breakthrough of holography was initiated by the development of the laser, which made available a powerful source of coherent light. This was accompanied by the solution of the twin-image problem encountered in Gabor's in-line arrangement. E. N. Leith and Y. Upatnieks [7–9] recognized the similarity of Gabor's holography to the synthetic aperture antenna problem of radar technology and introduced the off-axis reference beam technique. Y. N. Denisyuk combined the ideas of Gabor and Lippmann in his invention of the thick reflection hologram [10].

Now there was a working method for recording and reconstruction of complete wavefields with intensity and phase, and this also in the visible region of the spectrum. Besides the impressive display of three-dimensional scenes exhibiting effects like depth and parallax, moreover holography found numerous applications based on its unique features. Using the theory describing the formation of a hologram by interference of reference and object wave, holograms were created by calculation on a digital computer [11]. The result of this calculation was transferred to a transparency by printing or by printing on paper followed by a photographic process that might have included a reduction in scale. Now images of ideal objects not existing in reality could be generated, later on offering ways for interferometric comparison of, for example, optical components to be tested or for fabrication of diffracting elements with predescribed behavior [12,13]. The way holograms store information in a form of distributed memory has given incentive for research in holographic data storage [14]. Especially three-dimensional storage media, such as photorefractive crystals which are capable of providing Bragg selectivity became the focus of research, eventually yielding solutions to the always increasing demand for data storage capacity in the computer industry [15].

Perhaps the most important application of holography is in interferometric metrology, started by K. Stetson's discovery of holographic interferometry [16, 17]. In holographic interferometry, two or more wave fields are compared interferometrically, at least one of them must be holographically recorded and reconstructed [18]. This technique allows the measurement of changes of the phase of the wave field and thus the change of any physical quantity that affects the phase. The early applications ranged from the first measurement of vibration modes [16, 17], over deformation measurement [19–22], contour measurement [23–28], to the determination of refractive index changes [29, 30]. These developments were accompanied by rigorous investigations of the underlying principles, mainly performed by K. Stetson [31–35].

For certain arrangements of illumination and observation directions the resulting holographic interference fringes can be interpreted in a first approximation as contour lines of the amplitude of the change of the measured quantity. As an example, a locally higher deformation of a diffusely reflecting surface manifests in a locally higher fringe density. So such areas which give hints to possible material faults, risk of damage, or inadequate design, can easily be detected by applying a load of the same type and direction as the intended operational load, but of much less amplitude. This is the field of HNDT – holographic nondestructive testing.

Besides this qualitative evaluation of the holographic interference patterns there has been continuing work to use holographic interferometry for quantitative measurements. Beginning with manual fringe counting [36, 37], soon image processing computers were employed for quantitative evaluation, a process that consists of recording the reconstructed fringe pattern by TV camera, digitizing and quantizing it, calculating the interference phase distribution from the stored intensity values, using geometry data of the holographic arrangement to determine the distribution of the physical quantity to be measured, and the display of the results. The main one of these named tasks is the calculation of the interference phase. The first algorithms doing this resembled the former fringe counting [38]. A significant step forward in computerized fringe analysis was the introduction of the phase shifting methods of classic interferometric metrology [39, 40] into holographic interferometry [41, 42]. Now it was possible to measure – and not to estimate by numerical interpolation – the interference phase between the fringe intensity maxima and minima, and also the sign ambiguity was resolved. However, one had to pay for this increased accuracy by additional experimental effort. An alternative without the need for generating several phase shifted interferograms and also without requiring the introduction of a carrier [43] was presented by the author with the Fourier transform evaluation [44]. This is a flexible tool for fitting a linear combination of harmonic functions to the recorded interferogram, taking into account all intensity values even those between the fringe extrema.

1.2 Historical Developments

While the evaluation of holographic interferograms by computer was successfully developed, there was still the clumsy work of the fabrication of the interference pattern, which was not amenable to computer. The wet chemical processing of the photographic plates, photothermoplastic film, photorefractive crystals, and other recording media showed their typical drawbacks. So the endeavor to record the primary interfering optical fields by the camera of the image processing system and to perform their superposition and thus the generation of the interferogram in the computer generated two solutions: electronic (ESPI), resp. digital (DSPI) speckle pattern interferometry and digital holography (DH) resp. digital holographic interferometry (DHI).

The imaging of diffusely scattering objects with coherent light always produces speckles, the high-contrast granular structure with which the image of the object appears to be covered. If a mutually coherent reference field is superposed to the field scattered by an object, the resulting speckle fields before and after a variation of the object can be added on an intensity basis and yield correlation fringes of the same form as in holographic interferometry [45]. It was recognized that the speckle patterns have a structure easily recordable by existing image sensors, so this metrologic method was automated by computerized recording and processing [46–48]. Due to the analog TV cameras first employed the method was called TV-holography or electronic speckle pattern interferometry (ESPI); later emphasizing the digital recording and processing the name changed to digital speckle pattern interferometry (DSPI). Its big advance was the computerized real-time fringe generation, but the method suffered from the grainy appearance on the TV screen, i. e. severe speckle noise. In the meantime a number of improvements have been achieved with the result that DSPI now is a mature technique with numerous applications in science and technology. Perhaps the most important contribution was the introduction of phase stepping to speckle interferometry by K. Creath [49] and K. Stetson and W. R. Brohinsky [50], resulting in phase stepping digital speckle pattern interferometry (PSDSPI) where optical phase distributions are calculated and compared.

While in ESPI/DSPI the object is focused onto the recording target and the fringes are correlation fringes on an intensity basis, in digital holography a Fresnel or Fraunhofer hologram is recorded. In the following I will give an admittedly "biased" outline of its development. The earliest publication on digital holography that I have found is by J. W. Goodman and R. W. Lawrence [51] and dates back to 1967, so digital holography is older than ESPI/DSPI. In this classic paper Goodman and Lawrence record a wave field using the lensless Fourier transform geometry with a vidicon whose lens assembly was removed. They write: "The output of the vidicon is sampled in a 256×256 array, and quantized to eight grey levels. To avoid aliasing errors, the object-reference-detector geometry is specifically chosen to assure that the maximum spatial frequency in the pattern of interference [the microinterference constituting the hologram, T. K.] is sampled four times per period." The reconstruction was done on a PDP 6 computer, the squared modulus of the calculated complex distribution was displayed on a scope. The computation of the 256×256 pixel field lasted 5 minutes, "a time which compares favorably with the processing time generally required to obtain a photographic hologram in the conventional manner" as Goodman and Lawrence wrote in 1967.

A further classic paper [52] by T. S. Huang from 1971 treats both categories of digital holography: computer generated holography and computerized reconstruction from holograms. In this paper Fourier transform holograms as well as Fresnel holograms and their numerical reconstruction are discussed, also digitization and quantization effects are considered. In 1972 the work of a Soviet group around L. P. Yaroslavsky was presented [53], and in a paper published 1974 T. H. Demetrakopoulos and R. Mittra [54] consider the computer reconstruction of holograms which are recorded at acoustical or microwave frequencies. In 1980 the book of L. P. Yaroslavsky and N. S. Merzlyakov [55] was translated into English; in this book the theory of computer generation of holograms and of computer reconstruction of holograms is thoroughly treated, and the experiments performed worldwide up to this time are described, especially the work done in the Soviet Union is presented in great detail.

Then there began a long phase in which digital holography in the sense of this book was dormant. Computer generated holograms on the one hand and speckle interferometry on the other hand were fields of active research finding numerous applications. The dormancy of digital holography lasted until the beginning of the 1990s. The young scientist U. Schnars in the department led by the author in the institute (BIAS) directed by W. Jüptner was working towards his doctoral dissertation. Starting with the work of Yaroslavsky [55] and using modern CCD-cameras and computer facilities soon the first digital hologram was recorded and numerically reconstructed. All the time it was a known fact that numerically the whole complex wave field can be reconstructed from a digital hologram. But the emphasis in the first experiments [51–53] was on the intensity distribution. Now the potential lying in the numerically reconstructed phase distribution was recognized, leading in 1993 to digital holographic interferometry as a measurement tool [56]. After the first paper [56] of Schnars soon others followed [57–60] and not much later this approach to holographic metrology was taken up by other research groups. One of the first of these was the group around G. Pedrini [61–65], working during the early days of Schnars' development in a joint project with BIAS. In the context of this project [66], Schnars and I presented our first results. Other groups working in digital holography now can be found in Belgium [67–70], Brazil [71], Canada [72–79], China [80,81], Czech Republic [82], France [83–89], Hong Kong [90,91], Italy [92–95], Japan [96–110], Poland [111], Singapore [112–114], Sweden [115–119], Switzerland [120–129], Turkey [130–132], USA [81, 97, 98, 133–151], to name only some countries in alphabetical order.

Digital holography and digital holographic interferometry now are recognized metrologic methods which receive continuously increasing interest [152, 153]. This is indicated by the steadily increasing number of publications per year related to this topic or by the fact that to the knowledge of the author the "Conference on Interferometry in Speckle Light" in Lausanne/Switzerland in 2000 was the first conference with a session entitled and dedicated only to the topic "digital holography". There is reasonable hope that this technique will yield new interesting results and possibilities, maybe some which are not possible with optical reconstruction. I hope this book will contribute a little bit to the further advance of the promising techniques of digital holography and digital holographic interferometry.

1.3 Holographic Interferometry as a Measurement Tool

In holographic interferometry, two or more wave fields are compared interferometrically, at least one of them must be holographically recorded and reconstructed. The method gives rise to interference patterns whose fringes are determined by the geometry of the holographic

1.3 Holographic Interferometry as a Measurement Tool

setup via the sensitivity vectors and by the optical path length differences. Thus holographic interference patterns can be produced by keeping the optical path length difference constant and changing the sensitivity vectors, by holding the sensitivity vectors constant and varying the optical path length differences, or by altering both of them between the object states to be compared. Especially the path lengths can be modified by a number of physical parameters. The flexibility and the precision gained by comparing the optical path length changes with the wavelength of the laser light used, make holographic interferometry an ideal means for measuring a manifold of physical quantities [154–156]. The main advantages are:

- The measurements are contactless and noninvasive. In addition to an eventual loading for inducing the optical pathlength changes, the object is only impinged by light waves. The intensities of these waves are well below the level for causing any damage, even for the most delicate of biological objects.
- A reliable analysis can be performed at low loading intensities: the testing remains non-destructive.
- Not only may two states separated by a long time be compared, but furthermore the generation and evaluation of the holographic information can be separated both temporally and locally.
- Measurements can be made through transparent windows. We can therefore make measurements in pressure or vacuum chambers or protect against hostile environments. Due to the measurement of differences of the optical path lengths instead of absolute values, low quality windows do not disturb the results.
- Holographic interferometric measurements can be accomplished at moving surfaces: Short pulse illumination makes the method insensitive to a disturbing motion, vibrations can be investigated, the holographic setup can be made insensitive to specific motion components, and the rotation of spinning objects can be cancelled optically by using an image derotator.
- Deformation measurements can be performed at rough, diffusely reflecting surfaces, which occur frequently in engineering. No specular reflection of the object is required.
- The objects to be examined holographically may be of almost arbitrary shape. Using multiple illumination and observation directions or fiber optics, barely accessible areas can be studied.
- Holographic interferometry is nearly independent of the state of matter: Deformations of hard and soft materials can be measured. Refractive index variations in solids, fluids, gases and even plasmas can be determined.
- Lateral dimensions of the examined subjects may range from a few millimeters to several meters.
- The measurement range extends roughly speaking from a hundredth to several hundreds of a wavelength, for example displacements can be measured from about 0.005 μm to 500 μm.
- The achievable resolution and accuracy of a holographic interferometric displacement measurement permit subsequent numerical strain and stress calculations.

- Two-dimensional spatially continuous information is obtained: local singularities, for example local deformation extrema, cannot go undetected.
- Multiple viewing directions using a single hologram are possible, enabling the application of computerized tomography to obtain three-dimensional fields.

2 Optical Foundations of Holography

This chapter discusses the physical basis of holography and holographic interferometry. The primary phenomena constituting holography are interference and diffraction, which take place because of the wave nature of light. So this chapter begins with a description of the wave theory of light as far as it is required to understand the recording and reconstruction of holograms and the effect of holographic interferometry. In holographic interferometry the variation of a physical parameter is measured by its influence on the phase of an optical wave field. Therefore the dependence of the phase upon the geometry of the optical setup and the different parameters to be measured is outlined.

2.1 Light Waves

2.1.1 Solutions of the Wave Equation

Light is a transverse, electromagnetic wave characterized by time-varying electric and magnetic fields. Since electromagnetic waves obey the Maxwell equations, the propagation of light is described by the wave equation which follows from the Maxwell equations. The *wave equation* for propagation of light in vacuum is

$$\nabla^2 \boldsymbol{E} - \frac{1}{c^2}\frac{\partial^2 \boldsymbol{E}}{\partial t^2} = 0 \tag{2.1}$$

where $\boldsymbol{E}$ is the *electric field strength*, ∇^2 is the *Laplace operator*

$$\nabla^2 = \frac{\partial^2}{\partial x^2} + \frac{\partial^2}{\partial y^2} + \frac{\partial^2}{\partial z^2} \tag{2.2}$$

(x, y, z) are the Cartesian spatial coordinates, t denotes the temporal coordinate, the time, and c is the propagation speed of the wave. The *speed of light* in vacuum c_0 is a constant of nature

$$c_0 = 299\,792\,458 \text{ m s}^{-1} \quad \text{or almost exactly} \quad c_0 = 3 \times 10^8 \text{ m s}^{-1}. \tag{2.3}$$

Transverse waves vibrate at right angles to the direction of propagation and so they must be described in vector notation. The wave may vibrate horizontally, vertically, or in any direction combined of these. Such effects are called *polarization* effects. Fortunately for most applications it is not necessary to use the full vectorial description of the fields, so we can

Handbook of Holographic Interferometry: Optical and Digital Methods. Thomas Kreis
Copyright © 2005 Wiley-VCH Verlag GmbH & Co. KGaA, Weinheim
ISBN: 3-527-40546-1

assume a wave vibrating in a single plane. Such a wave is called *plane polarized*. For a plane polarized wave field propagating in the z-direction the *scalar wave equation* is sufficient

$$\frac{\partial^2 E}{\partial z^2} - \frac{1}{c^2}\frac{\partial^2 E}{\partial t^2} = 0. \tag{2.4}$$

It is easily verified that

$$E(z,t) = f(z - ct) \quad \text{or} \quad E(z,t) = g(z + ct) \tag{2.5}$$

are also solutions of this equation, which means that the wave field retains its form during propagation. Due to the linearity of (2.4)

$$E(z,t) = a\,f(z - ct) + b\,g(z + ct) \tag{2.6}$$

is likewise a solution to the wave equation. This *superposition principle* is valid for linear differential equations in general and thus for (2.1) also.

The most important solution of (2.4) is the *harmonic wave*, which in real notation is

$$E(z,t) = E_0 \cos(kz - \omega t). \tag{2.7}$$

E_0 is the *real amplitude* of the wave, the term $(kz - \omega t)$ gives the *phase* of the wave. The *wave number* k is associated to the *wavelength* λ by

$$k = \frac{2\pi}{\lambda}. \tag{2.8}$$

Typical figures of λ for visible light are 514.5 nm (green line of argon-ion laser) or 632.8 nm (red light of helium-neon laser). The *angular frequency* ω is related to the *frequency* ν of the wave by

$$\omega = 2\pi\nu \tag{2.9}$$

where ν is the number of periods per second, that means

$$\nu = \frac{c}{\lambda} \quad \text{or} \quad \nu\lambda = c. \tag{2.10}$$

If we have not the maximum amplitude at $x = 0$ and $t = 0$, we have to introduce the *relative phase* ϕ

$$E(z,t) = E_0 \cos(kz - \omega t + \phi). \tag{2.11}$$

With the *period* T, the time for a full 2π-cycle, we can write

$$E(z,t) = E_0 \cos\left(\frac{2\pi}{\lambda}z - \frac{2\pi}{T}t + \phi\right). \tag{2.12}$$

Figure 2.1 displays two aspects of this wave. Figure 2.1a shows the temporal distribution of the field at two points $z = 0$ and $z = z_1 > 0$, and Fig. 2.1b gives the spatial distribution of

2.1 Light Waves

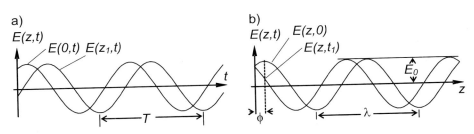

Figure 2.1: Spatial and temporal distribution of a scalar harmonic wave.

two periods for time instants $t = 0$ and $t = t_1 > 0$. We see that a point of constant phase moves with the so called *phase velocity*, the speed c.

The use of trigonometric functions leads to cumbersome calculations, which can be circumvented by using the complex exponential which is related to the trigonometric functions by *Euler's formula*

$$e^{i\alpha} = \cos \alpha + i \sin \alpha \qquad (2.13)$$

where $i = \sqrt{-1}$ is the imaginary unit. Since the cosine now is

$$\cos \alpha = \frac{1}{2}(e^{i\alpha} + e^{-i\alpha}) \qquad (2.14)$$

the harmonic wave (2.11) is

$$E(z,t) = \frac{1}{2} E_0 \, e^{i(kz - \omega t + \phi)} + \frac{1}{2} E_0 \, e^{-i(kz - \omega t + \phi)}. \qquad (2.15)$$

The second term on the right-hand side is the complex conjugate of the first term and can be omitted as long as it is understood that only the real part of $E(z,t)$ represents the physical wave. Thus the harmonic wave in complex notation is

$$E(z,t) = \frac{1}{2} E_0 \, e^{i(kz - \omega t + \phi)}. \qquad (2.16)$$

A *wavefront* refers to the spatial distribution of the maxima of the wave, or other surfaces of constant phase, as these surfaces propagate. The wavefronts are normal to the direction of propagation. A *plane wave* is a wave which has constant phase in all planes orthogonal to the propagation direction for a given time t. For describing the spatial distribution of the wave, we can assume $t = 0$ in an arbitrary time scale. Since

$$\boldsymbol{k} \cdot \boldsymbol{r} = \text{const} \qquad (2.17)$$

is the equation for a plane in three-dimensional space, with the *wave vector* $\boldsymbol{k} = (k_x, k_y, k_z)$ and the spatial vector $\boldsymbol{r} = (x, y, z)$, a plane harmonic wave at time $t = 0$ is

$$E(\boldsymbol{r}) = E_0 \, e^{i(\boldsymbol{k} \cdot \boldsymbol{r} + \phi)}. \qquad (2.18)$$

This wave repeats after the wavelength λ in direction $\boldsymbol{k}$, which can easily be proved using $|\boldsymbol{k}| = k = 2\pi/\lambda$ by

$$E\left(\boldsymbol{r} + \lambda \frac{\boldsymbol{k}}{k}\right) = E(\boldsymbol{r}). \tag{2.19}$$

The expression

$$E(\boldsymbol{r},t) = E_0\, e^{i(\boldsymbol{k}\cdot\boldsymbol{r} - \omega t + \phi)} \tag{2.20}$$

describes the temporal dependence of a plane harmonic wave propagating in the direction of the wavevector or

$$E(\boldsymbol{r},t) = E_0\, e^{i(\boldsymbol{k}\cdot\boldsymbol{r} + \omega t + \phi)} \tag{2.21}$$

if the wave propagates contrary to the direction of $\boldsymbol{k}$.

Another waveform often used is the *spherical wave* where the phase is constant on each spherical surface. The importance of spherical waves comes from the *Huygens principle* which states that each point on a propagating wavefront can be considered as radiating itself a spherical wavelet.

For a mathematical treatment of spherical waves the wave equation has to be described in polar coordinates (r, θ, ψ), transformed by $x = r\sin\theta\cos\psi$, $y = r\sin\theta\sin\psi$, $z = r\cos\theta$. Due to the spherical symmetry, a spherical wave is not dependent on θ and ψ. Then the scalar wave equation is

$$\frac{1}{r}\frac{\partial^2}{\partial r^2}(rE) - \frac{1}{c^2}\frac{\partial^2 E}{\partial t^2} = 0. \tag{2.22}$$

The solutions of main interest are the harmonic spherical waves

$$E(r,t) = \frac{E_0}{r}\, e^{i(kr - \omega t + \phi)}. \tag{2.23}$$

One observes that the amplitude E_0/r decreases proportionally to $1/r$. Furthermore at a long distance from the origin the spherical wave locally approximates a plane wave.

The complex amplitudes of wavefronts scattered by a surface are generally very complicated, but due to the superposition principle (2.6) they can be treated as the sum of plane waves or spherical waves. There are still other solutions to the wave equation. An example are the *Bessel waves* of the class of *nondiffracting beams* [157]. But up to now they have not found applications in holographic interferometry, so here we restrict ourselves on the plane and on the spherical waves.

2.1.2 Intensity

The only parameter of light which is directly amenable to sensors – eye, photodiode, CCD-target, etc. – is the *intensity* (and in a rough scale the frequency as color). Intensity is defined by the energy flux through an area per time. From the Maxwell equations we get

$$I = \varepsilon_0 c E^2 \tag{2.24}$$

2.2 Interference of Light

where we only use the proportionality of the intensity I to E^2

$$I \sim E^2. \tag{2.25}$$

It has to be recognized that the intensity has a nonlinear dependence on the electric field strength. Since there is no sensor which can follow the frequency of light, we have to integrate over a *measuring time* T_m, the momentary intensity is not measurable. So if $T_m \gg T = 2\pi/\omega$, omitting proportionality constants we define

$$I = E_0 E_0^* = |E_0|^2 \tag{2.26}$$

where * denotes the complex conjugate. The intensity of a general stationary wave field is

$$I(\boldsymbol{r}) = \langle E\, E^* \rangle = \lim_{T_m \to \infty} \frac{1}{T_m} \int_{-T_m/2}^{T_m/2} E(\boldsymbol{r}, t') E^*(\boldsymbol{r}, t')\, dt'. \tag{2.27}$$

This intensity is the limit of the *short time intensity*

$$I(\boldsymbol{r}, t, T_m) = \frac{1}{T_m} \int_{t-T_m/2}^{t+T_m/2} E(\boldsymbol{r}, t') E^*(\boldsymbol{r}, t')\, dt' \tag{2.28}$$

which is a sliding average of a temporal window centered around t with width T_m. The measuring time T_m always is large compared with the period of the light wave but has to be short in the time scale of the investigated process.

2.2 Interference of Light

2.2.1 Interference of Two Waves with Equal Frequency

The *interference* effect which occurs if two or more coherent light waves are superposed, is the basis of holography and holographic interferometry. So in this *coherent superposition* we consider two waves, emitted by the same source, which differ in the directions $\boldsymbol{k_1}$ and $\boldsymbol{k_2}$, and the phases ϕ_1 and ϕ_2, but for convenience have the same amplitude E_0 and frequency ω and are linearly polarized in the same direction. Then in scalar notation

$$\begin{aligned} E_1(\boldsymbol{r}, t) &= E_0\, e^{i(\boldsymbol{k_1} \cdot \boldsymbol{r} - \omega t + \phi_1)} \\ E_2(\boldsymbol{r}, t) &= E_0\, e^{i(\boldsymbol{k_2} \cdot \boldsymbol{r} - \omega t + \phi_2)}. \end{aligned} \tag{2.29}$$

For determination of the superposition of these waves we decompose the vectors $\boldsymbol{k_1}$ and $\boldsymbol{k_2}$ into components of equal and opposite directions, Fig. 2.2:, $\boldsymbol{k'} = (\boldsymbol{k_1} + \boldsymbol{k_2})/2$ and $\boldsymbol{k''} = (\boldsymbol{k_1} - \boldsymbol{k_2})/2$. If θ is the angle between $\boldsymbol{k_1}$ and $\boldsymbol{k_2}$ then

$$|\boldsymbol{k''}| = \frac{2\pi}{\lambda} \sin \frac{\theta}{2}. \tag{2.30}$$

[handwritten annotations: $= f(\text{spatial freq})$; $d = \dfrac{\lambda}{2 \sin \frac{\theta}{2}} \equiv \text{Fringe Spacing}$; Equation 2.30 in Digital Holography (Schnars)]

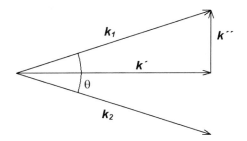

Figure 2.2: Decomposition of wave vectors.

In the same way we define the mean phase $\phi = (\phi_1 + \phi_2)/2$ and the half phase difference $\Delta\phi = (\phi_1 - \phi_2)/2$. Now the superposition gives the field

$$\begin{aligned}
(E_1 + E_2)(\boldsymbol{r}, t) &= E_0\, \mathrm{e}^{\mathrm{i}(\boldsymbol{k_1} \cdot \boldsymbol{r} - \omega t + \phi_1)} + E_0\, \mathrm{e}^{\mathrm{i}(\boldsymbol{k_2} \cdot \boldsymbol{r} - \omega t + \phi_2)} \\
&= E_0 \{ \mathrm{e}^{\mathrm{i}(\boldsymbol{k}' \cdot \boldsymbol{r} + \boldsymbol{k}'' \cdot \boldsymbol{r} - \omega t + \phi + \Delta\phi)} \\
&\quad + \mathrm{e}^{\mathrm{i}(\boldsymbol{k}' \cdot \boldsymbol{r} - \boldsymbol{k}'' \cdot \boldsymbol{r} - \omega t + \phi - \Delta\phi)} \} \\
&= E_0\, \mathrm{e}^{\mathrm{i}(\boldsymbol{k}' \cdot \boldsymbol{r} - \omega t + \phi)} \{ \mathrm{e}^{\mathrm{i}(\boldsymbol{k}'' \cdot \boldsymbol{r} + \Delta\phi)} + \mathrm{e}^{\mathrm{i}(-\boldsymbol{k}'' \cdot \boldsymbol{r} - \Delta\phi)} \} \\
&= 2 E_0\, \mathrm{e}^{\mathrm{i}(\boldsymbol{k}' \cdot \boldsymbol{r} - \omega t + \phi)} \cos(\boldsymbol{k}'' \cdot \boldsymbol{r} + \Delta\phi).
\end{aligned} \qquad (2.31)$$

In this field the exponential term is a temporally varying phase but the cosine term is independent of time. Thus we get the temporally constant intensity

$$\begin{aligned}
I(\boldsymbol{r}) &= (E_1 + E_2)(E_1 + E_2)^* \\
&= 4 E_0^2 \cos^2(\boldsymbol{k}'' \cdot \boldsymbol{r} + \Delta\phi).
\end{aligned} \qquad (2.32)$$

This means the intensity is minimal where $\cos^2(\boldsymbol{k}'' \cdot \boldsymbol{r} + \Delta\phi) = 0$. These are the loci where

$$\boldsymbol{k}'' \cdot \boldsymbol{r} + \Delta\phi = (2n+1)\frac{\pi}{2} \qquad n \in \mathbb{Z}. \qquad (2.33)$$

Here the wavefronts are said to be *anti-phase*, we speak of destructive interference. The intensity is maximal where

$$\boldsymbol{k}'' \cdot \boldsymbol{r} + \Delta\phi = n\pi \qquad n \in \mathbb{Z}. \qquad (2.34)$$

Here the wavefronts are *in-phase*, we have constructive interference.

The resulting time independent pattern is called an *interference pattern*, the fringes are called *interference fringes*. For plane waves they are oriented parallel to $\boldsymbol{k}'$ and have a distance of $\pi/|\boldsymbol{k}''|$ in the direction $\boldsymbol{k}''$. This is shown in moiré analogy in Fig. 2.3.

2.2.2 Interference of Two Waves with Different Frequencies

In the following we investigate the interference of two waves where not only the propagation directions and the phases but additionally the frequencies $\nu_i = \omega_i/(2\pi)$ are different.

$$\begin{aligned}
E_1(\boldsymbol{r}, t) &= E_0\, \mathrm{e}^{\mathrm{i}(\boldsymbol{k_1} \cdot \boldsymbol{r} - 2\pi\nu_1 t + \phi_1)} \\
E_2(\boldsymbol{r}, t) &= E_0\, \mathrm{e}^{\mathrm{i}(\boldsymbol{k_2} \cdot \boldsymbol{r} - 2\pi\nu_2 t + \phi_2)}.
\end{aligned} \qquad (2.35)$$

2.2 Interference of Light

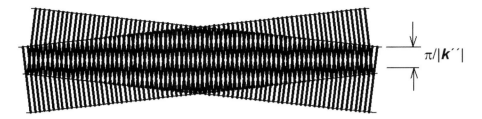

Figure 2.3: Interference fringes constant in time.

Besides the definitions of k', k'', ϕ and $\Delta\phi$ now let $\nu = (\nu_1 + \nu_2)/2$ and $\Delta\nu = (\nu_1 - \nu_2)/2$. Then we have

$$
\begin{aligned}
(E_1 + E_2)(\boldsymbol{r}, t) &= E_0 \{ e^{i(\boldsymbol{k'} \cdot \boldsymbol{r} + \boldsymbol{k''} \cdot \boldsymbol{r} - 2\pi\nu t - 2\pi\Delta\nu t + \phi + \Delta\phi)} \\
&\quad + e^{i(\boldsymbol{k'} \cdot \boldsymbol{r} - \boldsymbol{k''} \cdot \boldsymbol{r} - 2\pi\nu t + 2\pi\Delta\nu t + \phi - \Delta\phi)} \} \\
&= E_0 \, e^{i(\boldsymbol{k'} \cdot \boldsymbol{r} - 2\pi\nu t + \phi)} \{ e^{i(\boldsymbol{k''} \cdot \boldsymbol{r} - 2\pi\Delta\nu t + \Delta\phi)} \\
&\quad + e^{i(-\boldsymbol{k''} \cdot \boldsymbol{r} + 2\pi\Delta\nu t - \Delta\phi)} \} \\
&= 2\, E_0 \, e^{i(\boldsymbol{k'} \cdot \boldsymbol{r} - 2\pi\nu t + \phi)} \cos(\boldsymbol{k''} \cdot \boldsymbol{r} - 2\pi\Delta\nu t + \Delta\phi)
\end{aligned}
\tag{2.36}
$$

and the intensity is

$$
\begin{aligned}
I(\boldsymbol{r}, t) &= 4\, E_0^2 \, \cos^2(\boldsymbol{k''} \cdot \boldsymbol{r} + \Delta\phi - 2\pi\Delta\nu t) \\
&= 2\, E_0^2 \left[1 + \cos(2\boldsymbol{k''} \cdot \boldsymbol{r} + 2\Delta\phi - 4\pi\Delta\nu t) \right].
\end{aligned}
\tag{2.37}
$$

If the frequency difference is small enough, $\nu_1 \approx \nu_2$, a detector can register an intensity at $\boldsymbol{r}$ oscillating with the *beat frequency* $2\Delta\nu = \nu_1 - \nu_2$. The phase of this modulation is the phase difference $2\Delta\phi = \phi_1 - \phi_2$ of the superposed waves. Contrary to the frequencies of the optical waves the beat frequency can be measured electronically and further evaluated as long as it remains in the kHz or MHz range. The measurement of the beat frequency $\Delta\nu$ enables one to calculate the motion of a reflector via the *Doppler shift* or to determine the phase difference $\Delta\phi$ between different points of an object where the intensity oscillates with the same constant beat frequency.

2.2.3 Interference of Two Waves with Different Amplitudes

If we have plane linearly polarized waves of the same frequency, but different direction and phase and moreover different amplitudes

$$
\begin{aligned}
E_1(\boldsymbol{r}, t) &= E_{01} \, e^{i(\boldsymbol{k_1} \cdot \boldsymbol{r} - \omega t + \phi_1)} \\
E_2(\boldsymbol{r}, t) &= E_{02} \, e^{i(\boldsymbol{k_2} \cdot \boldsymbol{r} - \omega t + \phi_2)}
\end{aligned}
\tag{2.38}
$$

we get the intensity

$$I(r,t) = \left(E_{01}\, e^{i(k_1 \cdot r - \omega t + \phi_1)} + E_{02}\, e^{i(k_2 \cdot r - \omega t + \phi_2)}\right)$$
$$\times \left(E_{01}\, e^{-i(k_1 \cdot r - \omega t + \phi_1)} + E_{02}\, e^{-i(k_2 \cdot r - \omega t + \phi_2)}\right)$$
$$= E_{01}^2 + E_{02}^2 + E_{01}E_{02}\{e^{i(k_1 \cdot r - k_2 \cdot r + \phi_1 - \phi_2)} \quad (2.39)$$
$$+ e^{i(k_2 \cdot r - k_1 \cdot r + \phi_2 - \phi_1)}\}$$
$$= E_{01}^2 + E_{02}^2 + 2E_{01}E_{02}\cos(2k'' \cdot r + 2\Delta\phi).$$

This result can be written as

$$I = I_1 + I_2 + 2\sqrt{I_1 I_2}\cos(2k'' \cdot r + 2\Delta\phi) \qquad (2.40)$$

or using the identity $\cos\alpha = 2\cos^2(\alpha/2) - 1$ for comparison with (2.32) as

$$I = E_{01}^2 + E_{02}^2 + 4E_{01}E_{02}\cos^2(k'' \cdot r + \Delta\phi) - 2E_{01}E_{02}. \qquad (2.41)$$

The special case $E_{01} = E_{02} = E_0$ gives (2.32).

In general the result of superposing two waves consists of one part that is the addition of the intensities and another part, the interference term, (2.40). Up to now we only have investigated *parallelly polarized waves*. The other extreme are *orthogonally polarized waves*. These waves do not interfere, their superposition only consists of the addition of the intensities

$$I = I_1 + I_2. \qquad (2.42)$$

For other angles between the polarization directions the field vector has to be decomposed into components of parallel and orthogonal polarizations, the result contains interference parts as well as an addition of intensities.

Reasons for the additive intensity term not only may be mutually oblique polarization directions or different intensities, but also an insufficient coherence of the interfering waves. Because in the superposition of incoherent light we always observe a pure addition of the intensities but no interference, the additive term often is called the *incoherent part*, or we speak of *incoherent superposition*.

The *visibility* or *contrast* of the interference pattern is defined by

$$V = \frac{I_{\max} - I_{\min}}{I_{\max} + I_{\min}}. \qquad (2.43)$$

If two parallel polarized waves of the same intensity interfere, we have the maximal contrast of $V = 1$; we have minimal contrast $V = 0$ for incoherent superposition. For example, if the ratio of the intensities of interfering waves is 5:1, the contrast is 0.745.

2.3 Coherence

With sunlight or lamplight we rarely observe interference. Only light of sufficient coherence will exhibit this effect. Roughly speaking coherence means the ability of light waves to interfere. Precisely, coherence describes the correlation between individual light waves. The two aspects of the general spatio-temporal coherence are the temporal and the spatial coherence.

2.3.1 Temporal Coherence

Temporal coherence describes the correlation of a wave with itself as it behaves at different time instants [158, 159]. It is best explained with the help of a *Michelson interferometer*, Fig. 2.4. At a beam splitter the incoming wave field is divided into two parts, one being reflected into the orthogonal direction, one passing the splitter and maintaining the original direction. This type of wavefront division is called *amplitude division*. If we do not have a single beam, but plane waves collimated by properly placed lenses and if there is an extended screen instead of the point-detector, then we speak of a *Twyman-Green interferometer*.

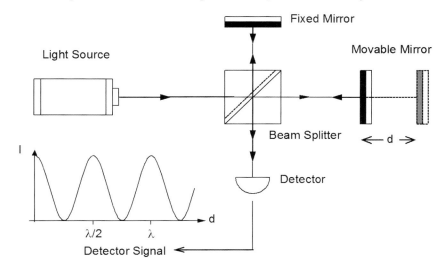

Figure 2.4: Michelson interferometer.

To keep the mathematics easy we assume that the beam splitter of the Michelson interferometer reflects 50 percent of the incident light and transmits the other 50 percent. The reflected wave travels to the fixed mirror, is reflected again and part of it hits the detector or screen for observation, where it meets the other part which was reflected at the movable mirror. At the detector mutually time shifted parts of the wave are superimposed, the time shift can be varied by changing the mirror shift d, Fig. 2.4. If in the Twyman-Green arrangement the mirrors are perfectly orthogonal we see a constant intensity over the screen, but with a minute tilt of one mirror around one axis, we observe fringes parallel to this axis.

Let the waves be E_1 and E_2. For a fixed point on the screen, or the position of the detector, we have

$$E_2(t) = E_1(t + \tau) \quad \text{or} \quad E_1(t) = E_2(t - \tau) \tag{2.44}$$

where

$$\tau = \frac{2d}{c}. \tag{2.45}$$

We have to recognize that the distance d is travelled forward and backward, therefore the factor 2 in (2.45).

At the observation point we have the superposition

$$E(t) = E_1(t) + E_2(t) = E_1(t) + E_1(t+\tau) \tag{2.46}$$

and see the intensity

$$\begin{aligned} I &= \langle E\, E^* \rangle \\ &= \langle E_1\, E_1^* \rangle + \langle E_2\, E_2^* \rangle + \langle E_2\, E_1^* \rangle + \langle E_1\, E_2^* \rangle \\ &= 2I_1 + 2\mathrm{Re}(\langle E_1\, E_2^* \rangle) \end{aligned} \tag{2.47}$$

due to our assumption of equal amplitudes.

According to (2.27) we define the complex *self coherence* $\Gamma(\tau)$ as

$$\begin{aligned} \Gamma(\tau) &= \langle E_1^*\, E_1(t+\tau) \rangle \\ &= \lim_{T_m \to \infty} \frac{1}{T_m} \int_{-T_m/2}^{T_m/2} E_1^*(t)\, E_1(t+\tau)\, dt \end{aligned} \tag{2.48}$$

which is the autocorrelation of $E_1(t)$. The normalized quantity

$$\gamma(\tau) = \frac{\Gamma(\tau)}{\Gamma(0)} \tag{2.49}$$

defines the *degree of coherence*. Since $\Gamma(0) = I_1$ is always real and the maximal value of $|\Gamma(\tau)|$, we have

$$|\gamma(\tau)| \leq 1 \tag{2.50}$$

and

$$I(\tau) = 2I_1(1 + \mathrm{Re}\,\gamma(\tau)). \tag{2.51}$$

The degree of coherence or the self coherence are not directly measurable, but it can be shown [159] that for the contrast V, which is easily measurable, we have

$$V(\tau) = |\gamma(\tau)|. \tag{2.52}$$

Now we can discriminate perfectly coherent light with $|\gamma(\tau)| = 1$, which nearly is emitted by a stabilized single-mode laser, incoherent light with $|\gamma(\tau)| = 0$ for all $\tau \neq 0$ where we have a statistically fluctuating phase, e. g. in sunlight, and partially coherent light, $0 \leq |\gamma(\tau)| \leq 1$. Often the contrast $V(\tau)$ decreases monotonically in τ. So we can introduce the *coherence time* τ_c as the time shift at which the contrast is reduced to $1/e$. The time shift is realized in interferometers by different optical pathlengths, so one can define the *coherence length*

$$l_c = c\tau_c. \tag{2.53}$$

If we have a periodic instead of a monotonically decreasing contrast function, e. g. from a two-mode laser, we take the time shift corresponding to the first minimum as the coherence time.

2.3 Coherence

2.3.2 Spatial Coherence

Spatial coherence describes the mutual correlation of different parts of the same wavefront [158, 159] and is explained using *Young's double aperture interferometer*, Fig. 2.5. This type

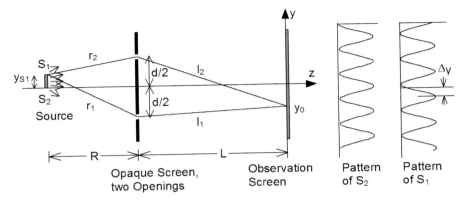

Figure 2.5: Young's interferometer.

of interferometer picks two geometrically different parts of the wavefront and brings them to interference, therefore it is called a *division of wavefront* interferometer. Let an opaque screen contain two small holes or parallel slits with a mutual distance d. For the moment we assume one point $S_1 = (0, y_{S1}, -R - L)$ of an extended source placed at the distance R behind that illuminates the opaque screen. Only the light passing through the holes forms an interference pattern on the observation screen placed some distance L in front of them. The distances of S_1 to the holes are r_1, r_2, those of the holes to the observation point are l_1, l_2. We can assume that the intensities of the two spherical waves leaving the holes are equal, therefore the intensity at the observation screen is (2.32)

$$I(x,y) = 4I_0(x,y) \cos^2 \Delta\phi(x,y). \tag{2.54}$$

The half phase difference $\Delta\phi$ is

$$\Delta\phi = \frac{1}{2}\left(\frac{2\pi}{\lambda}\Delta l\right) \tag{2.55}$$

where Δl is the difference in the pathlength of the light from the source S_1 to the observation point $B = (0, y_0, 0)$

$$\begin{aligned}\Delta l &= r_2 + l_2 - r_1 - l_1 \\ &= \sqrt{R^2 + \left(\frac{d}{2} - y_{S1}\right)^2} + \sqrt{L^2 + \left(\frac{d}{2} - y_0\right)^2} \\ &\quad - \sqrt{R^2 + \left(\frac{d}{2} + y_{S1}\right)^2} - \sqrt{L^2 + \left(\frac{d}{2} + y_0\right)^2}.\end{aligned} \tag{2.56}$$

Since y_0, y_{S1} and d are small compared with R and L, the square roots of the form $\sqrt{1+b}$ can be approximated by $1 + b/2$ and we obtain

$$\Delta l = -d\left(\frac{y_{S1}}{R} + \frac{y_0}{L}\right). \tag{2.57}$$

The resulting irradiance now is proportional to

$$I = I_0 \cos^2\left[\frac{\pi d}{\lambda}\left(\frac{y_{S1}}{R} + \frac{y_0}{L}\right)\right] \tag{2.58}$$

and describes a pattern of fringes parallel to the x-axis with a spacing of $\lambda L/d$ in the y-direction.

Next we investigate an extended source of perimeter l, Fig. 2.5. Source point S_2 on the optical axis emits a spherical wave which reaches the holes with equal phase, so we get an intensity maximum where the optical axis intersects the observation screen. The point S_1 gives rise to a fringe system which is shifted laterally because here r_1 and r_2 do not have equal length. Therefore the phase difference between the two spherical waves originating at the two holes is

$$\phi_1 - \phi_2 = \frac{2\pi}{\lambda}(r_1 - r_2) \tag{2.59}$$

which results in a lateral shift of the interference pattern by the amount

$$\Delta y = \frac{L}{d}(r_1 - r_2). \tag{2.60}$$

If there is a fixed phase relation between S_1 and S_2, simultaneous emission from S_1 and S_2 will produce an interference pattern similar to the one of one point source alone. If on the other hand we have a randomly fluctuating phase between S_1 and S_2, in the mean we get the sum of the intensities. As a condition for visibility of the fringes therefore we have to demand a lateral shift Δy less than a half fringe spacing

$$|\Delta y| < \frac{1}{2}\left(\frac{\lambda L}{d}\right) \tag{2.61}$$

or equivalently

$$|r_1 - r_2| < \lambda/2. \tag{2.62}$$

To express this in terms of the optical setup using the same arguments which led to (2.57) we get $|r_1 - r_2| = dl/R$ and thus

$$\frac{dl}{R} < \lambda/2. \qquad \text{Condition for Coherence} \tag{2.63}$$

The derivation was carried out for the two points S_1 and S_2. But since these are the furthest points of the extended source, the condition (2.62), if fulfilled, is valid for all points between.

2.4.3 Fraunhofer Approximation

In (2.71) r was approximated by

$$r \approx z + \frac{x^2 + y^2}{2z} - \frac{x\xi + y\eta}{z} + \frac{\xi^2 + \eta^2}{2z}. \tag{2.75}$$

The Fraunhofer approximation now also omits the term $(\xi^2 + \eta^2)/(2z)$, so that

$$\begin{aligned}
E(\nu, \mu) &= \frac{1}{i\lambda z} \int_{-\infty}^{\infty} \int_{-\infty}^{\infty} U(\xi, \eta) e^{ik\left[z + \frac{x^2+y^2}{2z} - \frac{x\xi+y\eta}{z}\right]} d\xi\, d\eta \\
&= \frac{1}{i\lambda z} \int_{-\infty}^{\infty} \int_{-\infty}^{\infty} U(\xi, \eta) e^{\frac{ik}{2z}\left[2z^2 + x^2 + y^2 - 2x\xi - 2y\eta\right]} d\xi\, d\eta \\
&= \frac{e^{ikz}}{i\lambda z} e^{i\pi z \lambda (\nu^2 + \mu^2)} \int_{-\infty}^{\infty} \int_{-\infty}^{\infty} U(\xi, \eta) e^{-2i\pi(\xi\nu + \eta\mu)} d\xi\, d\eta. \tag{2.76}
\end{aligned}$$

Now despite the phase factors and the intensity factor preceding the integrals the Fresnel-Kirchhoff formula is reduced to the Fourier transform.

It remains to analyze under which conditions the Fresnel approximation and when even the Fraunhofer approximation is permitted. The Fresnel approximation replaces the secondary spherical waves of the Huygens-Fresnel principle by waves with parabolic wavefronts, while the Fraunhofer approximation uses even plane wavefronts. The Fresnel approximation is applicable as long as the quadratic term in the binomial expansion, namely $\frac{1}{8}b^2$, leads to a phase change much less than 1 radian

$$kz\left(\frac{1}{8}b^2\right) \ll 1 \tag{2.77}$$

or equivalently $b^2 \ll 8/(kz)$. If we take for b again $((\xi - x)/z)^2 + ((\eta - y)/z)^2$ as in (2.71), we obtain

$$\left[\left(\frac{\xi - x}{z}\right)^2 + \left(\frac{\eta - y}{z}\right)^2\right]^2 \ll \frac{8}{kz} = \frac{4\lambda}{\pi z} \tag{2.78}$$

or

$$\left[(\xi - x)^2 + (\eta - y)^2\right]^2 \ll \frac{4\lambda}{\pi} z^3 \quad \text{for all} \quad x, y, \xi, \eta. \tag{2.79}$$

For a circular aperture of size 1 cm, a circular observation region of size 1 cm, and an assumed wavelength $\lambda = 0.5\ \mu\text{m}$, this condition indicates a distance $z \gg 25$ cm for accuracy. As long as the amplitude transmittance and the illumination of the diffracting aperture is smooth and slowly varying, a much less stringent condition can be obtained [160]. But this prerequisite does not hold if we investigate the diffraction at digitally recorded holograms.

To find the region where even the Fraunhofer approximation is valid we investigate the phase induced by the omitted term in (2.75) which should remain much less than 1 radian

$$k\left(\frac{\xi^2 + \eta^2}{2z}\right) \ll 1 \tag{2.80}$$

or $\xi^2 + \eta^2 \ll z\lambda/\pi$. Thus an aperture with width 1 cm and a wavelength $\lambda = 0.5$ μm requires a distance $z \gg 300$ m for validity of the Fraunhofer approximation. The Fresnel or near field region and the Fraunhofer or far field region are given in Fig. 2.8, but with the distances not exactly scaled.

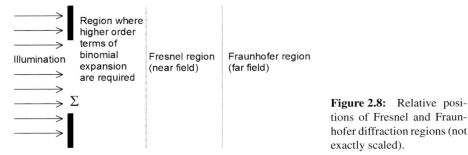

Figure 2.8: Relative positions of Fresnel and Fraunhofer diffraction regions (not exactly scaled).

An observation of the Fraunhofer diffraction patterns in practice is possible without going as far as implied by (2.80) from the diffracting aperture: One has to illuminate the aperture by a spherical wave converging to the observer or a positive lens is placed appropriately between aperture and observer. This last mentioned option deserves a more detailed analysis.

2.4.4 Thin Lens

A lens consists of a material optically more dense than air, usually glass with a refractive index of approximately between $n = 1.5$ and $n = 1.9$, in which the propagation velocity of an optical disturbance is less than in air. A lens is considered as a *thin lens* if a ray entering the lens at coordinates (x, y) at one face exits at nearly the same coordinates on the opposite face. Only a delay of the incident wavefront has occured but no translation of the ray is considered. The delay is proportional to the thickness $\Delta(x, y)$ of the thin lens at coordinates (x, y), Fig. 2.9. The phase delay $\Delta\phi(x, y)$ of a wave that passes the lens at (x, y) therefore is

$$\begin{aligned}\Delta\phi(x,y) &= kn\Delta(x,y) + k[\Delta_0 - \Delta(x,y)] \\ &= k\Delta_0 + k(n-1)\Delta(x,y)\end{aligned} \tag{2.81}$$

with Δ_0 denoting the maximum thickness of the lens and n the refractive index of the lens material. The multiplicative phase transformation of such a lens is of the form

$$t_l(x,y) = e^{ik\Delta_0} e^{ik(n-1)\Delta(x,y)}. \tag{2.82}$$

In practice the faces of the lens are spherical surfaces with radii of curvature R_1 and R_2. Again for practical applications the spheres are approximated by parabolic surfaces. This

2.4 Scalar Diffraction Theory

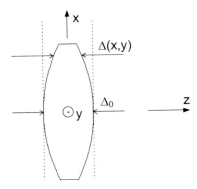

Figure 2.9: Thickness of a thin lens.

paraxial approximation restricts the analysis to portions of the wavefront that lie near the lens axis, the z-axis of Fig. 2.9. The thickness function $\Delta(x,y)$ now becomes [160]

$$\Delta(x,y) = \Delta_0 - \frac{x^2 + y^2}{2\left(\frac{1}{R_1} - \frac{1}{R_2}\right)}. \tag{2.83}$$

The lens parameters n, R_1, R_2 are combined with the *focal length* f by

$$f = \frac{1}{(n-1)\left(\frac{1}{R_1} - \frac{1}{R_2}\right)} \tag{2.84}$$

which yields a phase transformation

$$t_l(x,y) = \exp\left[-\frac{ik}{2f}(x^2 + y^2)\right] \tag{2.85}$$

where the constant factor has been omitted [160]. We adopt the convention that a positive focal length f produces a spherical wave converging towards a point on the z-axis a distance f behind the lens if it was illuminated with a plane wave. Such a lens is called a *positive lens* or *converging lens*. On the other hand if f is negative, the plane wave is transformed into a spherical wave diverging from a point on the lens axis a distance f in front of the lens, which is now called a *negative* or *diverging lens*, Fig. 2.10. We have to note that the phase transition (2.85) of a lens is described by a chirp function.

Now we can show that a thin positive lens produces a Fraunhofer diffraction pattern in a distance that is far less than that predicted by (2.80). This is equivalent to the statement that a lens performs a two-dimensional Fourier transform of a given input distribution. Let an input in the form of a transparency be pressed against a thin lens and be illuminated by a plane wave, Fig. 2.11. If the input image is represented by the function $U(\xi, \eta)$ then the distribution of light just behind the lens is

$$U_L(\xi, \eta) = U(\xi, \eta) \exp\left[-\frac{i\pi}{\lambda f}(\xi^2 + \eta^2)\right] \tag{2.86}$$

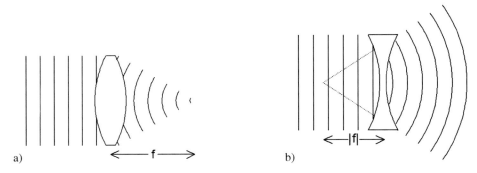

Figure 2.10: Positive (a) and negative (b) lens.

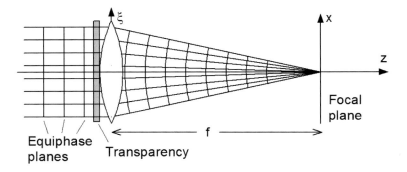

Figure 2.11: Optical Fourier transformation.

where the diameter of the lens is assumed infinite. To find the distribution $E(\nu, \mu, z = f)$ in the back focal plane of the lens, we apply the Fresnel diffraction formula (2.73) and obtain

$$E(\nu, \mu, f) = \frac{e^{ikf}}{i\lambda f} e^{i\pi f \lambda (\nu^2 + \mu^2)} \int_{-\infty}^{\infty} \int_{-\infty}^{\infty} \left[U(\xi, \eta) e^{-\frac{i\pi}{f\lambda}(\xi^2 + \eta^2)} \right]$$

$$\times e^{\frac{i\pi}{f\lambda}(\xi^2 + \eta^2)} e^{-2i\pi(\xi\nu + \eta\mu)} \, d\xi \, d\eta \qquad (2.87)$$

$$= \frac{e^{ikf} e^{i\pi f \lambda (\nu^2 + \mu^2)}}{i\lambda f} \int_{-\infty}^{\infty} \int_{-\infty}^{\infty} U(\xi, \eta) e^{-2i\pi(\xi\nu + \eta\mu)} \, d\xi \, d\eta.$$

Except for a factor that does not depend on the specific input function, the pattern in the focal plane is the Fourier transform of the input function, or in other words, we see the Fraunhofer diffraction pattern of the field incident on the lens in a distance equal to the focal length of the lens. This distance is significantly shorter than that demanded by (2.80). The cases of a lens having a finite aperture or those with inputs located at different distances behind and in front of the lens can be found in detailed treatments in [160, 161].

2.6 Holographic Recording and Optical Reconstruction

distribution, at least as long as the recording process, $\Delta\phi$ must be stationary, which means the wave fields must be mutually coherent.

It was D. Gabor [1–3] who has shown that by illuminating the recorded interference pattern by one of the two interfering wave fields we can reconstruct the other one: this reconstructed wave field then consists of amplitude and phase distributions, not only the intensity. Figures 2.18 and 2.19 show schematically two basic holographic setups, used for recording the wave field reflected from the object's surface.

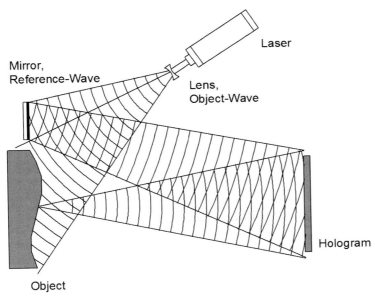

Figure 2.18: Basic holographic setup, wavefront division.

This field is called the object field or object wave, while the other field, necessary for producing the interference, is called the reference field or reference wave. To be mutually coherent, both waves must stem from the same source of coherent light, the laser. The division into object and reference wave can be performed by *wavefront division*, Fig. 2.18, or by *amplitude division*, Fig. 2.19.

The following description uses a point source which does not restrict the generality, because by the superposition principle (2.6) the results can be extended to all points of the object surface. Let the wave reflected by an object surface point P be the spherical wave (2.23), called the *object wave*

$$E_P = \frac{E_{0P}}{p} e^{i(kp + \phi)} \tag{2.120}$$

where p is the distance between the point P and the point $Q = (x, y, 0)$ on the photographic plate. The temporal factor ωt of (2.23) can be omitted. The *reference wave* is assumed to be

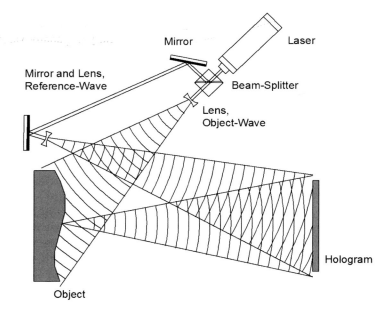

Figure 2.19: Basic holographic setup, amplitude division.

a spherical wave emitted at R

$$E_R = \frac{E_{0R}}{r} e^{i(kr + \psi)} \tag{2.121}$$

with the distance r between R and Q. The photographic plate registers the intensity

$$\begin{aligned} I(x,y) = |E_P + E_R|^2 &= E_P E_P^* + E_R E_R^* + E_P^* E_R + E_P E_R^* \\ &= \frac{E_{0P}^2}{p^2} + \frac{E_{0R}^2}{r^2} + \frac{E_{0P}}{p} e^{-i(kp+\phi)} \frac{E_{0R}}{r} e^{i(kr+\psi)} \\ &\quad + \frac{E_{0P}}{p} e^{i(kp+\phi)} \frac{E_{0R}}{r} e^{-i(kr+\psi)} \\ &= \frac{E_{0P}^2}{p^2} + \frac{E_{0R}^2}{r^2} + \frac{2 E_{0P} E_{0R}}{pr} \cos(k(r-p) + \psi - \phi). \end{aligned} \tag{2.122}$$

As long as E_P is a single spherical wave, this intensity distribution, which is spatially varying because $p = p(x,y)$ and $r = r(x,y)$, is the *hologram* of a point source, the phase ϕ of the object wave relative to the phase ψ of the reference wave is coded into the intensity variation. The same applies for a continuum of object surface points according to the superposition principle.

During the *time of exposure* t_B, the photographic plate receives the energy

$$B(x,y) = \int_0^{t_B} I(x,y,t)\, dt. \tag{2.123}$$

2.6 Holographic Recording and Optical Reconstruction

By processing, this energy is translated into a blackening and a change of the refractive index, summarized in the complex *degree of transmission* τ which is generally a spatially varying function

$$\tau = \tau(x,y) = T(x,y)e^{i\theta(x,y)}. \tag{2.124}$$

This contains the cases of the *amplitude hologram*, where $\theta = $ const, or the *phase hologram* with $T = $ const and θ varying with x and y.

If the exposed plate is processed to produce an amplitude hologram, the real transmission T depends on the received energy B as shown in Fig. 2.20. One has to work in the linear range, where the curve is approximated by the line

$$\begin{aligned} T &= \alpha - \beta B \\ &= \alpha - \beta t_B I \end{aligned} \tag{2.125}$$

for a temporally constant intensity I. α represents a uniform background transmittance and

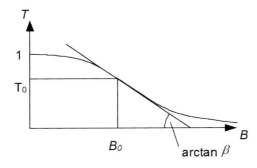

Figure 2.20: Amplitude transmittance versus received energy.

the positive value β is the slope of the amplitude transmittance. The working point B_0 is reached by adjusting the exposure time t_B. To keep the variation around B_0 small, the two wavefronts are given different amplitudes (2.40). The resulting real amplitude transmittance after processing is

$$\begin{aligned} T &= \alpha - \beta t_B (E_P E_P^* + E_R E_R^* + E_P^* E_R + E_P E_R^*) \\ &= \alpha - \beta t_B \left(\frac{E_{0P}^2}{p^2} + \frac{E_{0R}^2}{r^2} \right. \\ &\quad \left. + \frac{E_{0P} E_{0R}}{pr} e^{i(k(r-p)+\psi-\phi)} + \frac{E_{0P} E_{0R}}{pr} e^{-i(k(r-p)+\psi-\phi)} \right) \\ &= T_0 - \beta t_B \frac{E_{0P} E_{0R}}{pr} \left(e^{i(k(r-p)+\psi-\phi)} + e^{-i(k(r-p)+\psi-\phi)} \right) \\ &= T_0 - \frac{\beta t_B E_{0P} E_{0R}}{pr} \cos(k(r-p)+\psi-\phi) \end{aligned} \tag{2.126}$$

where T_0 is the mean transmittance $T_0 = \alpha - \beta t_B (E_{0P}^2/p^2 + E_{0R}^2/r^2)$.

If we produce a phase hologram, we must remain in the linear range of the curve describing the effective phase shift θ against exposure, Fig. 2.21.

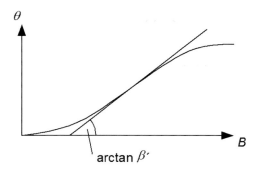

Figure 2.21: Phase shift versus received energy.

Again we have a range where we approximate by a line

$$\theta = \alpha' + \beta' t_B I. \tag{2.127}$$

The complex transmission after a series expansion of the exponential and neglecting higher than linear terms is

$$\tau = e^{i\theta(I)} \approx 1 + i\theta(I). \tag{2.128}$$

The resulting phase transmittance is

$$\tau = e^{i\theta(I)} \approx (1 + i\alpha') + i\beta' t_B (E_P E_P^* + E_R E_R^* + E_P^* E_R + E_P E_R^*) \tag{2.129}$$

analogous to (2.126). For convenience we have set $T = const = 1$.

The necessary *spatial resolution* of the recording media can be estimated by (2.30). If the angle between reference and object wave is θ, both waves for the moment being assumed as plane waves, the fringe distance is

$$\frac{\pi}{|k''|} = \frac{\lambda}{2 \sin \frac{\theta}{2}}. \tag{2.130}$$

If we assume a wavelength $\lambda = 0.5$ μm, for an angle of $\theta = 1°$ we need a spatial resolution of better than 35 LP/mm (line-pairs per millimeter), for $\theta = 10°$ already more than 350 LP/mm and for $30°$ more than 1035 LP/mm. As a consequence the spatial resolution puts an upper bound to the angular separation of the object points and thus to the object size, an issue becoming important when using CCD-arrays for recording the holograms (see Section 3.1.1).

2.6.2 Optical Reconstruction of a Wave Field

For optical *reconstruction* of the object wave we illuminate the processed photographic plate, also called a *hologram*, with the reference wave E_R, Fig. 2.22. This results in a modulation of the reference wave by the transmission $\tau(x, y)$.

2.6 Holographic Recording and Optical Reconstruction

We first investigate the distance p

$$p = \sqrt{(x - x_P)^2 + (y - y_P)^2 + z_P^2}$$
$$= \sqrt{x^2 + y^2 - 2xx_P - 2yy_P + p_0^2} \qquad (2.142)$$

where p_0 is the distance of P to the origin. We observe that p_0 is large compared to x, y, x_P and y_P, so we can expand p into a power series around p_0 [7–9, 21, 22, 166, 170, 171]

$$p = p_0 + \frac{x^2 + y^2}{2p_0} - \frac{2xx_P + 2yy_P}{2p_0} + - \ldots \qquad (2.143)$$

In the same way we expand r, r' and p'

$$r = r_0 + \frac{x^2 + y^2}{2r_0} - \frac{2xx_R + 2yy_R}{2r_0} + - \ldots, \qquad (2.144)$$

$$r' = r_0' + \frac{x^2 + y^2}{2r_0'} - \frac{2xx_R' + 2yy_R'}{2r_0'} + - \ldots, \qquad (2.145)$$

$$p' = p_0' + \frac{x^2 + y^2}{2p_0'} - \frac{2xx_P' + 2yy_P'}{2p_0'} + - \ldots \qquad (2.146)$$

From the definition of p' (2.141) we have

$$k'p' = k(p - r) + k'r'. \qquad (2.147)$$

Now we insert the approximations of p, r, r' and p' and observe that (2.147) is valid for all (x, y), so we equate the coefficients of similar terms and with the definition $\mu = \lambda'/\lambda$ we get

$$\frac{1}{p_0'} = \frac{\mu}{p_0} - \frac{\mu}{r_0} + \frac{1}{r_0'}, \qquad (2.148)$$

$$\frac{x_P'}{p_0'} = \frac{\mu x_P}{p_0} - \frac{\mu x_R}{r_0} + \frac{x_R'}{r_0'}, \qquad (2.149)$$

$$\frac{y_P'}{p_0'} = \frac{\mu y_P}{p_0} - \frac{\mu y_R}{r_0} + \frac{y_R'}{r_0'}. \qquad (2.150)$$

These are the *holographic imaging equations* for the direct image to a first order approximation. In a similar way one can derive the imaging equations of the conjugate image, the second term in (2.131). They have the same form, only the signs preceding μ are interchanged.

In the preceding derivation we have expanded about p_0, p_0', r_0 and r_0' [170], which gives a more accurate approximation than the expansion about the z-components z_P, z_P', z_R, z_R' [172]. This may be performed, if the x- and y-components are negligible compared to the z-components. In that case we would get the imaging equations

$$\frac{1}{z_P'} = \frac{\mu}{z_P} - \frac{\mu}{z_R} + \frac{1}{z_R'}, \qquad (2.151)$$

$$\frac{x_P'}{z_P'} = \frac{\mu x_P}{z_P} - \frac{\mu x_R}{z_R} + \frac{x_R'}{z_R'}, \qquad (2.152)$$

$$\frac{y_P'}{z_P'} = \frac{\mu y_P}{z_P} - \frac{\mu y_R}{z_R} + \frac{y_R'}{z_R'} \qquad (2.153)$$

which can be rearranged to yield the easier to handle and well known form [172, 173]

$$x'_P = \frac{\mu x_P z_R z'_R - \mu x_R z_P z'_R + x'_R z_P z_R}{\mu z_R z'_R - \mu z_P z'_R + z_P z_R}, \qquad (2.154)$$

$$y'_P = \frac{\mu y_P z_R z'_R - \mu y_R z_P z'_R + y'_R z_P z_R}{\mu z_R z'_R - \mu z_P z'_R + z_P z_R}, \qquad (2.155)$$

$$z'_P = \frac{z'_R z_P z_R}{\mu z_R z'_R - \mu z_P z'_R + z_P z_R}. \qquad (2.156)$$

We now consider some special cases. The easiest is the reconstruction with a reference wave identical to the one used for recording, even with the same wavelength. Then we have $p'_0 = p_0$ and $x'_P = x_P$, $y'_P = y_P$ and $z'_P = z_P$, as expected. Next we examine a plane reference wave impinging orthogonally onto the hologram, but during reconstruction with wavelength λ' instead of λ. Then we have $r_0 = z_R \to \infty$, $r'_0 = z'_R \to \infty$, $x_R/r_0 = 0$, $y_R/r_0 = 0$, $x'_R/r'_0 = 0$, $y'_R/r'_0 = 0$. From (2.148) we get $\mu p'_0/p_0 = 1$ and together with (2.149) and (2.150) resp.

$$x'_P = x_P, \qquad (2.157)$$

$$y'_P = y_P, \qquad (2.158)$$

$$z'_P = \frac{1}{\mu} z_P. \qquad (2.159)$$

This means the object is shifted and stretched or compressed in the z-direction, but the lateral dimensions remain unaffected.

The *lateral magnification* M_{lat} in the direct image can be defined [172]

$$M_{lat} = \frac{dx'_P}{dx_P} = \frac{dy'_P}{dy_P}. \qquad (2.160)$$

It is

$$M_{lat} = \frac{1}{1 + z_P \left(\frac{1}{\mu z'_R} - \frac{1}{z_R} \right)}. \qquad (2.161)$$

For the plane wave of the case above due to $z_R \to \infty$ and $z'_R \to \infty$ we get $M_{lat} = 1$.

Furthermore we may define the *angular magnification* by [172]

$$M_{ang} = \frac{d(x'_P/z'_P)}{d(x_P/z_P)} = \frac{d(y'_P/z'_P)}{d(y_P/z_P)} \qquad (2.162)$$

and calculate $|M_{ang}| = \mu$. For the *longitudinal magnification* M_{long} of the primary image we obtain

$$M_{long} = \frac{dz'_P}{dz_P} = \frac{1}{\mu} M_{lat}^2. \qquad (2.163)$$

The different magnifications for the conjugate image are derived analogously.

2.6 Holographic Recording and Optical Reconstruction

The holographic imaging equations can be used to compensate the effect of CW-reconstruction with a wavelength differing from that of the pulsed recording by an appropriate reference beam adjustment [143]. If the hologram is shifted during the real-time reconstruction, shearing fringes are produced which modify the deformation fringes in a controlled way [174]. If in the expansion (2.143) we retain higher order terms, we get the *aberrations* like spherical aberration, coma, astigmatism, field curvature and others [165, 166, 172, 175–178]. They give information about the quality of the imaged object, which means how the image of a point is washed out.

2.6.4 Types of Holograms

In Section 2.6.2 we already mentioned the distinction between Gabor's in-line holograms and the Leith-Upatnieks off-axis holograms. The *in-line arrangement* is still used for the analysis of transparent objects, or small particles like droplets, Fig. 2.25a.

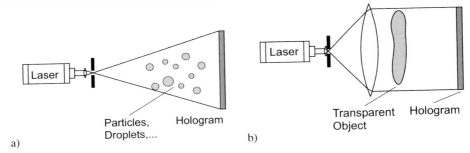

Figure 2.25: In-line holography with divergent (a) and collimated (b) light.

The reference wave is the light passing unaffected by the particles and the object wave is the wave field scattered by the particles. We may use divergent or collimated light, Fig. 2.25b. Reconstruction by the illuminating wave alone without the object gives the virtual image of the particles or the transparent object at its original position as well as the real image on the opposite side of the hologram. There is no beam splitting into reference and object beam, so the method sometimes is called *single beam holography*.

The main disadvantages of in-line holograms are the disturbed reconstruction due to the bright reference beam and the twin images: Virtual and real image are along the same line of sight, so while focusing on one of them we see it overlayed by an out of focus image of the other one. These drawbacks are avoided by the *off-axis arrangement*. Here the laser beam is split by wavefront division, Fig. 2.18, or by amplitude division, Fig. 2.19, into reference and object wave. This approach has been called *split beam holography* or *two beam holography*. If the offset angle between reference and object wave is large enough, we have no overlap between the virtual and the real reconstructed images nor do we stare into the directly transmitted reference beam.

If we have flat or nearly flat objects we may record the Fourier transform of the object and reference waves. Then we speak about *Fourier transform holography*. Imagine an object located in the front focal plane of a lens, Fig. 2.26a, illuminated by coherent light.

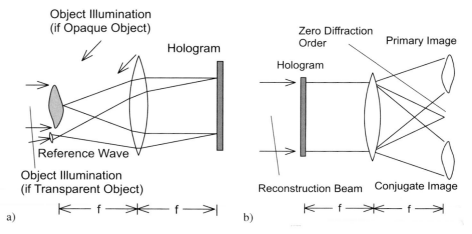

Figure 2.26: Recording (a) and reconstruction (b) of a Fourier transform hologram.

Let the complex amplitude leaving the object plane be $E_P(x, y)$, then the complex amplitude of the wave field at the holographic plate located in the back focal plane of the lens is the Fourier transform of $E_P(x, y)$

$$\mathcal{E}_P(\xi, \eta) = \mathcal{F}\{E_P(x, y)\}. \tag{2.164}$$

The reference wave is a spherical wave emitted from a point source at (x_0, y_0) in the front focal plane. Without loss of generality we can assume unit amplitude, so $E_R(x, y) = \delta(x - x_0, y - y_0)$. $\delta(x, y)$ describes the *Dirac delta impulse*, see (A.6). The complex amplitude of this reference wave in the hologram plane is

$$\mathcal{E}_R(\xi, \eta) = \mathcal{F}\{\delta(x - x_0, y - y_0)\} = e^{-i2\pi(\xi x_0 + \eta y_0)} \tag{2.165}$$

as follows from the shift theorem (A.15) and the forth line of Table A.1. In the hologram plane, the back focal plane of the lens, we obtain a hologram characterized by the transmission

$$\begin{aligned} T(\xi, \eta) &= \alpha - \beta t_B (\mathcal{E}_P + \mathcal{E}_R)(\mathcal{E}_P + \mathcal{E}_R)^* \\ &= \alpha - \beta t_B (\mathcal{E}_R \mathcal{E}_R^* + \mathcal{E}_P \mathcal{E}_P^* + \mathcal{E}_R^* \mathcal{E}_P + \mathcal{E}_R \mathcal{E}_P^*)(\xi, \eta). \end{aligned} \tag{2.166}$$

For reconstruction we illuminate this *Fourier transform hologram* with the plane wave of (2.165) and obtain

$$T\mathcal{E}_R = (\alpha - \beta t_B |\mathcal{E}_R|^2)\mathcal{E}_R - \beta t_B(\mathcal{E}_P \mathcal{E}_P^* \mathcal{E}_R + \mathcal{E}_R \mathcal{E}_R^* \mathcal{E}_P + \mathcal{E}_R \mathcal{E}_R \mathcal{E}_P^*). \tag{2.167}$$

The hologram plane now is the front focal plane of the reconstruction lens, see Fig. 2.26b,

2.6 Holographic Recording and Optical Reconstruction

which produces the Fourier transform of $T\mathcal{E}_R$ in its back focal plane, the (x', y')-plane.

$$\begin{aligned}
\mathcal{F}\{T(\xi,\eta)\}&(x',y') \\
&= (\alpha - \beta t_B |\mathcal{E}_R|^2)\mathcal{F}\{\mathcal{E}_R\}(x',y') - \beta t_B (\mathcal{F}\{\mathcal{E}_P \mathcal{E}_P^* \mathcal{E}_R\} \\
&\quad + \mathcal{F}\{\mathcal{E}_R \mathcal{E}_R^* \mathcal{E}_P\} + \mathcal{F}\{\mathcal{E}_R \mathcal{E}_R \mathcal{E}_P^*\})(x',y') \\
&= \text{const.}\, \delta(x'+x_0, y'+y_0) - \text{const.}\,[(E_P \otimes E_P)(x'+x_0, y'+y_0) \\
&\quad + E_P(-x',-y') + E_P^*(x'+2x_0, y'+2y_0)]
\end{aligned} \qquad (2.168)$$

according to the rules for manipulating Fourier transforms. The $\otimes$ denotes correlation and the sign change in the arguments is due to the fact that the lens performs a direct Fourier transform instead of an inverse one. The first term of this reconstructed wave field represents a focus that is the pointwise dc-term. The second term constitutes a halo around this focus. The third term is proportional to the original object wave field but inverted, while the forth term is a conjugate of the original wave shifted by $(-2x_0, -2y_0)$. Both images are real and sharp and can be registered by film or TV camera in the back focal plane. The conjugated image in the registered intensity distribution is identified only by its geometric inversion. The main advantage of such a Fourier hologram is the stationary reconstructed image even when the hologram is translated in its own plane, due to the shift invariance of the Fourier transform intensity (see Appendix A.4). Furthermore the resolution requirements on the recording medium are less.

Collimation by the lens in Fourier transform holography means that the object points and the reference source are at infinity. *Lensless Fourier transform holography* is possible if object point and reference source are at a finite but the same distance from the holographic plate. For proving this statement, let the wave leaving the object be $E_P(x,y)$, Fig. 2.27.

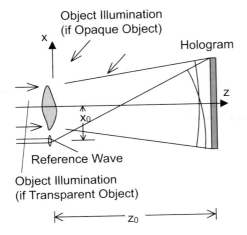

Figure 2.27: Recording of a lensless Fourier transform hologram.

The Fresnel-Kirchhoff integral (2.73) implies that the complex amplitude in the hologram plane is

$$\mathcal{E}_P(\xi,\eta) = \frac{e^{ikz}}{i\lambda z} e^{i\pi z\lambda(\xi^2+\eta^2)}$$
$$\times \int_{-\infty}^{\infty}\int_{-\infty}^{\infty} E_P(x,y) e^{\frac{i\pi}{z\lambda}(x^2+y^2)} e^{-2i\pi(\xi x+\eta y)}\,dx\,dy \quad (2.169)$$
$$= \text{const.}\, e^{i\pi z\lambda(\xi^2+\eta^2)} \mathcal{F}\{E_P(x,y) e^{\frac{i\pi}{z\lambda}(x^2+y^2)}\}(\xi,\eta)$$
$$= \text{const.}\, e^{i\pi z\lambda(\xi^2+\eta^2)} \mathcal{E}'_P(\xi,\eta)$$

with $\mathcal{E}'_P = \mathcal{F}\{E_P(x,y) e^{\frac{i\pi}{z\lambda}(x^2+y^2)}\}$. The same formalism holds for the reference wave $E_R(x,y) = \delta(x-x_0, y-y_0)$

$$\mathcal{E}_R(\xi,\eta) = \frac{e^{ikz}}{i\lambda z} e^{i\pi z\lambda(\xi^2+\eta^2)}$$
$$\times \int_{-\infty}^{\infty}\int_{-\infty}^{\infty} \delta(x-x_0,y-y_0) e^{\frac{i\pi}{z\lambda}(x^2+y^2)} e^{-2i\pi(\xi x+\eta y)}\,dx\,dy \quad (2.170)$$
$$= \text{const.}\, e^{i\pi z\lambda(\xi^2+\eta^2)} \mathcal{E}'_R(\xi,\eta)$$

with $\mathcal{E}'_R = e^{\frac{i\pi}{z\lambda}(x^2+y^2)} e^{-2i\pi(\xi x+\eta y)}$. The resulting hologram transmission is

$$T = \alpha - \beta t_B(|\mathcal{E}_R|^2 + |\mathcal{E}_P|^2 + \mathcal{E}_R^*\mathcal{E}_P + \mathcal{E}_R\mathcal{E}_P^*) \quad (2.171)$$
$$= \alpha - \beta t_B(|\mathcal{E}'_R|^2 + |\mathcal{E}'_P|^2 + \mathcal{E}'^*_R\mathcal{E}'_P + \mathcal{E}'_R\mathcal{E}'^*_P)$$

which has the same form as (2.166). The crucial fact is that the z in (2.169) and in (2.170) is the same, so the factors $e^{i\pi z\lambda(\xi^2+\eta^2)}$ and its conjugate in each term multiply to unity. The effect of the spherical phase factor associated with the near-field Fresnel diffraction pattern here is eliminated by the spherical reference wave with the same curvature.

If the object is focused into the plane of the hologram, Fig. 2.28, we speak of an *image hologram*. Here the real image of the object is recorded instead of the wave field reflected or scattered by the object. The advantage of image holograms is that they can be reconstructed by an incoherent light source of appreciable size and spectral bandwidth and they still produce acceptably sharp images. Also the image luminance is increased. But we have to pay with an observation angle limited by the angular aperture of the used lens [172].

For the above mentioned image hologram the distance between object and hologram appears to be zero. The most general case is the holographic plate in the *near-field* or *Fresnel diffraction region*, Fig. 2.29, then we speak of a *Fresnel hologram*. Fresnel holograms are the most generated in holographic interferometry.

As the distance between object and hologram increases we reach the *far-field* or *Fraunhofer diffraction region* and get a so-called *Fraunhofer hologram*. In this case either the object

2.6 Holographic Recording and Optical Reconstruction

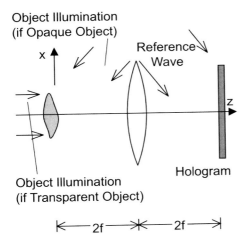

Figure 2.28: Recording of an image hologram.

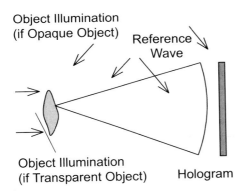

Figure 2.29: Recording of a Fresnel hologram.

has to be small compared to the dimensions of the holographic arrangement

$$z_0 \gg \frac{x_0^2 + y_0^2}{\lambda} \tag{2.172}$$

where x_0 and y_0 are the maximal lateral dimensions of the object, or the object must be in the focus of a lens, Fig. 2.30.

Of course holographers have attempted early to get rid of the coherence requirements, at least in the reconstruction stage. If a usual hologram is illuminated with white light, all the frequencies of the white light are diffracted in different directions. At each point different frequencies stemming from different points in the hologram are superimposed, so normally we will not recognize a reconstruction of an object wave field. The quality may be improved slightly by a color filter, but then the brightness decreases drastically.

The so-called *white light holograms*, which can be reconstructed with white light, use the finite thickness of the photographic emulsion in the hologram [179]. Up to now we have only considered the lateral distribution of the phase or transmittance of the hologram. If the thickness of the sensitive layer is much greater than the distance between adjacent surfaces

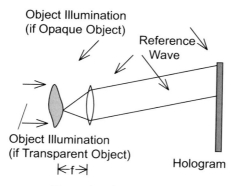

Figure 2.30: Recording of a Fraunhofer hologram.

of the interference maxima, then the hologram should be considered as a *volume* or *three-dimensional hologram*. Generally we may speak of a volume hologram, if the thickness d_H of the layer is

$$d_H > \frac{1.6 d^2}{\lambda} \tag{2.173}$$

where d is the distance between adjacent interference planes [180].

If now the coherent object and reference waves impinge onto the hologram from opposite sides we get interference layers nearly parallel to the hologram surface. The distance between subsequent interference layers is $\lambda/(2\sin\frac{\theta}{2})$ according to (2.30) and Fig. 2.3. For reconstruction this thick hologram is illuminated with white light which is reflected at the layers. Dependent on the wavelength the reflected waves interfere constructively in defined directions, an effect called *Bragg reflection*. Let the distance of the interference layers be d and illuminate by an angle α, then we find the n-th diffraction order for the wavelength λ in the observation direction of angle β

$$d = \frac{n\lambda}{\sin\alpha + \sin\beta}. \tag{2.174}$$

The most intense wave is that in the first diffraction order $n = 1$. So if we look under the angle β onto the hologram illuminated under the angle α, we see a clear image with color corresponding to λ. The parallel layers modulated by the information about the image react like an interference filter for the specific wavelength λ.

An arrangement for recording such a white light hologram is given in Fig. 2.31. The expanded and collimated laser beam is directed through the hologram plate onto the object. The reference wave is given by the light coming directly from the laser, the wave passing through the plate and reflected by the object is the object wave. For good results we need strongly reflecting objects and a hologram plate close to the object.

Another way to prevent the different diffracted colors from overlaying during the reconstruction with white light is the exchange of the variation of the vertical parallax against a variation of the wavelength [159]. For this task first a master hologram of the object is produced as usual, Fig. 2.32a.

2.7 Elements of the Holographic Setup

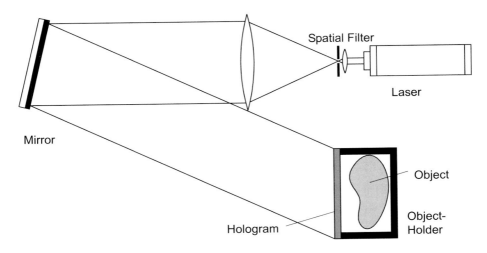

Figure 2.31: Recording of a white light hologram.

By reversing the direction of the reference beam or by turning the hologram by 180° a real pseudoscopic image is reconstructed. In front of the hologram a horizontal slit aperture is now placed and a hologram of the wave field passing through this slit is recorded, Fig. 2.32b. This has the first effect that the vertical parallax is lost, but this is not recognized immediately, as long as the eyes of the observer are horizontally arranged. The second effect is that the different colors reconstructed from the second hologram still overlap in space. But the colors converge to different reconstructed slits. Although neighboring colors are overlayed in the reconstructed slits, the range from blue to red can be stretched over a broad area so that each reconstructed slit produces a sharp image. The eyes of the observer are placed in one reconstructed slit and see the object in one color. If the head is moved in the vertical direction the object is seen in another spectral color than before, Fig. 2.32c. Since in this way the object can be observed in the successive colors of the rainbow, the secondary hologram is called a *rainbow hologram* and we speak of *rainbow holography*.

2.7 Elements of the Holographic Setup

2.7.1 Laser

Optical holography and holographic interferometry in the visible range of the spectrum became possible with the invention of a source radiating coherent light, the *laser*. The basic principle behind the laser is the *stimulated emission* of radiation. Contrary to the ubiquitous *spontaneous emission* here the emission of photons is triggered by an electromagnetic wave. All photons generated this way have the same frequency, polarization, phase, and direction as the stimulating wave.

Normally in a collection of atoms each one tends to hold the lowest energy configuration. Therefore in thermal equilibrium or even when excited the majority of all atoms are in the

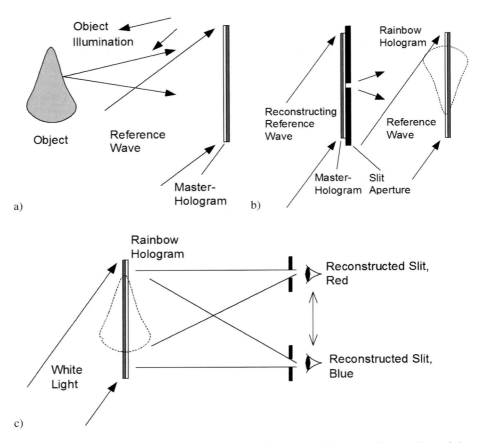

Figure 2.32: Rainbow holography: (a) recording of a master hologram, (b) recording of the rainbow hologram, (c) reconstruction with white light.

ground state. Only if one succeeds in bringing a larger part of the atoms into a higher excited state than remain in a lower state, which may be the ground state or itself an excited state, then an impinging wave can stimulate the emission of an avalanche of waves, all with equal phase and propagating in the same direction. This stimulated emission takes place as long as the *population inversion* between the lower and the higher energy states is maintained. To achieve the inversion, energy must be provided to the system by a process called *pumping*. So the laser can be regarded as an amplifier since an impinging wave generates a manifold of waves of the same direction, frequency, and phase.

To prevent this amplifier from amplifying only noise, a feedback is introduced by installing two mirrors on opposite sides of the active medium. If plane mirrors are adjusted exactly parallel, photons are reflected back and forth, and what we get is an oscillator of high quality. Standing waves will be formed between the two mirrors. If one of the mirrors is semi-transparent, some of the photons can leave the laser as the so called *laser beam* of coherent

2.7 Elements of the Holographic Setup

radiation. The resonant frequencies possible in a *cavity* of length L, the separation between the mirrors, are   *These, I believe, are the longitudinal modes*

$$\nu_n = \frac{nc}{2L} \tag{2.175}$$

with n an integer and c the speed of light.

Since the active medium possesses a gain curve around a central line, amplification takes place only at those frequencies where the gain is higher than the cavity losses, Fig. 2.33a.

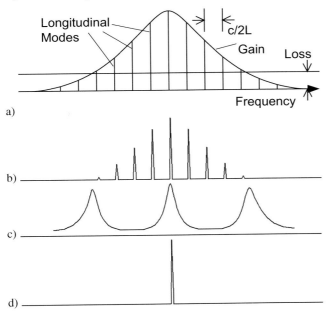

Figure 2.33: Single frequency selection: (a) oscillation frequencies, gain, and loss, (b) multi-frequency output, (c) etalon transmittance, (d) single frequency output.

If the medium allows for several central lines in the spectrum, these are selected by a *wavelength selector prism*, Fig. 2.34.

The existence of several frequencies ν_n under the gain curve – one speaks of the *longitudinal modes* – results in a coherence length too short for most holographic applications. On the other hand short cavities would produce the desired single longitudinal mode under the gain curve, but with extremely low power. The solution to this contradiction is an intracavity *etalon*, Fig. 2.34, which is basically a *Fabry-Perot interferometer*. Only those modes are allowed to oscillate that match into the long laser cavity as well as the short cavity of the etalon. The result is a single frequency radiation of high coherence length, Figs. 2.33c and d.

Although principally a laser may emit in various *transverse modes*, which describe the intensity variation across the diameter of the laser beam, in holography only the TEM_{00} mode, rendering the best spatial coherence, has to be used. This mode is achieved by inserting an aperture of small diameter into the resonator, the pinhole in Fig. 2.35. In the TEM_{00}-mode

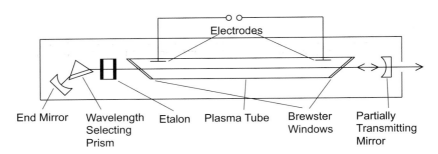

Figure 2.34: Typical gas laser.

the intensity is Gaussian distributed, which constitutes the Gaussian background illumination of most holographic interferograms.

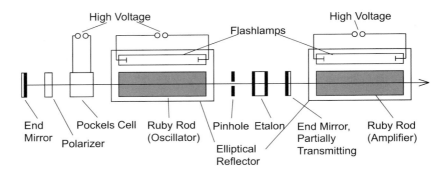

Figure 2.35: Q-switched ruby laser.

There is a vast number of materials showing laser activity, pumped in various ways. But only some of them have gained importance for holographic applications, Table 2.1.

The *ruby laser*, Fig. 2.35, is the most widely used *pulsed laser* in optical holography [181], mainly because of its output energy of up to about 10 J per pulse. Pumping is performed with xenon flashlamps. To achieve a single pulse of short duration an intracavity *Q-switch* is employed. This is a fast-acting optical shutter, normally realized by a *Pockels cell*. The Pockels cell is a crystal showing birefringence when an electric field is applied. If its principal axes are oriented at $45°$ to the direction of polarization of the laser beam, it produces a mutual phase shift of $90°$ between the two polarization components when the voltage is applied. A wave transmitted by the activated Pockels cell after reflection at the end mirror is polarized $90°$ to its original direction and thus blocked by a polarizer. As soon as the flashlamp is fired, the Q-switch is closed. A large population inversion is built up. At the end of the flashlamp pulse, the voltage is switched off, Pockels cell and polarizer transmit freely, and the oscillation can take place. A short pulse, typically of 10 to 20 ns duration, is emitted. The wavelength of the ruby emission is 694 nm, which fits well to the sensitivity of photographic emulsions relative polarization of the object wavefront. Barnhart et al. report on its employment in digital holographic velocimetry [192].

TV camera-tubes and *CCD-targets* generally do not have the resolution required by off-axis holography. This on the one hand led to the development of the *ESPI methods*, Section 7.2, and on the other hand to *digital holography*, Chapter 3. Digital holography utilizes all the advantages the CCD-targets are offering: Fast acquisition of the primary holograms, rapid digital storage, numerical evaluation instead of optical reconstruction, thus not suffering from optical imperfections or limited diffraction efficiency. Because of their importance for digital holography CCD and CMOS image sensors are treated in more detail in Section 2.8.

2.7.3 Optical Components

Here we have not the room to treat the fundamentals of all optical components used in holographic interferometry; these can be found in every optics textbook [159, 193, 194]. Only some hints and practical considerations are given. To produce good quality holograms, the mechanical and optical components always have to be fixed in a way to prevent spurious motions or vibrations. Sources of vibrations like transformers, ventilators, cooling with streaming water, etc. should be kept away from the holographic arrangement. A mechanical shutter should not be mounted on the same vibration-isolated table as the other components. If motions cannot be avoided, a pulsed laser or one of the methods described in Section 6.1.4 have to be applied.

The quality of the expanded laser beam illuminating the object can be significantly improved by positioning a *pinhole* in the focal point of the magnifying lens. Dust and scratches on optical surfaces produce spatial noise. The pinhole acts as a *spatial filter* that allows to pass only the dc-term of the spectrum. Thus a clean illumination of nearly Gaussian characteristic will result. Magnifying the beam by a positive lens and spatial filtering is not possible when using a pulsed ruby laser. Here the power concentration in the focus would be high enough to ionize the air, the resulting plasma is opaque to light. Therefore with pulsed ruby lasers only negative lenses should be used. Employing adequate optics, objects of varying size and form can be investigated. Even the interior walls of pipes have been holographically analyzed by utilizing panoramic annular lenses [195].

Polarization plays a minor role in holographic interferometry. Of course one has to guarantee that object and reference wave are polarized at identical angles to obtain good contrast holographic fringes. Holographic setups most often have optical axes only in one horizontal plane, vertical deflections are rather rare. It is recommended to use the lasers in such an orientation that the normals to the Brewster windows of the laser are in a vertical plane. This prevents any reflecting mirror from being accidentally positioned at the Brewster angle, thus cutting off the reflection [196].

Optical fibers are a means to conduct the laser light along paths differing from the straight propagation [197, 198]. Thus the laser, the splitting of the primary beam into object and reference beam, as well as parts of the paths of these beams can be decoupled from the rest of the holographic arrangement. Areas which are unaccessible by straight rays become accessible to holographic interferometry if optical fibers are used. Furthermore an optical fiber can be used as a sensor for pressure or temperature, since any expansion of the fiber due to the pressure or temperature variations influences the measurable optical path length.

Generally it can be stated that the influence of optical fibers on the temporal coherence can be neglected. But the spatial coherence is affected significantly if *multimode fibers* are used. Many modes can propagate in the fiber which may interfere and produce speckles. This may be tolerated for object illumination, since the diffusely scattering object also will degrade the spatial coherence. But a good spatial coherence in the reference wave during recording as well as for reconstruction is crucial. *Monomode fibers* with a core diameter less than 50 μm should be employed [199]. Also results are reported on the use of coherent multimode fiber bundles to transmit the image of the test object from the test site to the holographic plate [200].

Optical fibers are already used in holographic interferometry in combination with CW lasers. The use of Q-switched lasers with fibers is under investigation [201]. The problems are that the high energies can destroy the faces of the fiber and may stimulate Brillouin scattering in the fiber volume.

2.7.4 Beam Modulating Components

A number of holographic interferometric methods require the modulation of the laser beam, like the *phase sampling* methods which employ a *phase shifting* device or the *heterodyne methods* which make use of a *frequency shift*. The fast shutter realized by a Pockels cell was already introduced in the context of Q-switched lasers.

A *phase shift* in a beam can be introduced by rotating a half-wave plate, moving a grating or employing an acoustooptical modulator, tilting a glass plate, [94] shifting a mirror [202], or stretching an optical fiber, Fig. 2.37.

The effect of a $\lambda/2$-plate on circularly polarized light, Fig. 2.37a, is best described in the formalism of *Jones matrices*: Let a left-hand circularly polarized wave E_{circ} be

$$E_{\text{circ}} = \begin{pmatrix} 1/\sqrt{2} \\ i/\sqrt{2} \end{pmatrix} = \frac{1}{\sqrt{2}} \begin{pmatrix} e^{i0} \\ e^{i\frac{\pi}{2}} \end{pmatrix}. \quad (2.176)$$

Ordinary and extraordinary rays are mutually shifted by $\pi/2$. The $\lambda/2$-plate oriented with the ordinary ray in the x-direction and the extraordinary ray in the y-direction is described by the matrix

$$M_{\lambda/2} = \begin{pmatrix} 1 & 0 \\ 0 & -1 \end{pmatrix} = \begin{pmatrix} e^{i0} & 0 \\ 0 & e^{i\pi} \end{pmatrix}. \quad (2.177)$$

Wave E_{circ} after passing the $\lambda/2$-plate is

$$M_{\lambda/2}\, E_{\text{circ}} = \begin{pmatrix} e^{i0} & 0 \\ 0 & e^{i\pi} \end{pmatrix} \frac{1}{\sqrt{2}} \begin{pmatrix} e^{i0} \\ e^{i\frac{\pi}{2}} \end{pmatrix} = \frac{1}{\sqrt{2}} \begin{pmatrix} e^{i0} \\ e^{i\frac{3\pi}{2}} \end{pmatrix}. \quad (2.178)$$

The $\lambda/2$-plate after undergoing a rotation R_{45} of $45°$ has the matrix

$$R_{45}^{-1} M_{\lambda/2} R_{45} = \frac{1}{\sqrt{2}} \begin{pmatrix} 1 & -1 \\ 1 & 1 \end{pmatrix} \begin{pmatrix} e^{i0} & 0 \\ 0 & e^{i\pi} \end{pmatrix} \frac{1}{\sqrt{2}} \begin{pmatrix} 1 & 1 \\ -1 & 1 \end{pmatrix}$$
$$= \begin{pmatrix} 0 & 1 \\ 1 & 0 \end{pmatrix}. \quad (2.179)$$

2.7 Elements of the Holographic Setup

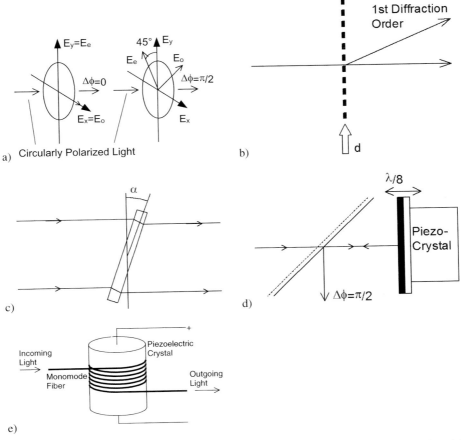

Figure 2.37: Phase shifting by (a) rotating half-wave plate, (b) shifted diffraction grating, (c) tilted glass plate, (d) translated mirror, (e) elongation of fiber.

The wave E_{circ} after passing the rotated half-wave plate is

$$R_{45}^{-1} M_{\lambda/2} R_{45} E_{\text{circ}} = \begin{pmatrix} 0 & 1 \\ 1 & 0 \end{pmatrix} \frac{1}{\sqrt{2}} \begin{pmatrix} e^{i0} \\ e^{i\frac{\pi}{2}} \end{pmatrix} = \frac{1}{\sqrt{2}} \begin{pmatrix} e^{i\frac{\pi}{2}} \\ e^{i0} \end{pmatrix}$$

$$= \frac{1}{\sqrt{2}} \begin{pmatrix} e^{i(0+\frac{\pi}{2})} \\ e^{i(\frac{3\pi}{2}+\frac{\pi}{2})} \end{pmatrix} \quad (2.180)$$

which is the wave after going through the unrotated plate (2.178) but now shifted by $\pi/2$ [203].

In an analogous way e. g. three consecutive phase shifts by $\pi/2$ can be obtained. This is frequently used in phase shifting holographic interferometry, see Section 5.5, or phase shifting digital holography, see Section 3.4.1. The $\lambda/2$-plate is accompanied by a $\lambda/4$-plate as shown in Fig. 2.38.

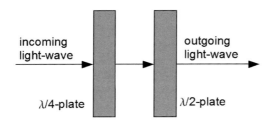

Figure 2.38: Phase shifting using quarter and half wave plate.

The $\lambda/4$-plate is described by the matrix

$$M_{\lambda/4} = \begin{pmatrix} 1 & 0 \\ 0 & -i \end{pmatrix} = \begin{pmatrix} e^{i0} & 0 \\ 0 & e^{i\frac{\pi}{2}} \end{pmatrix}. \qquad (2.181)$$

Take as the incoming wave a linearly polarized wave with polarization direction oriented $45°$ with respect to the x-axis of a coordinate system in which the ordinary axes of the $\lambda/4$- and the $\lambda/2$-plates coincide with the x-axis. This wave of unit amplitude is

$$E_{\text{lin}} = \frac{1}{\sqrt{2}} \begin{pmatrix} e^{i0} \\ e^{i0} \end{pmatrix}. \qquad (2.182)$$

After passing through the two plates it is

$$M_{\lambda/2} M_{\lambda/4} E_{\text{lin}} = \begin{pmatrix} 1 & 0 \\ 0 & -1 \end{pmatrix} \begin{pmatrix} 1 & 0 \\ 0 & -i \end{pmatrix} \frac{1}{\sqrt{2}} \begin{pmatrix} e^{i0} \\ e^{i0} \end{pmatrix} = \frac{1}{\sqrt{2}} \begin{pmatrix} e^{i0} \\ e^{i\frac{\pi}{2}} \end{pmatrix} \qquad (2.183)$$

which is a left hand circularly polarized wave. Now we rotate the plate by $+45°$ and $-45°$ using the rotation matrices

$$R_{45} = R_{-45}^{-1} = \begin{pmatrix} \frac{1}{\sqrt{2}} & -\frac{1}{\sqrt{2}} \\ \frac{1}{\sqrt{2}} & \frac{1}{\sqrt{2}} \end{pmatrix} \quad \text{and} \quad R_{45}^{-1} = R_{-45} = \begin{pmatrix} \frac{1}{\sqrt{2}} & \frac{1}{\sqrt{2}} \\ -\frac{1}{\sqrt{2}} & \frac{1}{\sqrt{2}} \end{pmatrix}. \qquad (2.184)$$

First we rotate both plates by $-45°$

$$R_{-45}^{-1} M_{\lambda/2} R_{-45} R_{-45}^{-1} M_{\lambda/4} R_{-45} E_{\text{lin}} = \frac{1}{\sqrt{2}} \begin{pmatrix} e^{i0} \\ e^{i0} \end{pmatrix}. \qquad (2.185)$$

Next both plates are rotated by $+45°$ with respect to the original orientation

$$R_{45}^{-1} M_{\lambda/2} R_{45} R_{45}^{-1} M_{\lambda/4} R_{45} E_{\text{lin}} = \frac{1}{\sqrt{2}} \begin{pmatrix} e^{i\frac{\pi}{2}} \\ e^{i\frac{\pi}{2}} \end{pmatrix}. \qquad (2.186)$$

Now we rotate the half wave plate by $+45°$ and the quarter wave plate by $-45°$

$$R_{45}^{-1} M_{\lambda/2} R_{45} R_{-45}^{-1} M_{\lambda/4} R_{-45} E_{\text{lin}}$$

$$= \frac{1}{4\sqrt{2}} \begin{pmatrix} 1 & 1 \\ -1 & 1 \end{pmatrix} \begin{pmatrix} 1 & 0 \\ 0 & -1 \end{pmatrix} \begin{pmatrix} 1 & -1 \\ 1 & 1 \end{pmatrix} \begin{pmatrix} 1 & -1 \\ 1 & 1 \end{pmatrix} \begin{pmatrix} 1 & 0 \\ 0 & -i \end{pmatrix} \begin{pmatrix} 1 & 1 \\ -1 & 1 \end{pmatrix} \begin{pmatrix} 1 \\ 1 \end{pmatrix} \qquad (2.187)$$

$$= \frac{1}{2\sqrt{2}} \begin{pmatrix} -1-i & -1+i \\ 1+i & -1-i \end{pmatrix} \begin{pmatrix} 1 \\ 1 \end{pmatrix} = \frac{1}{\sqrt{2}} \begin{pmatrix} e^{i\pi} \\ e^{i\pi} \end{pmatrix}.$$

The last rotation is $-45°$ of the $\lambda/2$-plate and $+45°$ of the $\lambda/4$-plate

$$R_{-45}^{-1} M_{\lambda/2} R_{-45} R_{45}^{-1} M_{\lambda/4} R_{45} E_{\text{lin}}$$
$$= \frac{1}{4\sqrt{2}} \begin{pmatrix} 1 & -1 \\ 1 & 1 \end{pmatrix} \begin{pmatrix} 1 & 0 \\ 0 & -1 \end{pmatrix} \begin{pmatrix} 1 & 1 \\ -1 & 1 \end{pmatrix} \begin{pmatrix} 1 & 1 \\ -1 & 1 \end{pmatrix} \begin{pmatrix} 1 & 0 \\ 0 & -i \end{pmatrix} \begin{pmatrix} 1 & -1 \\ 1 & 1 \end{pmatrix} \begin{pmatrix} 1 \\ 1 \end{pmatrix} \quad (2.188)$$
$$= \frac{1}{2\sqrt{2}} \begin{pmatrix} -1-i & 1-i \\ 1-i & -1-i \end{pmatrix} \begin{pmatrix} 1 \\ 1 \end{pmatrix} = \frac{1}{\sqrt{2}} \begin{pmatrix} e^{i\frac{3\pi}{2}} \\ e^{i\frac{3\pi}{2}} \end{pmatrix}.$$

We see that (2.185), (2.186), (2.187), and (2.188) generate four waves with consecutive mutual phase shifts of $\pi/2$.

A lateral displacement d of a *diffraction grating* [204, 205] shifts the phase of the n-th diffraction order by $\Delta\phi = 2\pi n d f$, where f is the spatial frequency of the grating, Fig. 2.37b. A moving diffraction grating also is realized in an *acoustooptical modulator* (AOM) that can be employed for phase shifting likewise.

If a plane parallel glass plate is tilted, the path through the plate changes and a phase shift is produced that depends on the thickness of the plate, its refractive index, the wavelength and the tilt angle, Fig. 2.37c. Because the exact phase shift is strongly influenced by the quality of the plate, this approach is not frequently used.

Most often the phase is shifted by a reflecting mirror mounted on a *piezoelectric transducer*, Fig. 2.37d. The piezo-crystal can be controlled electrically with high precision. A mirror shift of $\lambda/8$ corresponds to a pathlength change of $2\pi/8$ and due to the double pass results in a phase shift of $\pi/2$ in the reflected wave. If the light for illumination and/or the reference wave is transmitted through optical fibers, a common way to perform the phase shift is to wrap a portion of the fiber firmly around a piezoelectric cylinder that expands when a voltage is applied [206]. The stretching of the fiber results in a phase shift, Fig. 2.37e.

A *frequency shift* in principle can be produced the same way as the phase shift with the only difference that a continuous motion of the phase shifting component is required instead of a single step. Practically the frequency shift is realized by a diffraction grating moving with continuous velocity. This may be achieved by a rotating *radial grating*, or an AOM. An AOM or *Bragg cell* is a quartz material through which an ultrasonic wave propagates. Because this is a longitudinal or compression wave, the index of refraction of the material varies sinusoidally with the same wavelength. Incident light that is diffracted or deflected into the Bragg angle gets a Doppler shift that is equal to the sound frequency. Wavelength shifts also can be induced by applying a ramp current to a laser diode [207].

2.8 CCD- and CMOS-Arrays

Digital holography in the context treated here is based on the digital recording of holograms using CCD-arrays instead of holographic film, photorefractive crystals, photothermoplast, or others. In recent decades the development of CCDs has reached maturity, the arrays needed for our specific purposes now can be ordered off the shelf. CCD-cameras have more and more replaced conventional tube cameras in the vast majority of applications, ranging from entertainment to science and engineering.

In this chapter the fundamental operational principle of CCDs is introduced and the main performance parameters of CCD-arrays are defined. This should help to find the optimally suited CCD for the special application of digital holography. The concept of modulation transfer functions (MTF) is introduced to explain spatial sampling with CCD-arrays. Also CMOS technology and consumer cams, the color still cams, which meanwhile reached a technological standard enabling their use in digital holography, are described.

2.8.1 CCD Concept

The essential idea behind the *charge coupled devices* (CCDs) is the way the image is read off the array. Contrary to their forerunners like the microchannel plate (MCP), where individual narrow electron multiplier tubes are fused together into an imaging array, or the SIDAC using an electron beam to read the image off a photodiode array, CCDs actually transfer stored images through the detector array itself [208].

The basic building block of this technique, first proposed in [209], is the metal-oxide-semiconductor (MOS) capacitor, Fig. 2.39. A MOS capacitor typically consists of an intrinsic silicon substrate with an insulation layer of silicon dioxide on it. A thin metal electrode, called the gate, is deposited on to this layer. Applying a positive voltage across this structure causes the holes in the p-type silicon to move away from the Si-SiO$_2$ interface below the gate, leaving a region depleted of positive charges. This depletion region, that of course has not the sharp boundaries of Fig. 2.39 but a gradual shape, now can act as a potential energy well for mobile electrons.

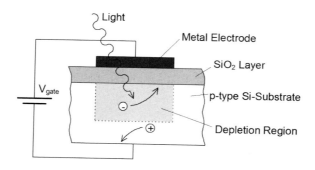

Figure 2.39: MOS capacitor.

The higher the gate voltage, the deeper the potential-energy well. If a photon of sufficient energy is absorbed in the depletion region, while voltage is applied, an electron-hole pair is generated. While the hole moves to the ground electrode the electron stays in the depletion region. The more light is absorbed by the silicon the more electrons are collected in the potential-energy well under the gate electrode. This continues until the well becomes saturated or the voltage is removed. The amount of negative charge, the electrons, that can be collected is proportional to the applied voltage, oxide thickness, and gate electrode area. The total number of electrons that can be stored is called the *well capacity* [210]. A two-dimensional array of such MOS capacitors can therefore store images in the form of trapped charge carriers beneath the gates.

2.8 CCD- and CMOS-Arrays

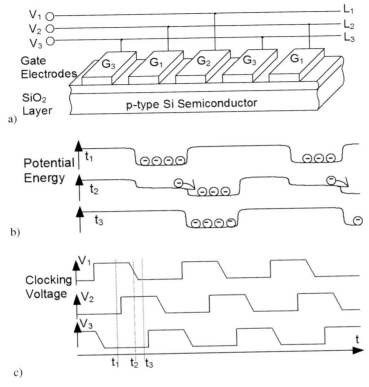

Figure 2.40: Charge transfer in three-phase CCD.

The CCD solution to move the accumulated charges off the array is to adjust the voltages to each gate of the array in a way that transfers the charges from one potential well to the neighboring one. This process is illustrated for three-phase clocking in Fig. 2.40. While a CCD register consists of a series of gates, a CCD array is a series of registers, say column registers. For three-phase clocking the column gates are connected to separate voltage lines L_1, L_2, L_3 in contiguous groups of three gates G_1, G_2, G_3, Fig. 2.40a. Initially, say at time instant t_1, a high voltage V_1 is applied to gate G_1 so that photoelectrons are collected in the wells below all the gates G_1, Fig. 2.40b. The period with which the high voltage is applied is known as the *integration time*. After integration V_1 is reduced to zero while V_2 is increased. That occurs at time instant t_2 according to Fig. 2.40c. When V_1 drops below V_2, the G_2 gates will have deeper potential wells, which draws the electrons from the G_1 wells to the G_2 wells. The same voltage sequence is repeated for lines L_2 and L_3, now the electrons are transferred from all G_2 wells to the G_3 wells. This V_1, V_2, V_3 voltage sequence is repeated to move all electrons off the array. They are read out as a serial signal by a linear MOS register called the readout register, Fig. 2.41.

In this method of charge transfer, called *line address transfer*, one row at a time exits the readout register. But whenever the high voltage level is applied to the gate to move the

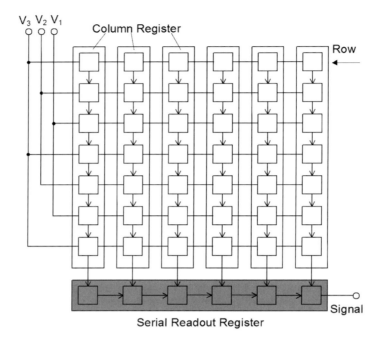

Figure 2.41: Three-phase CCD array, line address transfer.

charges, new charges can be produced by light shining on the array. Existing and new charges accumulate and cause image smearing. A reasonable solution is to mechanically shutter the array between the integration times. A more efficient solution is the *interline transfer* CCD. Here columns of MOS capacitors, the vertical shift registers, are interlaced between the light collecting columns of pixels, Fig. 2.42. These shift registers are covered by opaque metal shields. While the stored charges are transfered off the array over the light protected shift and readout registers, the light sensitive pixels may integrate a new image.

A third charge transfer method is *frame transfer*. Frame transfer CCDs quickly move all the accumulated charges into an adjacent shielded storage array the same size as the light sensitive pixel array, Fig. 2.43. While the pixels integrate the next image, the stored image is clocked off to the readout register.

Frame transfer CCDs as well as the line-address transfer CCDs are slower than interline transfer CCDs. Image smear is least with interline transfer. But the interline transfer concept leaves less room at the array area for active sensors. The shields obscure half the information that is available in the scene. The area fill factor may be as low as 20 percent. In this case the output voltage is only 20 percent of a detector that would completely fill the pixel area. With the help of microlenses concentrating light of a larger area onto the sensitive area, the fill factor may be optically increased [210].

Holography of moving scenes normally demands the employment of pulsed lasers. Traditional camera tubes had a readout time of 19 ms on a total readout cycle of 20 ms according

2.8 CCD- and CMOS-Arrays

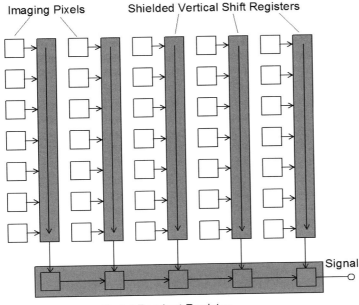

Figure 2.42: Interline transfer CCD array.

to the CCIR standard. So the laser pulses had to be triggered onto the 1 ms blanking period just before the readout of an even field [211]. With CCD arrays the image is read out much faster. In the frame transfer arrays the image is transferred from the photosensitive area to the shielded area in typically 300 µs. With interline transfer arrays the charge transfer to the shielded shift registers is a single step process and thus even shorter: typically 1 µs, which makes pulse timing even more flexible. It may be advantageous to reset the array prior to the generation of the laser pulse to get rid of the ghost image problem [211]. A ghost image may occur if some of the incident photons are detected by the supposedly light-shielded shift registers, because of diffraction, refraction, or multiple reflections.

The ghost image problem should not be confused with *blooming*. This effect occurs if very bright parts of scenes, e. g. specular reflections, oversaturate some of the pixels and cause spill-over of charge to neighboring pixels. This results in a local loss of spatial information. Some CCD arrays are equipped with antiblooming drains, where excess carriers produced by local overload are disposed so that the neighboring pixels are not affected by the oversaturation.

Besides the three-phase clocking scheme described here there exist other clocking schemes, like two-phase clocking, four-phase clocking and even more specific schemes which can be used for electronic shuttering, iris control, or image-scanning functions such as interlaced video fields.

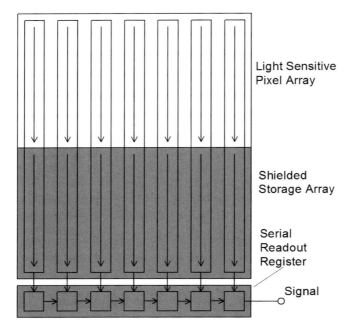

Figure 2.43: Frame transfer CCD array.

CCDs can be illuminated either from the front or the back. In front-illuminated CCDs, Fig. 2.44a, the light enters through the gate electrodes, which must therefore be produced of transparent material such as polysilicon. Back-illuminated CCDs avoid interference and absorption effects caused by the gates but must be thinned for imaging in visible or near-infrared regions of the spectrum, Fig. 2.44b.

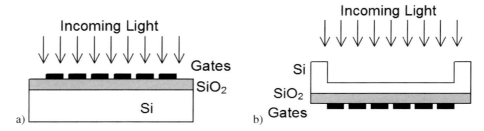

Figure 2.44: Front-illuminated (a) and back-illuminated (b) CCD arrays.

2.8.2 CCD Array Performance Parameters

CCD arrays exhibit a number of features which influence their performance. To select the best array for a specific application, e. g. digital holography, one has to know these features

2.8 CCD- and CMOS-Arrays

and their consequences on the application. A number of parameters precisely quantify these features.

The rate of conversion of the incident photons into electrons is the *quantum efficiency*. Assuming an ideal material and a photon energy higher than the semiconductor band gap energy, then each photon would produce one electron-hole pair. We would have a quantum efficiency of one. However, the absorption coefficient is wavelength dependent and decreases with increasing wavelengths. Therefore long wavelength photons may even pass through the CCD without being absorbed. Any photon absorbed within the depletion region will yield a quantum efficiency near unity. For photons absorbed in the bulk material their contribution depends on the diffusion length. A diffusion length of zero implies immediate recombination of electrons created in the bulk material and a quantum efficiency going to zero for these wavelengths. On the other hand for a diffusion length of nearly infinity the electrons eventually reach the charge well and contribute to a quantum efficiency approaching one. Since the quantum efficiency depends upon the gate voltage and material thickness, the *spectral responsivity* differs between front-illuminated and back-illuminated CCD arrays. Typical curves are shown in Fig. 2.45.

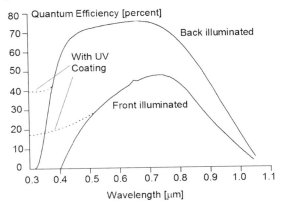

Figure 2.45: Typical spectral response of front- and back-illuminated CCD arrays.

The quantum efficiency of the back-illuminated CCD generally is higher, mainly since the photons do not have to pass the polysilicon gate. The response in the ultraviolet may be enhanced with UV fluorescent phosphors deposited onto the array. When they are excited by light of wavelengths between 0.12 μm and 0.45 μm these phosphors emit light at approximately 0.54 μm to 0.58 μm, which is efficiently absorbed, .

Ideally the CCD output is proportional to the exposure. But in practice for long exposure times as used for low-light-level operation, the *dark current* may become a problem. Sources of dark current are thermal generation and diffusion of electrons in the neutral bulk material and in the depletion region or due to surface states. Typical dark current densities vary between 0.1 nA/cm^2 and 10 nA/cm^2 in silicon CCDs. As an example, in a pixel of dimension 24 μm × 24 μm a dark current of 1000 pA/cm^2 produces 36,000 electrons/(pixel·sec). If the well capacity is 360,000 electrons, the well fills in 10 seconds [210]. Dark current due to thermally generated electrons can be reduced by cooling the device. Some arrays contain shielded extra pixels at the end of the array. These establish a reference dark current level, that can be subtracted from each active pixel value as a first crude correction.

Besides the dark current there are other noise sources influencing the performance of the CCD array. *Shot noise* is due to the discrete nature of electrons. It always occurs when photoelectrons are created and when dark current electrons are present. Shot noise is a random process that can be modeled by a Poisson distribution. Additional noise is added when reading the charge (*reset noise*) or may be introduced by the amplifier (*1/f noise* and *white noise*). If the output is digitized, *quantization noise* has to be regarded. *Fixed pattern noise* refers to the pixel-to-pixel variation that occurs when the array is in the dark. It is a signal-independent additive noise. On the other hand *photoresponse nonuniformity*, due to differences in the responsivity, is a signal-dependent multiplicative noise. For a more thorough discussion of these noise sources the reader is refered to [210].

The dependence of the output signal voltage on exposure is shown in an idealized sketch in Fig. 2.46. The slope of this transformation in the linear range between dark current and saturation is the average *responsivity* of the array. It has the units of volts/(joule cm^2). Cameras intended for consumer video indicate their responsivity normalized to photometric units. The units of radiation are normalized to the spectral sensitivity of the human eye. In this case the responsivity is given in volts/lux.

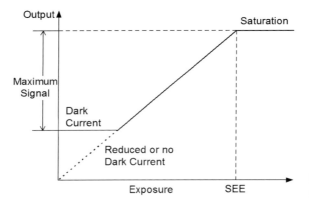

Figure 2.46: Dependence of output voltage on exposure.

The *noise equivalent exposure* (NEE) is the exposure that produces a signal-to-noise ratio of one. For CCD arrays, NEE is the noise value in rms electrons. The minimum noise level is the noise floor, this value is used most often for the NEE. The *saturation equivalent exposure* (SEE) is that input that fills the charge wells completely, Fig. 2.46. With these definitions of NEE and SEE the *dynamic range* DR is defined as the maximum signal divided by the rms noise

$$DR = \frac{SEE}{NEE} = \frac{N_{\text{well}}}{\langle n_{\text{sys}} \rangle} = \frac{V_{\text{max}}}{V_{\text{noise}}}. \tag{2.189}$$

Instead of SEE and NEE we can use the well capacity N_{well} in electrons divided by the system noise $\langle n_{\text{sys}} \rangle$ measured in electrons rms or the voltages corresponding to the maximum and the noise signal. With $DR = 20\log(SEE/NEE)$ dB the dynamic range is expressed in decibels.

The exposure can be varied by changing the integration time, therefore an *integration time control* improves the application of CCDs in many areas. Reduction of the integration time is

2.8 CCD- and CMOS-Arrays

interesting when objects are studied with a cw laser under unstable experimental conditions, provided that enough laser intensity is available. When pulsed lasers are applied, shuttering synchronously to the laser pulses is a way to remove background light, which otherwise would destroy the dynamic range for the signal of interest [211].

Both interline transfer and frame transfer CCD cameras are offered with integration times adjustable from about 1/50,000 s to arbitrarily long exposures, while retaining the TV standard readout. The latter is an advantage because standard video equipment then still can be used to transmit and record the camera signal. For extended integration this means that the integration time always is an integer times the standard integration time, empty television frames are presented at the output while integrating [211].

The standard video synchronization pulses from a commercial television camera often are not precisely locked to the pixel clocks, which control the readout of the CCD array. This effect is called *jitter*. Jitter can cause substantial errors when the video signal is being digitized. Especially when two digital holograms have to be compared in holographic interferometry it is extremely important that pixels have the same positions in subsequent digitized frames. This can be ensured by synchronizing the frame grabber to a pixel clock signal from the camera, which is often offered as an option.

A useful measure of performance of CCD arrays, especially if they are compared to hologram recording emulsions, see Section 2.7.2, is the *space-bandwidth product*. According to (A.61) the space-bandwidth product of a CCD device is defined as the product of the dimensions of the device and the pixel frequency. Therefore a conventional hologram with, say, 3000 linepairs/mm and a size of 20 cm^2 has a space-bandwidth product in the range of 10^{10}, while CCDs with 1000 – 2000 pixels in each direction and pixel sizes down to 7 μm exhibit a space-bandwidth product in the range of 10^6 to 10^7 [212]. The consequences using CCDs as recording media in holography are described in Section 3.1.1.

2.8.3 CMOS Image Sensors

In recent years CMOS image sensors have entered the market. The CMOS (*complementary metal oxide semiconductor*) devices have complementary pairs of p- and n-type transistors. While CCDs are specialized chips, used only for image capture, they are manufactured by only a handful of very specialized fabrication facilities. On the other hand CMOS technology is used in the vast majority of electronic devices. Because CMOS tends to be produced using standard processes in high volume facilities, the economies of scale are such that CMOS image sensors are considerably less expensive to manufacture [213].

In the past CMOS sensors exhibited less light sensitivity than CCDs and significantly more noise. But progress in technology has closed this gap, so now CMOS image sensors have found their way into consumer and professional digital video and still cameras.

At its most basic, an image sensor needs to achieve five key tasks: absorb photons, generate a charge from the photons, collect the charge, transfer the charge, and convert it to a voltage. Both CCD and CMOS sensors perform all five tasks. The first three tasks are performed similarly, but they diverge in their methods of charge transfer and voltage conversion. A key characteristic of CMOS devices is that they consume negligible power when they are doing nothing other than storing a one or zero; significant power consumption is confined to the times when a CMOS is switching from one state to another. CMOS image sensors can have much more functionality on-chip than CCDs. In addition to converting photons to electrons and transferring them, the CMOS sensor might also perform image processing, edge detection, noise reduction, or analog to digital conversion, to name just a few. This functional integration onto a single chip is the CMOS's main advantage over the CCD. It also reduces the number of external components needed. In addition, because CMOS devices consume less power than CCDs, there is less heat, so thermal noise can be reduced.

As far as digital holograms are recorded, there is no principal difference whether we use a CCD or a CMOS image sensor, so in the following the term CCD chip refers to both CCD and CMOS image sensors.

2.8.4 Spatial Sampling with CCD-Arrays

Spatial resolution describes the ability of an imaging device to resolve image details. The main parameters which determine spatial resolution are *pixel size* and *pixel number*. Normally we have rectangular arrays of light sensitive pixels organized in N lines and M rows. Often $N = M$ holds. The center-to-center spacing of the pixels, the pixel pitch, is $\Delta\xi$ and $\Delta\eta$ in the two orthogonal directions. If there is a gap between consecutive pixels, the effective pixel dimensions are $\alpha\Delta\xi$ and $\beta\Delta\eta$ with $\alpha \leq 1.0$ and $\beta \leq 1.0$ in the two directions, see also Fig. 3.41. The α and β are the *fill-factors* in the ξ- and η-directions. For square pixels $\alpha\Delta\xi = \beta\Delta\eta$ and with identical spacing in vertical and in horizontal directions $\Delta\xi = \Delta\eta$, the fill factor is the square root of the light sensitive detector area divided by the whole array area. According to the sampling theorem the signals can be reproduced faithfully from data recorded by the array up to spatial frequencies

$$f_{N\xi} = \frac{1}{2\Delta\xi} \quad \text{and} \quad f_{N\eta} = \frac{1}{2\Delta\eta} \qquad (2.190)$$

the Nyquist frequencies in both directions.

A tool for analyzing the capability of imaging devices to record and reproduce spatially varying signals is the *modulation transfer function* (MTF). The MTF is defined as the ratio of output and input modulation at each frequency

$$MTF = \frac{M_{\text{output}}}{M_{\text{input}}} \qquad (2.191)$$

where the modulation M is given by

$$M = \frac{V_{\max} - V_{\min}}{V_{\max} + V_{\min}} \qquad (2.192)$$

with $V_{\max}$ and $V_{\min}$ denoting the maximum and minimum signal levels. This concept is presented in Fig. 2.47 for three input signals [210]. The MTF of a system indicates how the various frequencies are affected by the system.

The influence of the spatial integration on the modulation is illustrated in Fig. 2.48. In Fig. 2.48a we have a very small, ideally a pointwise, sampling detector. The maximum and minimum input and output amplitudes agree, and we obtain an $MTF = 1$. Figure 2.48b shows the sampling with a CCD array whose pixel width is one-half of the center-to-center

2.8 CCD- and CMOS-Arrays

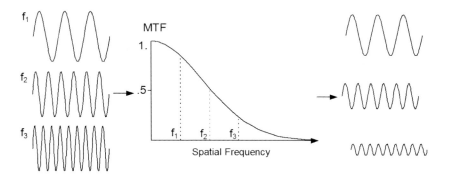

Figure 2.47: Modulation transfer function applied to three frequencies.

spacing. The heavy lines indicate the pixel output. The maximum output amplitude is less than the maximum input, and we have an $MTF < 1$. For an array with a fill-factor of one, Fig. 2.48c, the MTF is even lower. We see that with increasing pixel width Δ the MTF decreases.

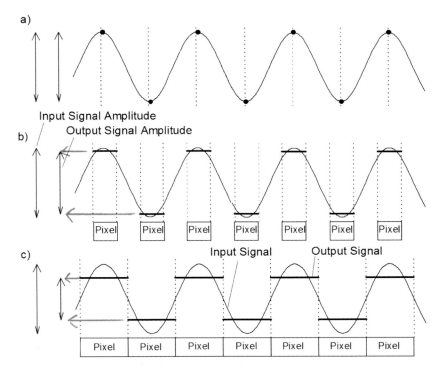

Figure 2.48: Sampling with different spatial integration.

Furthermore each individual pixel has an optical MTF, which in one dimension is [210]

$$MTF_{\text{pixel}}(f) = \frac{\sin(\pi \Delta f)}{\pi \Delta f}. \qquad (2.193)$$

This is shown for normalized spatial frequencies Δf in Fig. 2.49. The MTF is zero when $f = k/\Delta$, $k \in \mathbb{N}$. The first zero ($k = 1$) defines the pixel cutoff frequency f_C; higher frequencies cannot be resolved by such a pixel.

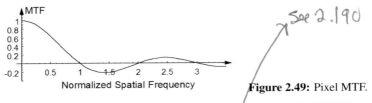

Figure 2.49: Pixel MTF.

A CCD-array will sample with a sampling rate d_{CC} denoting the center-to-center distance between the individual pixels of width Δ. The fill-factor for square detectors with equal spacing in horizontal and vertical directions is $(\Delta/d_{CC})^2$. According to the sampling theorem signals up to the Nyquist frequency $f_N = 1/(2d_{CC})$ can be faithfully reproduced. Figure 2.50 gives a plot of the MTF up to the sampling frequency $f_S = 1/(d_{CC})$ (here normalized to one) which agrees with the pixel cutoff frequency in the case $\Delta = d_{CC}$. For fill-factors less than one the sampling frequency is below the pixel cutoff frequency but still the Nyquist frequency is half the sampling frequency. Summarizing, we can state that in each case the Nyquist frequency defines the upper limit for faithful reconstruction.

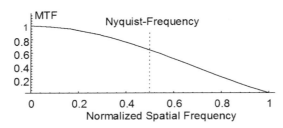

Figure 2.50: MTF for Nyquist and sampling frequencies.

2.8.5 Color Still Cameras

The biggest market for digital image sensors is digital photo-cameras and digital video-cameras. Mass production enables large numbers of items and thus low prices. The sensors needed in this realm generally are *color sensors*. The monochrome sensors, although less complex, have only a limited market, therefore their significantly higher price.

Basically all image sensors are grayscale devices that record the intensity of light from black to white with the appropriate intervening gray. To sensitize the sensors to color, a layer of color filters is bonded to the silicon using a photolithography process to apply color dyes. Image sensors employing micro lenses have the color filter between the micro lens and the photodetector. Some expensive high-end digital cameras use three array image sensors with

2.8 CCD- and CMOS-Arrays

the incoming light being split by properly arranged prisms. For this approach it is easy to coat each of the three sensors with a separate color. But nearly all consumer or professional digital still cameras today are single sensor devices, which use a color filter array (CFA).

Theoretically color is described in terms of what are called *tristimulus values* X, Y, Z:

$$\begin{align}
X &= k \sum \phi(\lambda) \bar{x}(\lambda) \, \Delta\lambda \\
Y &= k \sum \phi(\lambda) \bar{y}(\lambda) \, \Delta\lambda \\
Z &= k \sum \phi(\lambda) \bar{z}(\lambda) \, \Delta\lambda.
\end{align} \qquad (2.194)$$

Here $\bar{x}, \bar{y}$, and $\bar{z}$ are the *color matching functions*, and they represent the sensitivities of human visual systems' three color channels, Fig. 2.51.

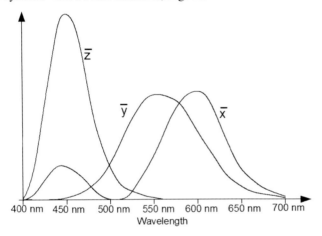

Figure 2.51: Color matching functions.

The term $\phi(\lambda)$ is the stimulus function, which is $\phi(\lambda) = S(\lambda)R(\lambda)$ for opaque objects, $\phi(\lambda) = S(\lambda)T(\lambda)$ for transparent objects, and $\phi(\lambda) = S(\lambda)$ for light sources. In these formulas $S(\lambda)$ is the spectral power distribution of the light source, $R(\lambda)$ is the spectral reflectance of the opaque object, and $T(\lambda)$ is the spectral transmittance of the transparent object. $\Delta\lambda$ is the wavelength resolution of the color-matching and stimulus functions and k is a normalizing constant [214]. In a similar way image detectors sense three primary colors, mostly red, green, and blue (RGB), Fig. 2.52, or yellow, cyan, and magenta (YCM).

There are principally two distinct categories of color filter design: the relatively new *vertical color filter detector* and the *lateral color filter detector* [215]. The vertical color filter detector features three separate layers of photodetectors embedded in silicon, Fig. 2.53a.

Each layer captures a different color, since slab silicon absorbs different colors of light at different depths, Fig. 2.53b. Stacked together, full-color pixels capture red, green, and blue light at every lateral pixel location, Fig. 2.53c. The more conventional lateral color filter detector uses a single layer of photodetectors with the color filters arranged in a tiled mosaic pattern, Fig. 2.54a.

Here each filter lets only one color of light – e. g. red, green, or blue – pass through, allowing the pixel to record only one color, Fig. 2.54b. The most common arrangement of colors in the CFA is that of the *Bayer array* [216]. In the Bayer array the sensor captures a mosaic of

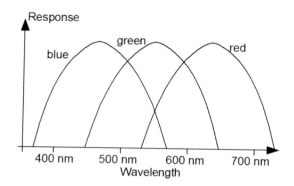

Figure 2.52: Color filter response.

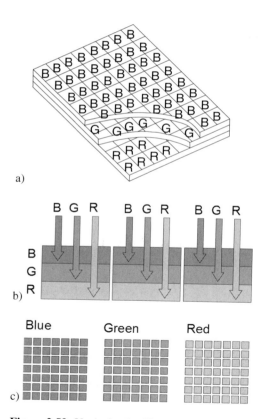

Figure 2.53: Vertical color filter detector.

2.8 CCD- and CMOS-Arrays

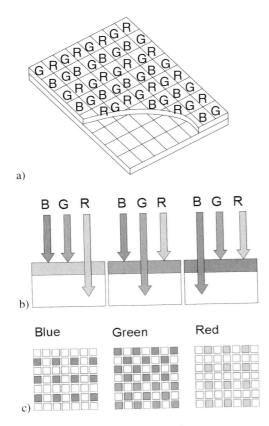

Figure 2.54: Lateral color filter detector.

25% of the red, 25% of the blue, and 50% of the green light, Fig. 2.54 c. This accords with the principle that the luminance channel (here the color green) needs to be sampled at a higher rate than the chrominance channels (red and blue). The choice of green as the representative for the luminance is due to the fact that the luminance response curve of the human eye peaks at around the frequency of green light (around 550 nm) [217]. Generally *luminance* is defined as any positive weighted sum of the primary colors R, G, and B. The weights are chosen so that luminance approximates the photopic function $V(\lambda)$ which represents the human luminosity sensation [216]. The luminance record contains most of the spatial information of the image. On the other hand *chrominance* contains the color information.

Since in single layer sensors with lateral CFAs like the Bayer array at each pixel only one spectral measurement is made, the other colors must be estimated using information from all the color planes in order to obtain a high resolution color image. This process is often referred to as *demosaicking* [217]. Demosaicking algorithms can be categorized into three different groups:

- Simple straightforward interpolation algorithms using surrounding cells of the same color channel, like linear filtering, nearest neighbor interpolation, and median filtering.

- Advanced interpolation algorithms that use surrounding cells of all three channels and take into account spatial information about the image data.
- Reconstruction techniques that use information about the capturing device itself and that make certain assumptions about the captured scene.

The goal of the demosaicking process can be to reconstruct a color image that comes as close as possible to a full resolution image, or aims at maximum perceptual quality [218]. However, in the context of this book, using monochromatic laser light, only luminance is the interesting measure.

3 Digital Recording and Numerical Reconstruction of Wave Fields

Here an introduction to the technique called digital holography is given. In this book the term 'digital holography' is understood as the digital recording of the hologram field by CCD- or CMOS-arrays and the numerical reconstruction of the wave fields in a computer. This should not be confused with the computer generation of holograms which then are printed onto physical media and from which the wave fields are reconstructed optically – an approach also called digital holography occasionally [219]. In this chapter the preliminaries as well as the main methods of numerical reconstruction are presented, forming the basis for digital holographic interferometry to be introduced in more detail in Section 5.8.

3.1 Digital Recording of Holograms

This section presents the way to record Fresnel holograms as well as Fraunhofer holograms with common CCD- or CMOS-arrays. In the discussion we will address only CCD-arrays, although also CMOS-arrays are meant. It is the angle between object wave and reference wave that must be controlled carefully in order to produce holograms which can be resolved by a given CCD-array. Optical methods for reducing the wave fields emanating from large objects are described which enable also the recording of such wave fields. A variety of possible reference waves and their numerical description are revealed. Based on the given number and spacing of the pixels of the CCD and based on the size of the object to be recorded, this section supplies the necessary information to design the optical arrangement for recording successfully digital holograms.

3.1.1 CCD Recording and Sampling

The aim of *digital holography* is the employment of CCD-arrays for recording holograms which are then stored in a computer memory and can be reconstructed numerically. As presented in detail in Section 2.6 the hologram is the microscopically fine interference pattern generated by the coherent superposition of an object and a reference wave field. The spatial frequency of this interference pattern is defined mainly by the angle between these two wave fields, (2.30) and (2.32). A typical geometry for recording a digital hologram is shown in Fig. 3.1.

Let the CCD-array have $N \times M$ light sensitive pixels with distances $\Delta \xi$ and $\Delta \eta$ between the pixel centers in the x- and y-directions, respectively. Without restriction of generality we

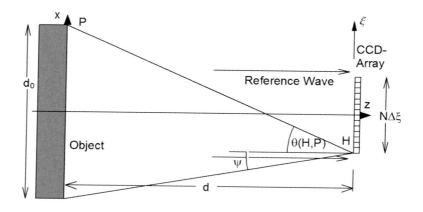

Figure 3.1: Geometry for recording a digital Fresnel hologram.

can assume that $N = M$ and $\Delta\xi = \Delta\eta$. Since the x- and y-directions, and in the same way the ξ- and η-directions, are equivalent, it is sufficient to perform the analysis only in the x-z-plane. For our analysis we assume a plane reference wave impinging normally onto the CCD, which constitutes the most frequently used arrangement, Fig. 3.1. θ is the angle at point H between the reference and object wave emitted from an object point P. The distance p between two consecutive interference fringes forming the hologram at H was seen in (2.30) and (2.32) to be

$$p = \frac{\lambda}{2\sin(\frac{\theta}{2})}. \tag{3.1}$$

A meaningful sampling of the intensity distribution constituting the hologram is only guaranteed if the *sampling theorem* (A.50) is obeyed. The sampling theorem requires that the period p must be sampled with more than two pixels, implying more than two pixels fitting into the distance p

$$p > 2\Delta\xi. \tag{3.2}$$

If one pleases, this can also be expressed using the spatial frequency f of the holographic fringes

$$f < \frac{1}{2\Delta\xi}. \tag{3.3}$$

In practical applications the CCD-array with parameters $N, M, \Delta\xi, \Delta\eta$ is given, so it sets a limit to the angle θ. Because θ in all practical cases remains small, we can use $\sin\theta/2 = \tan\theta/2 = \theta/2$ in the calculations. From (3.1) and (3.2) we obtain an upper limit to the angle θ

$$\theta < \frac{\lambda}{2\Delta\xi} \tag{3.4}$$

3.1 Digital Recording of Holograms

or, with the definition $\theta_{\max} = \max\{\theta(H,P) : H,P\}$, see Fig. 3.1

$$f < \frac{2}{\lambda} \sin\left(\frac{\theta_{\max}}{2}\right). \tag{3.5}$$

The following table lists the spatial resolution of some frequently used holographic recording media. These are holofilm on the base of silver halides, photothermoplastic film, and the CCD-arrays of the frequently used videocameras Kodak Megaplus 1.4 and Megaplus 4.2. The maximum angle θ is given for these materials assuming the employment of a helium-neon laser with a wavelength $\lambda = 0.6328$ µm.

Recording material	Resolution	Maximum angle
Holofilm (Silver halide)	up to 7000 line pairs/mm	arbitrary
Photothermoplast	750 – 1250 line pairs/mm	27° to 47°
Megaplus 1.4 ($\Delta\xi = 6.8$ µm)	73 line pairs/mm	2.67°
Megaplus 4.2 ($\Delta\xi = 9.0$ µm)	55 line pairs/mm	2.01°

The conclusion is that we can employ CCD-arrays to record holograms as long as the angle between reference wave and object wave remains small enough so that the sampling theorem is fulfilled. The restricted angles are obtained either by objects of small lateral dimensions or by objects placed far away from the CCD-target. A further solution to this problem is presented in the next section.

For the typical geometry of Fig. 3.1 with the plane object of lateral extension d_0 in the x-direction placed symmetrically to the optical axis, and a plane reference wave travelling along the optical axis and impinging orthogonally onto the CCD, we can calculate the maximum object width d_0 for each distance d. According to Fig. 3.1 we have

$$\tan\theta = \frac{\frac{d_0}{2} + \frac{N\Delta\xi}{2}}{d}. \tag{3.6}$$

Together with the maximum angle of (3.4) we obtain [since $\tan\theta \approx \theta$]

$$\frac{\frac{d_0}{2} + \frac{N\Delta\xi}{2}}{d} < \frac{\lambda}{2\Delta\xi} \qquad \text{[To fit in CCD]} \tag{3.7}$$

which is resolved with respect to d_0 to yield the limit to the lateral extension [Permissible object Size at a distance]

$$d_0(d) < \frac{\lambda d}{\Delta\xi} - N\Delta\xi. \tag{3.8}$$

In its other form this inequality defines the minimum distance between an object of given lateral width d_0 and the recording target [Object Distance as a function of object size]

$$d(d_0) > \frac{(d_0 + N\Delta\xi)\Delta\xi}{\lambda}. \tag{3.9}$$

In this derivation we have claimed that all pixels of the CCD-array together with all object points fulfill the sampling theorem. A weaker requirement is that for each point of the object at least one pixel fulfills the sampling theorem. This leads to the angle ψ, see Fig. 3.1, with

$$\tan\psi = \frac{\frac{d_0}{2} - \frac{N\Delta\xi}{2}}{d} \tag{3.10}$$

and instead of (3.8) to

$$d_0(d) = \frac{\lambda d}{\Delta\xi} + N\Delta\xi \tag{3.11}$$

or conversely

$$d(d_0) > \frac{(d_0 - N\Delta\xi)\Delta\xi}{\lambda}. \tag{3.12}$$

In this case slightly larger objects are admitted. But as a consequence the information about marginal points of the object which make full use of these limits are faithfully stored only in a few hologram points. As a result they appear noisy, with weak contrast, and unreliable in the reconstruction. It must be thought over carefully whether one is willing to pay this price for having an object of 5.3 cm width in 50 cm distance instead of one 4 cm wide (assuming $\lambda = 0.6328$ μm, $N = 1024$, $\Delta\xi = 6.8$ μm) while on the other hand the larger object also can be recorded with full resolution in all pixels if it stands 65 cm apart from the CCD instead of 50 cm. As a rule of thumb the distance between object and CCD must be roughly at least

$$d > d_0 \frac{\Delta\xi}{\lambda}. \tag{3.13}$$

We have seen here that the pixel size $\Delta\xi \times \Delta\eta$ plays a crucial role in determining the maximum allowable angle between object and reference waves and we will see later on that spatial resolution and speckle size also depend on pixel size. Although CCD- or CMOS-arrays with very small pixels – comparable to the resolution of holographic plates – are not offered on the market today, Jacquot et al. [120, 121] have found a technical solution. In the recording plane they position an opaque screen containing an array of 128 transparent apertures each of 2 μm × 2 μm size with a pitch of 8 μm in both directions, Fig. 3.2. The light transmitted by the transparent apertures is imaged onto the CCD by a magnification objective. In a sequential acquisition process 4×4 digital holograms are recorded with the mask shifted by a piezo translation stage in 2 μm steps in the ξ- and η-direction. The final image is composed from these 16 sub-images and acts like a digital $N\Delta\xi \times M\Delta\eta$-hologram with $\Delta\xi = \Delta\eta = 2$ μm and $N = M = 4 \times 128 = 512$.

3.1.2 Reduction of the Imaging Angle

In practical applications of holographic metrology we often have objects with large surfaces. A surface of 50 cm lateral dimension would require a distance of at least 5.4 m between object and CCD if we use $\lambda = 0.6328$ μm, $N = 1024$, $\Delta\xi = 6.8$ μm. This is not a feasible distance to record holograms, since the refractive index of air may vary significantly during recording

3.1 Digital Recording of Holograms

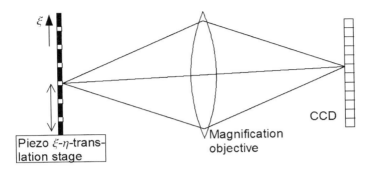

Figure 3.2: Decreasing the pixel size by a mask.

or between the records to be compared interferometrically, vibration isolation may become more difficult, or simply the room in the laboratory is restricted. But in such cases the wave field reflected from the object's surface can be drastically reduced by using a lens [59, 60]. Figure 3.3 displays this for a concave (negative) lens. The object wave field impinging onto

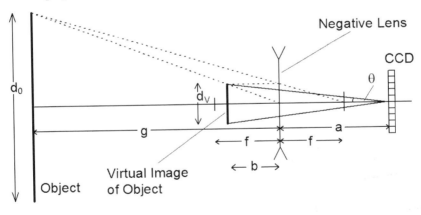

Figure 3.3: Reduction of the imaging angle by a concave lens.

the CCD target seems to come from the small virtual image of the object and not from the large object itself. Although the angle between rays from edge points of the large object and the normally impinging plane reference wave is too large for producing holographic interferences which fulfill the sampling theorem, the angle β between rays coming from the virtual object and the reference wave is much smaller so that it meets requirement (3.4).

In typical applications the object is given; let its lateral dimension in the direction investigated here be d_O. The results in the same way are valid for the other lateral coordinate direction. The CCD-array with its specified pixel distance $\Delta\xi$ and pixel number N together with the wavelength λ of the available laser defines the angle θ. Furthermore the focal length f of the lens is provided, which is negative for the concave lens. From these quantities we can calculate the distance a the lens must be from the CCD and the distance g of the object from the lens, see Fig. 3.3. The calculation is based on the *lens formula*

$$\frac{1}{f} = \frac{1}{g} - \frac{1}{b} \tag{3.14}$$

where b is the distance of the virtual image from the lens and the *magnification formula*

$$M_T = \frac{d_V}{d_O} = -\frac{f}{g-f}. \tag{3.15}$$

Here M_T denotes the transversal magnification and d_V is the lateral extension of the virtual image. With $\tan\theta = d_V/[2(a+b)]$, $d_V = -d_O f/(g-f)$, and $b = gf/(f-g)$ we obtain

$$a = \frac{-d_O f}{(g-f)2\tan\theta} + \frac{fg}{g-f}. \tag{3.16}$$

In the reconstruction stage of digital holograms recorded in this way, of course one has to consider the distance between the CCD-array and the small virtual image of the object which now is

$$d = a + b \tag{3.17}$$

instead of the object's distance $g + a$.

Equation (3.16) shows that different distances g between object and lens correspond to different distances a between lens and CCD. This is illustrated for three object positions in Fig. 3.4. We see that the object size is reduced by the magnification factor M_T but the distances b are not reduced by the same factor. For them the longitudinal magnification factor M_L applies, which is a consequence of the lens formula. We have $M_L = -M_T^2$ [220]. The minus sign accounts for the reversal of the position order. The difference between lateral or transverse magnification and longitudinal magnification is important for holographic contour measurements when the object size is reduced by a lens.

For practical applications we have to fix the arrangement of object, lens, and CCD by selecting an optimal choice of g and z. It is often helpful to calculate some tables. For some available lenses and a given object dimension d_O we list the width of the virtual image d_V and the distance b of the virtual image from the lens in dependence on some typical lens–object distances g. The underlying equations are $d_V = -d_O f/(g-f)$, and $b = gf/(f-g)$.

3.1 Digital Recording of Holograms

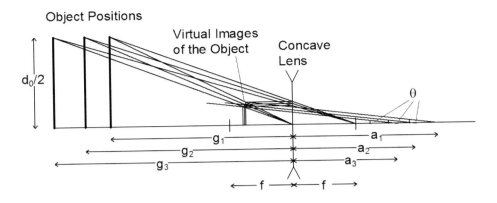

Figure 3.4: Influence of the object distance.

$f = -50$ mm $d_0 = 130$ mm			$f = -100$ mm $d_0 = 130$ mm			$f = -200$ mm $d_0 = 130$ mm		
g [mm]	d_V [mm]	b [mm]	g [mm]	d_V [mm]	b [mm]	g [mm]	d_V [mm]	b [mm]
20	92.9	14.3	20	108.3	16.6	20	118.2	18.2
50	65.0	25.0	50	86.7	33.3	50	104.0	40.0
100	43.3	33.3	100	65.0	50.0	100	86.7	66.7
200	26.0	40.0	200	43.3	66.7	200	65.0	100.0
300	18.6	42.9	300	32.5	75.0	300	52.0	120.0
400	14.4	44.4	400	26.0	80.0	400	43.3	133.3
500	11.8	45.5	500	21.7	83.0	500	37.1	142.9
600			600	18.6	85.7	600	32.5	150
700			700	16.2	87.5	700	28.9	155.6
800			800	14.4	88.9	800	26.0	160.0

Equations (3.8) and (3.11) now are used to calculate $d_V = d_0(d)$ to enter one of these tables. The distance d of the virtual image from the CCD here is composed of the individual distances $d = b + a$.

Instead of a concave lens we also may use a convex lens, Fig. 3.5. Equation (3.16) now changes to

$$a = \frac{+d_0 f}{(g-f) 2 \tan \theta} + \frac{fg}{g-f}. \tag{3.18}$$

The focal length f now is positive, the image is real and appears upside down. Generally if we use a concave lens the total length $g + a$ of the arrangement is shorter. Therefore the use of concave lenses for reducing the object wave field is recommended. On the other hand if we have to magnify very small objects in holographic microscopy applications it is recommended to produce a magnified virtual image using a convex lens.

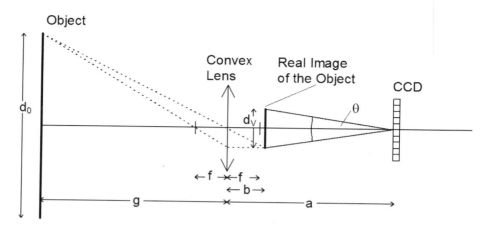

Figure 3.5: Reduction of the imaging angle by a convex lens.

Another way for reducing a large imaging angle caused by extended objects is to introduce an aperture of appropriate dimensions between object and CCD, Fig. 3.6 [221, 222].

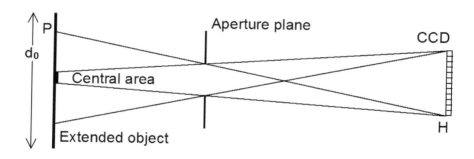

Figure 3.6: Aperture reducing the imaging angle.

This aperture limits the angles of the light rays scattered from the object points to within acceptable values. The aperture, by blocking those ray bundles making relatively large angles with the reference beam, serves to restrict the maximum angle between object and reference beams. A digital hologram of a large object could thus be recorded. But one must notice that not all points of the imaged area of the object surface contribute equally to any CCD-pixel. In Fig. 3.6 we recognize that the central area contributes to all hologram points, while e. g. object point P illuminates only point H of the hologram. Moreover the aperture may cause vignetting with the result that not the whole object wave field is stored in the recorded digital hologram. Multiple recordings with shifted positions of the aperture in its plane and a

3.2 Numerical Reconstruction by the Fresnel Transform

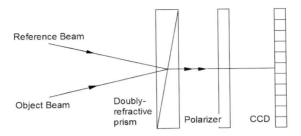

Figure 3.12: Beam combination by doubly refractive prism.

3.2 Numerical Reconstruction by the Fresnel Transform

In this chapter the *numerical reconstruction* of digitally recorded holograms is addressed. Although the captured digital hologram can be fed to a spatial light modulator enabling an optical reconstruction of the recorded wave field [226, 227], here we concentrate on the numerical calculation of the wave field. The general coordinate system for the description of the generation of holograms as well as the reconstruction of the real and virtual images is introduced. The theoretical tool for the numerical reconstruction is the scalar diffraction theory that was already introduced in Section 2.4. In this section we confine ourselves to the Fresnel approximation of the diffraction integral which is feasible due to the sufficient distance between object and CCD-array, and which on the other hand enables one easily to handle reconstruction procedures. The continuous formulas are transferred into finite discrete algorithms which can be implemented in digital image processing systems. The parameters of the resulting numerically expressed wave fields and the corresponding digital images are specified, the roles of the real and virtual images are explained, and the influence of the various possible reference waves is studied. The d.c.-term in the reconstructed image is analyzed and a method to suppress the d.c.-term is presented. At this point we will encounter the concept of negative intensities, which becomes possible numerically, but has no physical counterpart in real-world optics.

3.2.1 Wave Field Reconstruction by the Finite Discrete Fresnel Transform

Let us assume an opaque diffusely reflecting object that is illuminated by a coherent wave field. The object surface contour may be denoted by $F(x, y, z) = 0$, it is illuminated by the wave field $E(x, y, z) = |E(x, y, z)|e^{i\alpha(x, y, z)}$. The reflection of the surface is described by the complex reflection coefficient $b(x, y, z)$

$$b(x, y, z) = |b(x, y, z)|e^{i\beta(x, y, z)} \tag{3.27}$$

where b and β indicate the variation of amplitude and phase by the surface. However, for our purposes it is sufficient to consider the object surface as a self-luminescent object, where each surface point (x, y, z) emits a spherical wavelet $b(x, y, z)$. Since we assumed a diffusely scattering object, the phases $\beta(x, y, z)$ are random; this characteristic is not changed by the illuminating wave $E(x, y, z)$. Therefore it is sufficient to describe the object surface by the complex amplitude $b(x, y, z)$ of (3.27).

Let the geometry for the numerical description be as in Fig. 3.13. To keep the analysis simple the object surface is approximately flat, that means the z in $b(x, y, z)$ is constant. The microstructure of the rough surface is only contained in the stochastic phase β. A distance d apart from the object surface we have the recording medium, the hologram plate or in the digital case the CCD target. The plane of the recording medium has the coordinates (ξ, η).

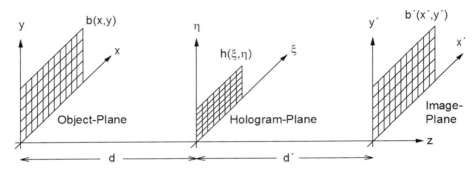

Figure 3.13: Geometry for digital holography.

At distance d' from this hologram plane we have the image plane also named the observation plane where the real image can be reconstructed. The coordinates of this plane are chosen as (x', y'). While the object is denoted by $b(x, y)$, the corresponding field in the hologram plane is $B(\xi, \eta)$. Its superposition with the reference wave field $r(\xi, \eta)$ by interference produces a wave field whose intensity distribution $h(\xi, \eta)$ is recorded by the CCD-array, see (2.122). After recording this real distribution, $h(\xi, \eta)$ is stored in a quantized and digitized form in the computer. Numerical evaluation then produces the complex distribution $b'(x', y')$ which represents the reconstructed image.

Let us start our analysis with the complex amplitude distribution $b(x, y)$ in the object plane. In most holographic applications its distance from the hologram plane is far enough that the Fresnel approximation can be applied. Therefore according to (2.73) in the hologram plane we obtain by diffraction the complex field

$$B(\nu, \mu) = \frac{e^{ikd}}{i\lambda d} e^{i\pi d\lambda(\nu^2 + \mu^2)} \iint b(x, y) e^{\frac{i\pi}{d\lambda}(x^2 + y^2)} e^{-2i\pi(x\nu + y\mu)} \, dx \, dy \quad (3.28)$$

with the correspondence

$$\nu = \frac{\xi}{d\lambda} \qquad \mu = \frac{\eta}{d\lambda} \quad (3.29)$$

between the coordinates (ξ, η) in the hologram plane and the spatial frequencies (ν, μ). The constant factor $e^{ikd}/(i\lambda d)$ does not depend on the spatial frequency coordinates nor the special object, so it will be omitted in the following.

3.2 Numerical Reconstruction by the Fresnel Transform

In practical applications the pixel numbers N, M and the pixel distances $\Delta\xi$, $\Delta\eta$ are given by the CCD array at hand. So in the following discussion these are the fundamental parameters to which the others are related. In the hologram plane we have the discrete coordinates

$$\begin{aligned} \xi &= n\,\Delta\xi & n &= 1,\ldots,N \\ \eta &= m\,\Delta\eta & m &= 1,\ldots,M. \end{aligned} \qquad (3.30)$$

With the given parameters in the object plane we have the step widths

$$\Delta x = \frac{1}{N\Delta\nu} = \frac{d\lambda}{N\Delta\xi} \qquad \Delta y = \frac{1}{M\Delta\mu} = \frac{d\lambda}{M\Delta\eta}. \qquad (3.31)$$

In the (ξ,η)-plane the discrete reference wave model $r(n\Delta\xi, m\Delta\eta)$ is superposed, the resulting intensity $h(n\Delta\xi, m\Delta\eta)$ is recorded as the digital hologram

$$h = (B+r)(B+r)^*. \qquad (3.32)$$

The reconstruction of the real and virtual images in optics would require the illumination of the hologram by the reference wave. This process now is modeled numerically by a multiplication of the digital hologram h with the reference wave. Since we want to obtain a real image we have to multiply the digital hologram with the conjugate $r^*(\xi,\eta)$ of the reference wave, see (2.137). The real image in the (x',y')-plane is determined by the diffraction formula that is approximated by the inverse Fresnel transform (2.73). To yield a sharp image we must choose $d' = d$. The z of (2.73) is replaced directly by d and not by d' to avoid confusion.

$$\begin{aligned} b'(\delta,\varepsilon) &= e^{\frac{i\pi}{d\lambda}(x'^2+y'^2)} \iint h(\xi,\eta) r^*(\xi,\eta) e^{\frac{i\pi}{d\lambda}(\xi^2+\eta^2)} e^{\frac{-2i\pi}{d\lambda}(x'\xi+y'\eta)} d\xi\,d\eta \\ &= e^{i\pi d\lambda(\delta^2+\varepsilon^2)} \iint h(\xi,\eta) r^*(\xi,\eta) e^{\frac{i\pi}{d\lambda}(\xi^2+\eta^2)} e^{-2i\pi(\xi\delta+\eta\varepsilon)} d\xi\,d\eta \end{aligned} \qquad (3.33)$$

where we used the substitutions

$$\delta = \frac{x'}{d\lambda} \qquad \varepsilon = \frac{y'}{d\lambda}. \qquad (3.34)$$

With these substitutions we can clearly recognize that despite of a phase factor which does not depend on the specific hologram the field is calculated by a Fourier transform of the digital hologram $h(\xi,\eta)$ multiplied with the reference wave $r^*(\xi,\eta)$ and also multiplied with the chirp function $\exp[i\pi(\xi^2+\eta^2)/(d\lambda)]$.

The discrete version of (3.33) is

$$b'(n\Delta\delta, m\Delta\varepsilon) = e^{i\pi d\lambda(n^2\Delta\delta^2 + m^2\Delta\varepsilon^2)}$$
$$\cdot \sum_{k=0}^{N-1}\sum_{l=0}^{M-1} h(k\Delta\xi, l\Delta\eta) r^*(k\Delta\xi, l\Delta\eta) e^{\frac{i\pi}{d\lambda}(k^2\Delta\xi^2 + l^2\Delta\eta^2)} e^{-2i\pi(\frac{kn}{N} + \frac{lm}{M})}. \qquad (3.35)$$

Due to (3.31) and

$$\Delta\delta = \frac{1}{N\Delta\xi} = \frac{\Delta x'}{d\lambda} \qquad \Delta\varepsilon = \frac{1}{M\Delta\eta} = \frac{\Delta y'}{d\lambda} \qquad (3.36)$$

the pixel spacing in the real image coincides with the pixel spacing in the object plane

$$\Delta x' = \Delta x \quad \text{and} \quad \Delta y' = \Delta y. \tag{3.37}$$

The reconstruction formula (3.35) now expressed with $\Delta \xi$ and $\Delta \eta$, the fundamental parameters, is the *central reconstruction formula of digital holography*

$$b'(n\Delta x', m\Delta y') = e^{i\pi d\lambda (\frac{n^2}{N^2 \Delta \xi^2} + \frac{m^2}{M^2 \Delta \eta^2})}$$
$$\cdot \sum_{k=0}^{N-1} \sum_{l=0}^{M-1} h(k\Delta \xi, l\Delta \eta) r^*(k\Delta \xi, l\Delta \eta) e^{\frac{i\pi}{d\lambda}(k^2 \Delta \xi^2 + l^2 \Delta \eta^2)} e^{-2i\pi(\frac{kn}{N} + \frac{lm}{M})}. \tag{3.38}$$

If the plane normally impinging reference wave field $r(\xi, \eta) = 1.0$ is employed, then the multiplication with $r^*(k\Delta \xi, l\Delta \eta)$ in (3.38) can be omitted.

Formula (3.38) constitutes the practical discrete finite calculation method on the basis of the Fresnel transform for the reconstruction of the wave field coded in a digital hologram. The result $b'(n\Delta x', m\Delta y')$ is a numerical representation of a complex optical wave field from which by

$$I(n\Delta x', m\Delta y') = |b'(n\Delta x', m\Delta y')|^2 \tag{3.39}$$

$$\text{and} \quad \phi(n\Delta x', m\Delta y') = \arctan \frac{\text{Im}\{b'(n\Delta x', m\Delta y')\}}{\text{Re}\{b'(n\Delta x', m\Delta y')\}} \tag{3.40}$$

intensity and phase distributions can be determined. This is a real advantage compared to the optical reconstruction, because in the optical case we only obtain the intensity distribution. In the digital case we also have access to the phase modulo 2π, which at first glance seems to be of no concern, because for rough object surfaces it varies stochastically. But we will see that the phase access makes up a real advantage when we come to applications in digital holographic interferometry [228], see Section 5.8.

From (3.36) we get the pixel size

$$\Delta x' = \frac{d\lambda}{N\Delta \xi} \quad \Delta y' = \frac{d\lambda}{M\Delta \eta}. \tag{3.41}$$

For this special relation it is possible to have the last exponential of (3.35) and (3.38) in the form $\exp\{-2i\pi(kn/N + lm/M)\}$. This form enables the use of the FFT-algorithm. If we want another pixel size we have to use (3.33) and find the discrete version

$$b'(n\Delta \delta, m\Delta \varepsilon) = e^{i\pi d\lambda (n^2 \Delta \delta^2 + m^2 \Delta \varepsilon^2)} \sum_{k=0}^{N-1} \sum_{l=0}^{M-1} h(k\Delta \xi, l\Delta \eta) r^*(k\Delta \xi, l\Delta \eta)$$
$$\times e^{\frac{i\pi}{d\lambda}(k^2 \Delta \xi^2 + l^2 \Delta \eta^2)} e^{-2i\pi(k\Delta \xi n\Delta \delta + l\Delta \eta m\Delta \varepsilon)} \tag{3.42}$$

but now we have to perform all the complex multiplications under the double sum without the effective acceleration by the FFT-algorithm.

3.2 Numerical Reconstruction by the Fresnel Transform

We obtain the maximum angle θ at any hologram point H for the object point P furthest away from to the source point R. The angle θ is $\theta = \theta_1 + \theta_2$ with

$$\theta_1 \approx \tan\theta_1 = \frac{\frac{d_0}{2} + \frac{N\Delta\xi}{2}}{d}$$

$$\theta_2 \approx \tan\theta_2 = \frac{\frac{d_0}{2} - \frac{N\Delta\xi}{2}}{d} \qquad (3.48)$$

resulting in

$$\theta = \theta_1 + \theta_2 = \frac{d_0}{d}. \qquad (3.49)$$

Now the sampling theorem demands

$$\frac{d_0}{d} < \frac{\lambda}{2\Delta\xi} \qquad (3.50)$$

which gives a minimum distance d for a given lateral width d_0 of the object

$$d(d_0) > \frac{2d_0\Delta\xi}{\lambda} \qquad (3.51)$$

or a maximum width of the object d_0 for a given distance d

$$d_0(d) < \frac{d\lambda}{2\Delta\xi}. \qquad (3.52)$$

Now we compare the minimum distance d_N for a normally impinging plane reference wave as given in (3.12) with the minimum distance between object and CCD-array d_{LFTH} in lensless Fourier transform holography, which is given in (3.51). We see that the limit $2d_0\Delta\xi/\lambda$ is less than $(d_0\Delta\xi + N\Delta\xi^2)/\lambda$ if $d_0 < N\Delta\xi$. This means that for objects with lateral dimensions less than those of the CCD-array, a smaller distance between object and CCD-array is allowed in lensless Fourier transform holography. This is particularly important in digital holographic microscopy: While for the plane reference wave the distance is always larger than $N\Delta\xi^2$ for any d_0, in the case of lensless Fourier transform holography this limit decreases linearly with decreasing object size.

In lensless Fourier transform holography each object point is encoded in a fringe system with a spatial frequency that is proportional to the distance of that point from the reference point [160]. This is shown in Fig. 3.17a: For each point P the angle $\theta(P, H)$ remains nearly constant over the hologram points H. We have seen, Section 2.8.4, that the MTF results in an increasing attenuation for increasing spatial frequencies. Therefore now this attenuation will increase for points P further from the reference points, because their $\theta(P, H)$ is larger producing higher spatial frequencies. Thus the MTF can be interpreted as an attenuating mask placed over the object, or equivalently over the real and virtual image [160]. The intensity decrease at the borders of the reconstructed picture due to a 100% fill factor of the CCD pixels is analyzed in [229]. Nevertheless a vignetting is not expected, because the Nyquist frequency f_N is always less than the frequency f_{DC}, where the MTF falls to zero, see Figs. 3.17a or b.

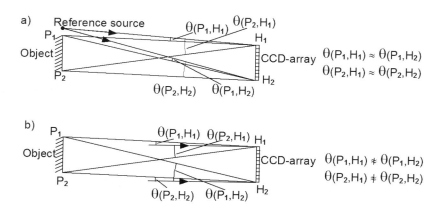

Figure 3.17: Divergent (a) and plane (b) reference waves.

These facts which hold for lensless Fourier transform holography are not valid for collimated reference waves, because there the angle $\theta(P,H)$ for each object point varies significantly with H, Fig. 3.17b. There the influence of the MTF on the image cannot be described in an easy fashion. The introduction of an aperture for reducing the imaging angle – see Section 3.1.2 – in lensless digital Fourier transform holography is treated in [221–223].

3.2.4 The D.C.-Term of the Fresnel Transform

In the intensity displays of the holographic reconstructions of Fig. 3.14b and Fig. 3.15 we recognize a bright central square. It is much brighter than the reconstructed real or virtual images, for the display its intensity has been clipped to enhance the eligibility of the overall pattern. The physical meaning of the bright central square is that it represents the zero-order diffraction of the reference wave or equivalently it is the projection of the illuminated CCD array. From the optical point of view it is the undiffracted part of the reconstructing reference wave; from the computational standpoint it is the *d.c.-term* of the Fresnel transform, as can be seen in the following.

If we neglect the factors before the integrals of (3.33) or the sums of (3.38), which only affect the phase in a way independent of the specific hologram, the Fresnel transform is the Fourier transform of a product, the factors being the hologram times the reference wave $h \cdot r^*$ and the chirp function. According to the convolution theorem this gives the same result as the convolution of the Fourier transforms of the individual factors. The Fourier transform H of the hologram multiplied with the reference wave $h(k\Delta\xi, l\Delta\eta) \cdot r^*(k\Delta\xi, l\Delta\eta)$ generally is trimodal with a high-amplitude peak at the spatial frequency $(0,0)$. This d.c.-term $H(0,0)$, whose value is calculated by

$$H(0,0) = \sum_{k=0}^{N-1}\sum_{l=0}^{M-1} h(k\Delta\xi, l\Delta\eta) r^*(k\Delta\xi, l\Delta\eta), \qquad (3.53)$$

can be modeled by a Dirac delta function. The d.c.-term of the Fresnel transform now is the d.c.-term of the Fourier transform of the digital hologram multiplied by the reference

3.2 Numerical Reconstruction by the Fresnel Transform

wave convolved with the Fourier transform of the two-dimensional chirp function. Since we assume a Dirac delta function for the former, the d.c.-term of the whole Fresnel transform is the Fourier transform of the finite chirp function

$$e^{\frac{i\pi}{\lambda d}(k^2 \Delta \xi^2 + l^2 \Delta \eta^2)} = e^{\frac{i\pi}{\lambda d} k^2 \Delta \xi^2} e^{\frac{i\pi}{\lambda d} l^2 \Delta \eta^2} \tag{3.54}$$

restricted to the finite extent of the hologram.

The Fourier transform $G(x')$ of the finite chirp function $g(\xi)$ is investigated in Appendix A.13. Only the coordinates in the results of Appendix A.13 must be modified. For the one-dimensional function $\exp[(i\pi/d\lambda)k^2 \Delta \xi^2]$, $k = 0, \ldots, N-1$ we translate $\beta = \Delta \xi^2/(d\lambda)$, $\xi = k$, and $2L = N$. Then the d.c.-term has a width of $2\pi N \Delta \xi^2/(d\lambda)$. If we count the spatial frequencies $2\pi n/N$ by the corresponding pixel numbers $n = 0, \ldots, N-1$, we obtain the width of the d.c.-term expressed in the experimental parameters extending over $N^2 \Delta \xi^2/(d\lambda)$ pixels. In two dimensions this gives the area of the d.c.-term as

$$\frac{N^2 \Delta \xi^2}{d\lambda} \times \frac{M^2 \Delta \eta^2}{d\lambda} \tag{3.55}$$

with $N^2 \Delta \xi^2/(d\lambda)$ being the width in the x'-direction and $M^2 \Delta \eta^2/(d\lambda)$ that in the y'-direction. The width of the d.c.-term increases with increasing pixel dimensions and pixel number of the CCD target and it decreases with increasing distance d. This effect can be seen in Fig. 3.15, where different reconstruction distances d have been used. For the limiting case of infinite d we have the d.c.-term of the Fourier transform which covers only a single pixel.

Also in Section A.13 it is investigated how a shift of the finite chirp function influences the location of its Fourier spectrum and thus the d.c.-term. If we reconstruct using (3.38), the hologram is defined in $[0, N\Delta\xi] \times [0, M\Delta\eta]$, so the chirp function $\exp\left(\frac{i\pi}{d\lambda} k^2 \Delta \xi^2\right) \exp\left(\frac{i\pi}{d\lambda} l^2 \Delta \eta^2\right)$ carries local frequencies from 0 to $N\Delta\xi/(d\lambda\pi)$ in the ξ-direction and from 0 to $M\Delta\eta/(d\lambda\pi)$ in the η-direction. The square d.c.-term is located totally in the first quadrant starting at $(0,0)$. The display normally is reordered so that the d.c.-term appears in the center of the pattern as we are used to from optics, e. g. with holographic reconstructions or with Fraunhofer diffraction patterns.

As can be seen in Fig. 3.15 after reordering the edge point of the square d.c.-term coming from $(0,0)$ is shifted for all reconstruction distances to the central point while the other edge points reflect the varying size of the d.c.-term due to the different d. If we shift the finite chirp function by say $k_0 \Delta \xi$ in the ξ-direction and $l_0 \Delta \eta$ in the η-direction, we obtain

$$\exp\left[\frac{i\pi}{d\lambda}(k-k_0)^2 \Delta \xi^2\right] \exp\left[\frac{i\pi}{d\lambda}(l-l_0)^2 \Delta \eta^2\right] \tag{3.56}$$

which carries local frequencies from $-k_0\Delta\xi/(d\lambda\pi)$ to $(N\Delta\xi - k_0\Delta\xi)/(d\lambda\pi)$ and from $-l_0\Delta\eta/(d\lambda\pi)$ to $(M\Delta\eta - l_0\Delta\eta)/(d\lambda\pi)$ in the two directions. If k_0 is between $-N$ and $+N$ (l_0 between $-M$ and $+M$) the d.c.-term in the reconstruction is divided to the four edges of the display before reordering. The location of the d.c.-term is demonstrated in Fig. 3.18.

The real part of the unshifted, $k_0 = l_0 = 0$, chirp function is shown in Fig. 3.18a, the resulting virtual image has a d.c.-term totally in the upper right corner before reordering, Fig. 3.18b. It is in the third quadrant after reordering, Fig. 3.18c. The real part of the shifted chirp function with $k_0 = N/2$, $l_0 = M/2$ is given in Fig. 3.18d. Here the d.c.-term is divided into equally sized parts at all four corners before reordering, Fig. 3.18e. It appears in the center of the image after reordering, Fig. 3.18f.

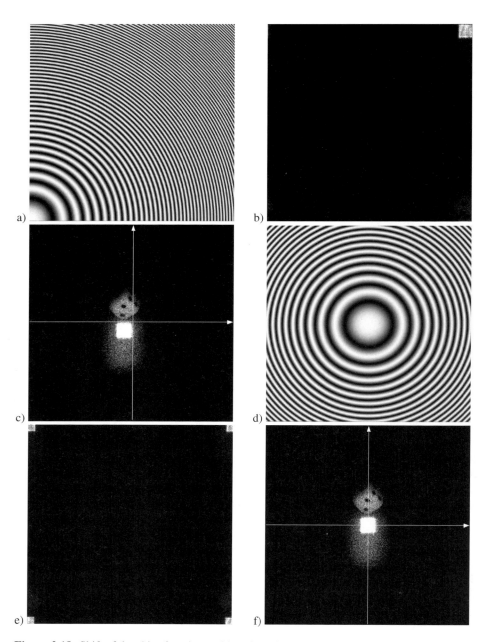

Figure 3.18: Shift of the chirp function and location of the d.c.-term.

3.2 Numerical Reconstruction by the Fresnel Transform

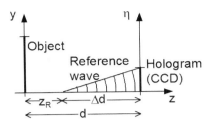

Figure 3.24: Spherical reference wave.

for the influence of any non-vanishing x_R and y_R is analogous to that of the inclined plane reference wave treated above. Thus we have

$$r(k,l) = E_r \, e^{\frac{2i\pi \Delta d}{\lambda}} \, e^{\frac{i\pi}{\Delta d \lambda}(k^2 \Delta \xi^2 + l^2 \Delta \eta)}. \tag{3.68}$$

With the conjugate of such a reference wave the correct reconstruction would be

$$b'(n,m) = e^{i\alpha} \, e^{\frac{2i\pi \Delta d}{\lambda}} \sum_{k=0}^{N-1} \sum_{l=0}^{M-1} h(k,l) r^*(k,l) e^{\frac{i\pi}{d\lambda}(k^2 \Delta \xi^2 + l^2 \Delta \eta^2)} e^{-2i\pi(\frac{kn}{N} + \frac{lm}{M})}$$

$$= E_r \, e^{i\beta}$$
$$\cdot \sum_{k=0}^{N-1} \sum_{l=0}^{M-1} h(k,l) e^{-\frac{i\pi}{\Delta d \lambda}(k^2 \Delta \xi^2 + l^2 \Delta \eta^2)} e^{\frac{i\pi}{d\lambda}(k^2 \Delta \xi^2 + l^2 \Delta \eta^2)} e^{-2i\pi(\frac{kn}{N} + \frac{lm}{M})}$$

$$= E_r \, e^{i\beta}$$
$$\cdot \sum_{k=0}^{N-1} \sum_{l=0}^{M-1} h(k,l) e^{-\frac{i\pi}{\lambda}\left(\frac{1}{d} - \frac{1}{\Delta d}\right)(k^2 \Delta \xi^2 + l^2 \Delta \eta^2)} e^{-2i\pi(\frac{kn}{N} + \frac{lm}{M})}. \tag{3.69}$$

Up to the phase term before the double sum this is the same as a reconstruction with a plane normally impinging reference wave but a reconstruction distance changed from d to d', which depend on each other by

$$\frac{1}{d'} = \frac{1}{d} - \frac{1}{\Delta d}. \tag{3.70}$$

An experiment where a plane reference wave is assumed but in reality a spherical one is existent will give an unsharp reconstructed intensity distribution when the exact distance d between object surface and CCD array is used, but a sharp reconstruction when one employs the d' of (3.70). On the other hand one may examine different reconstruction distances in conjunction with the plane normally impinging reconstruction wave, the one giving the sharpest result is d'. If d' differs from the measured d, this indicates that a spherical reference wave was arranged for the recording and furthermore informs us about the location of the source point of this wave. The special case $z_R = 0$ implies $\Delta d = d$. So $1/d' = 0$ and we have again the geometry of lensless Fourier transform holography.

In Fig. 3.21e an example was shown of a hologram recorded with a spherical reference wave and reconstructed with a conjugated replica of this wave. The object's distance from the

CCD was $d = 1.0$ m and the z-coordinate of the reference source point was $z_R = 0.4$ m. If we reconstruct with a plane normally impinging reference wave, but retain the distance d, we obtain the unsharp reconstruction of Fig. 3.25a. The reconstruction with $d' = 1.5$ m and a plane reconstruction wave on the other hand yields the intensity distribution of Fig. 3.25b, which agrees with that of Fig. 3.21e.

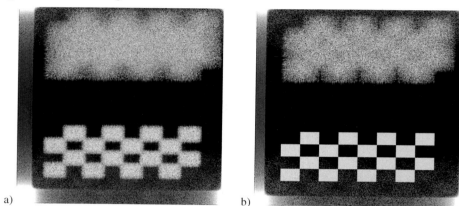

Figure 3.25: Reconstructed intensities from hologram produced with spherical reference wave and plane reconstruction wave, different image distances.

3.2.8 Anamorphic Correction

Digital holography allows the modification of the reconstructed wave field by a variation of the numerical process. We have recognized already the suppression of the d.c.-term, Section 3.2.5, and the suppression of the twin image, Section 3.2.6. In digital holographic microscopy the aberration induced by the magnifying objective is compensated by an additional phase factor introduced into (3.35) during reconstruction. The compensation of spherical aberrations is demonstrated e. g. in [236]. In the same way abberations which manifest as *anamorphism* or *astigmatism* in the reconstructed image can be compensated [94, 95].

Especially when using a *reflective grating interferometer* [95] strong anamorphism is produced that prevents correct imaging of the object. Without loss of generality let us assume that decorrelation of the reconstructed object field's phase occurs along the horizontal direction without disturbing the field along the vertical direction. It is possible to compensate for this anamorphism by substituting the η in (3.33) by $\eta \cos \alpha$ and the ε by $\varepsilon \cos \alpha$ with a suitable α. In reflective grating interferometry performed by digital holography this α is the angle between the object wave's incidence direction and the grating normal. As a result of this substitution the pixel size in the reconstructed image is differently affected in both directions. While according to (3.34) and (3.36) $\Delta \delta = \Delta x'/(d\lambda) = 1/(N \Delta \xi)$ resp. $\Delta x' = d\lambda/(N\Delta\xi)$ remains unchanged, now $\Delta \varepsilon = \Delta y' \cos \alpha/(d\lambda) = \cos \alpha/(N \Delta \eta)$ giving $\Delta y' = d\lambda \cos \alpha /(N \Delta \eta)$. An originally square object will be reconstructed as a rectangular one but the anamorphism is corrected. Examples of this procedure working with reflective grating interferometry and with phase shifting digital holography can be found in [94, 95].

Astigmatism is a third-order wavefront aberration that results in different focal lengths of the imaging systems for structures oriented in horizontal direction compared to those oriented in vertical direction. Such astigmatism can be compensated by replacing the chirp function $\exp\{\frac{i\pi}{d\lambda}(\xi^2 + \eta^2)\}$ in (3.33), (3.35), or (3.38) with

$$\exp\left\{\frac{i\pi}{\lambda}\left(\frac{\xi^2}{d_x} + \frac{\eta^2}{d_y}\right)\right\} \tag{3.71}$$

where d_x and d_y are the different distances from hologram plane to the plane of the sharply focused horizontal and vertical object structures.

3.3 Numerical Reconstruction by the Convolution Approach

The numerical reconstruction of digital holograms consists in a multiplication of the stored hologram with a wave field representing the reference wave followed by the calculation of the field diffracted from this product distribution into the image plane. In Section 3.2.1 the diffraction formula used for this calculation was approximated by the Fresnel transform. An effective reconstruction method resulted. However, in Section 2.4.5 we have seen that the diffraction formula is a superposition integral of a linear shift invariant system, so it can be realized as a convolution. In Section A.5 the convolution theorem is introduced which offers an effective way to calculate convolutions. Therefore in this chapter reconstruction procedures based on that concept, which in the following will be called the *convolution approach*, will be introduced, some consequences will be treated and the methods will be compared to reconstruction by the Fresnel transform, because there are more than only conceptual differences between the two approaches.

3.3.1 The Diffraction Integral as a Convolution

As we have seen in Section. 2.4.5 the field diffracted at the distribution $h(\xi, \eta) \cdot r^*(\xi, \eta)$ in the hologram plane can be calculated in any (x', y')-plane by the convolution integral

$$b'(x', y') = \iint h(\xi, \eta) \cdot r^*(\xi, \eta) g(x' - \xi, y' - \eta) d\xi \, d\eta \tag{3.72}$$

where g is the impulse response of free space propagation as introduced in (2.90)

$$g(x' - \xi, y' - \eta) = \frac{i}{\lambda} \frac{e^{ik\sqrt{(x'-\xi)^2 + (y'-\eta)^2 + d^2}}}{\sqrt{(x'-\xi)^2 + (y'-\eta)^2 + d^2}}. \tag{3.73}$$

In short notation (3.72) becomes

$$b'(x', y') = [h(\xi, \eta) \cdot r^*(\xi, \eta)] \star g(\xi, \eta) \tag{3.74}$$

where $\star$ denotes the convolution operation. The convolution theorem allows us to reduce the computational effort drastically by replacing the convolution in the spatial domain by a multiplication of the complex spectra in the spatial frequency domain followed by an inverse Fourier transform of this product back into the spatial domain. Now the effective FFT algorithm is repeatedly used to calculate the forward Fourier transform $\mathcal{F}$ as well as the inverse Fourier transform $\mathcal{F}^{-1}$:

$$b' = \mathcal{F}^{-1}\left\{\mathcal{F}\{h \cdot r^*\} \cdot \mathcal{F}\{g\}\right\}. \tag{3.75}$$

Of these three Fourier transforms one may be saved at least theoretically if we do not define the impulse response g and calculate $G = \mathcal{F}\{g\}$, but directly define the free space transfer function G that was introduced in its continuous form in (2.92).

For the numerical realization of this approach the continuous coordinates (ξ, η) are replaced by the discrete values $k\Delta\xi$ and $l\Delta\eta$, so that the impulse response now reads

$$g(k,l) = \frac{\mathrm{i}}{\lambda} \frac{\mathrm{e}^{\frac{2\pi\mathrm{i}}{\lambda}\sqrt{d^2 + (k-1)^2\Delta\xi^2 + (l-1)^2\Delta\eta^2}}}{\sqrt{d^2 + (k-1)^2\Delta\xi^2 + (l-1)^2\Delta\eta^2}}. \tag{3.76}$$

For programming we use $i e^{i\phi} = -\sin\phi + i\cos\phi$. The discrete impulse response $g(k,l)$; $k = \{1,\ldots,N\}$, $l = \{1,\ldots,M\}$ now has to be transformed using the FFT algorithm to obtain the discrete finite transfer function G.

The direct calculation of the transfer function G proceeds as in (2.92)

$$G(\nu,\mu) = \begin{cases} \exp\left[-\frac{\mathrm{i}2\pi d}{\lambda}\sqrt{1-(\lambda\nu)^2-(\lambda\mu)^2}\right] & : \ (\lambda\nu)^2 + (\lambda\mu)^2 \le 1 \\ 0 & : \ \text{otherwise.} \end{cases} \tag{3.77}$$

The discrete values of ν and μ are

$$\nu = \frac{n-1}{N\Delta\xi} \quad \text{and} \quad \mu = \frac{m-1}{M\Delta\eta} \quad n = 1,\ldots,N; \ m = 1,\ldots,M \tag{3.78}$$

so that the finite discrete transfer function becomes

$$\begin{aligned} G(n,m) &= \mathrm{e}^{\frac{2\pi\mathrm{i}d}{\lambda}\sqrt{1-\left(\frac{\lambda(n-1)}{N\Delta\xi}\right)^2-\left(\frac{\lambda(m-1)}{M\Delta\eta}\right)^2}} \\ &= \mathrm{e}^{\frac{2\pi\mathrm{i}d}{N\Delta\xi}\sqrt{\frac{N^2\Delta\xi^2}{\lambda^2}-(n-1)^2-(m-1)^2}} \end{aligned} \tag{3.79}$$

where the last expression holds in the frequently occurring case $N = M$ and $\Delta\xi = \Delta\eta$. As soon as the argument under the square root becomes negative we set $G(n,m) = 0$.

If we take a closer look at the Fresnel approximation (2.72) we recognize that it also has the form of a convolution with the convolution kernel

$$g_F(\xi - x, \eta - y) = \frac{\mathrm{e}^{\mathrm{i}kd}}{\mathrm{i}\lambda d} \mathrm{e}^{\frac{\mathrm{i}k}{2d}[(\xi-x)^2 + (\eta-y)^2]}. \tag{3.80}$$

The finite discrete version is

$$g_F(k,l) = \frac{\mathrm{e}^{\mathrm{i}kd}}{\mathrm{i}\lambda d} \mathrm{e}^{\frac{\mathrm{i}k}{2d}[(k-1)^2\Delta\xi^2 + (l-1)^2\Delta\eta^2]}. \tag{3.81}$$

3.3 Numerical Reconstruction by the Convolution Approach

applied. It would be advantageous if we could control the size of the reconstructed image field to fit the size of the object.

A variation of d or λ in the Fresnel reconstruction process would only result in unsharp wave fields. If we analyze the Fresnel reconstruction formulas, e. g. (3.38), we recognize that d and λ always appear as a product, therefore a change of one of them has always the same effect as if d was varied away from the optimal distance, so that the reconstructed field only becomes blurred.

Let us now turn to reconstruction by the convolution approach. If we try to rescale the image field without changing the scale of the hologram field, the Fresnel-Kirchhoff formula (2.69) or the related superposition integral (2.88) will exhibit coordinate differences $\xi - ax$ and/or $\eta - by$ with $a \neq 1$ and $b \neq 1$. As a consequence the system characterized by the superposition integral (2.88) is no longer shift-invariant, so the convolution theorem cannot be applied and none of the four methods introduced in Section 3.3.1 works.

The only way out of this dilemma is to rescale the hologram. Consequently the image plane is rescaled simultaneously in the same way. In the following we treat this by doubling the resolution to show the principles in an easy way. Nevertheless we have to keep in mind that factors other than 2 also apply. But the holographer should not be guided into a wrong direction: The way to perform practical measurements should not be to record the hologram with any geometry and to find the best scaling and resolution in the reconstruction stage. It is best practice to carefully plan the experiment and the holographic arrangement's geometry to obtain objects covering nearly the full frame of the reconstructed field without the need for rescaling.

The first way to scale the fields is to expand the $N \times M$-pixel hologram to, say, a $2N \times 2M$-pixel hologram. The original $N \times M$-pixel hologram is surrounded by pixels of intensity 0. Now the pixel numbers in each direction are doubled, $N' = 2N$, $M' = 2M$, the pixel size $\Delta\xi \times \Delta\eta$ remains unchanged. The reconstruction now must proceed with N' and M' and we get a reconstructed image of size $N'\Delta\xi \times M'\Delta\eta = 4NM\Delta\xi\Delta\eta$. In Fig. 3.28a we see the hologram of Fig. 3.14a augmented by black pixels around it. The intensity distribution of

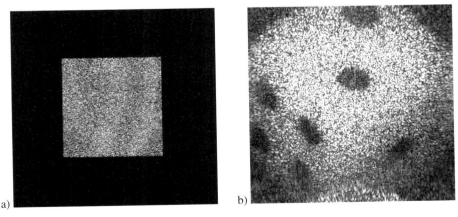

Figure 3.28: Reconstruction after quadrupling the pixel number of the hologram.

the field reconstructed from this augmented hologram is shown in Fig. 3.28b. It consists of 2048×2048 pixels of size 6.8×6.8 μm² each. The degradation of the reconstructed fields by the overlayed information from outside is much less severe than in the 1024×1024-pixel case, as can be seen by comparing with Fig. 3.27.

If we want to keep the numbers N and M while doubling the image size in each direction, we can average the intensities of all contiguous 2×2-pixel neighborhoods and in this way build a new hologram of $N/2 \times M/2$ pixels with a pixel size of $\Delta\xi' = 2\Delta\xi$ and $\Delta\eta' = 2\Delta\eta$. This new hologram forms the central part of an $N \times M$-pixel hologram with black pixels surrounding it as described before. The numerical reconstruction proceeds with $\Delta\xi'$, $\Delta\eta'$, N, M. Again we obtain a field of size $4NM\Delta\xi\Delta\eta = N\Delta\xi' \times M\Delta\eta'$. The differences in the results from those of the aforementioned approach are below the resolution of printing, the results of the experimental example look like Figs. 3.28a and b.

Both approaches to scale the reconstructed image field can be combined: The original $(\Delta\xi, \Delta\eta, N, M)$-hologram is reduced to a $(2\Delta\xi, 2\Delta\eta, N/2, M/2)$-hologram, and this new hologram is embedded into surrounding black pixels which fill up to a $(2\Delta\xi, 2\Delta\eta, 2N, 2M)$-hologram. The result will be a reconstructed image field of size $16NM\Delta\xi\Delta\eta = 2N2\Delta\xi \times 2M2\Delta\eta$. Figure 3.29a displays the averaged and augmented digital hologram now containing

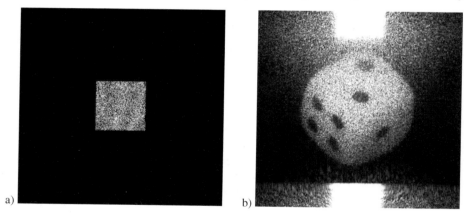

Figure 3.29: Reconstruction after combined quadrupling the pixel number and changing the pixel size of the hologram.

2048×2048 pixels of size 13.6×13.6 μm² each, where only the central 512×512 pixels contain the averaged recorded digital hologram. The intensity distribution of the resulting reconstructed wave field is given in Fig. 3.29b. Factors for scaling other than 2 are feasible as well. If non-integer factors should be used, an interpolation of hologram intensity values may be necessary. Furthermore scaling factors leading to pixel numbers N and M, which are $N \neq 2^n$ and $M \neq 2^m$ require discrete finite Fourier transforms which are generally calculated by (A.45) without the possibility of simplification by the FFT algorithm.

Now let us return to the experiment concerning particle analysis by digital in-line holography. The first results have been illustrated in Fig. 3.26. If we now embed the partial 1024×1024-pixel hologram into the 2048×2048-pixel field by augmenting with zero-intensity

3.3 Numerical Reconstruction by the Convolution Approach

pixels, Fig. 3.30a, then the 2048×2048-pixel reconstruction displays the particles at their correct location, Fig. 3.30b. Here the reconstructed particles can be seen outside the d.c.-term which in size corresponds to the non-vanishing part of the hologram.

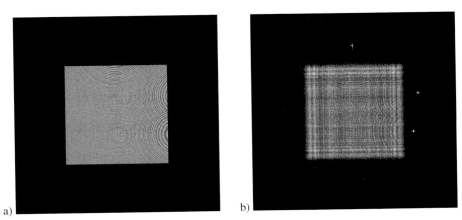

Figure 3.30: Reconstruction from partial hologram.

These facts represent a digital analogy to an effect known from optical holography. If we break a hologram into parts, the whole scene can be reconstructed from each single fragment, only the noise increases when the size of the used hologram decreases.

Another way to rescale the image field is to use a reference wave R' during reconstruction which is different from R, the reference wave employed for recording the digital hologram. We have seen in Section 2.6.3 that a change of the reference wave geometry and a change in reference wave wavelength produces a lateral magnification M_{lat} described in (2.161). The sharp reconstruction of the image in this case is found in a modified distance. This distance change is given in the holographic imaging equations (2.156).

The scaling of the image by reference wave variation will not succeed when using the Fresnel reconstruction. This is easily seen when we check the pixel size. Let the hologram be recorded with a reference wave having a z-coordinate z_R and wavelength λ. The reconstruction is performed with z'_R and wavelength λ'. The object point originally in distance $d = z_P$ now is reconstructed at distance

$$d' = z'_P = \frac{dz'_R z_R}{\mu z_R z'_R - \mu d z'_R + d z_R} \tag{3.89}$$

according to (2.156). The pixel size $d\lambda/(N\Delta\xi)$ if using the original reference wave now is changed to

$$\frac{d'\lambda'}{N\Delta\xi} = \frac{d\lambda}{N\Delta\xi}\left(\frac{\mu z'_R z_R}{\mu z_R z'_R - \mu d z'_R + d z_R}\right). \tag{3.90}$$

On the other hand the lateral magnification resulting from the reference wave variation according to (2.161) is

$$M_{lat} = \frac{1}{1 + d\left(\frac{1}{\mu z'_R} - \frac{1}{\mu z_R}\right)} = \frac{\mu z'_R z_R}{\mu z_R z'_R - \mu d z'_R + d z_R} \qquad (3.91)$$

which obviously is the same as the pixel size magnification calculated before. So the reconstructed image appears in the same size as before in the reconstructed frame due to the corresponding pixel size modification.

We have already seen that the pixel size remains constant for any distance or wavelength if we use the convolution approach. And indeed now we can scale the image field to any desired lateral magnification. This is demonstrated again for the well known example of the die: Its digital hologram was recorded with a plane reference wave R assumed to diverge from $(0., 0., \infty)m$. Thus in (2.151) the term μ/z_R vanishes. The distance between the CCD and the die was $d = z_P = 1.054$ m. We now reconstruct with spherical reference waves diverging from $(0., 0., z'_R)$. The distances d' where the images are sharp are given in Table 3.1 together with the resulting lateral magnifications – in our case reductions of the size – and the used wavelengths.

Table 3.1: Data for scaling experiment.

z'_R	λ'	d'	M_{lat}	Display
$\infty = z_R$	0.6328 μm = λ	1.054 m = d	1	Fig. 3.27
0.2 m	0.6328 μm = λ	0.1681 m	0.16	Fig. 3.31a
0.3 m	0.6328 μm = λ	0.2335 m	0.22	Fig. 3.31b
0.5 m	0.6328 μm = λ	0.339 m	0.32	Fig. 3.31c
0.5 m	0.3164 μm = λ	0.404 m	0.19	Fig. 3.31d

The other parameters are the pixel size 6.8 μm × 6.8 μm and $(s_K, s_l) = (768, 1536)$. The resulting intensities are shown in Fig. 3.31. So starting with a digital hologram with known recording reference wave and object distance we begin with (2.161) to calculate the plane from where the reconstruction wave has to diverge to obtain a desired image size magnification or reduction. Then we calculate by (2.151) the plane where the image is sharply reconstructed and with these data we perform the reconstruction by convolution. Scaling of the reconstructed image field is a frequent option in digital holography software packages [232].

3.4 Further Numerical Reconstruction Methods

3.4.1 Phase-Shifting Digital Holography

In the preceding sections several methods have been introduced for calculating the object wave field from a single digital hologram. A complex field in the hologram plane was generated

3.4 Further Numerical Reconstruction Methods

Reconstruction by the convolution approach with $\Delta x' = \Delta \xi$ defines a focal depth from

$$a_v = d\frac{N}{N+1} \quad \text{to} \quad a_h = d\frac{N}{N-1}. \tag{3.121}$$

The different depths are elucidated by an example. Let $d = 1$ m, $N = 1024$, $\Delta \xi = 6.8$ μm, and $\lambda = 0.6328$ μm. The reconstruction by one of the convolution methods gives a focal range $[a_v, a_h] = [d - 976 \text{ μm}, d + 978 \text{ μm}]$ while the application of the Fresnel transform defines $[a_v, a_h] = [d - 12.9 \text{ mm}, d + 13.2 \text{ mm}]$. In the Fresnel case we have a focal range more than 13 times wider than in the convolution case.

If we perform a digital holographic experiment with optical reduction of the object size according to the methods presented in Section 3.1.2, the depth of focus considered above applies to the reduced virtual image of the object. Furthermore in Section 3.1.2 we considered the transverse magnification M_T, while for the depth resolution the longitudinal magnification M_L is valid. It is $M_L = M_T^2$. Since here we have a reduction of the wave field, the depth dimensions of the object are reduced more than the lateral dimensions. Therefore large depth ranges are simultaneously sharp and even for strongly curved object surfaces all surface points can be reconstructed in focus with a single reconstruction distance d.

3.4.5 Hologram Recording Using Consumer Cameras

The rapid progress in the sector of *consumer still cameras* leading to higher pixel numbers and falling prices makes these cameras interesting for digital holography. Investigations on the applicability of common color photo-cameras have been performed by Sekanina and Pospisil [82] (submitted for publication in June 2001) and the author (in 2002).

A necessary condition for a successful application in digital holography is the access of raw-image data, which means the hologram data must not be compressed by a lossy data compression method, e. g. the JPEG format. The TIFF mode provides the uncompressed color data at each pixel after the demosaicking process due to the Bayer array, see Section 2.8.5. It is advantageous if the camera objective is detachable; however this is only the case for some reflex cameras but not an attribute of the most common cheap digital still cameras. For the reflex cameras then it is necessary that the mirror can be manually fixed before recording to prevent minute vibrations.

The experiments presented in [82] employ a camera with a CCD-chip of 2048×1536 pixels having a pixel pitch of 5.51 μm in both directions. Unfortunately this cam has an undetachable objective, therefore an optical arrangement as shown in Fig. 3.39 has been used.

We recognize the point source of a spherical reference wave in the plane of the object, thus here we have lensless Fourier transform holography. The microinterference field present in aperture A_1 is imaged onto the CCD via the lenses L_1, L_2, and L_3. The apertures A_2 and A_3 are the images of A_1 created by L_1 and L_2, respectively. Lens L_3 is the given objective of the camera.

The experiments performed by the author employed a reflex camera with a 3072×2048 pixel CMOS image sensor. Its pixel pitch is 7.4 μm in each direction. The holograms are recorded with a plane reference wave, and green laser light has been used. The reconstructed field was calculated by the Fresnel transform from the green color TIFF image, which also can be interpreted as the luminance signal. Effective FFT calculation as described in Section A.11

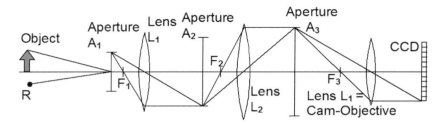

Figure 3.39: Digital holography using a camera with undetachable objective.

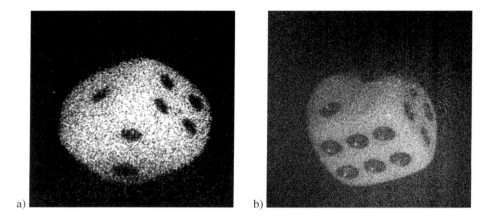

Figure 3.40: Digitally reconstructed intensity of a die. Hologram recorded with (a) 1024×1024 pixels, (b) 3072×2048 pixels.

was performed. The result of one such experiment is given in Fig. 3.40b. For comparison in Fig. 3.40a we have the identical object but whose digital hologram was recorded with a 1024×1024 pixel monochromic CCD array. It is an excerpt from the reconstructed intensity distribution already presented in Fig. 3.14. We recognize the smaller speckles and the better resolution obtained with the 6 times higher pixel number. The applicability of consumer still cams offers to holographers all the advantages of these machines like flexibility, automatic exposure control, in-cam storage of images, and no need for extra frame-grabber cards, and this for a lower price.

3.5 Wave-Optics Analysis of Digital Holography

The process of digital holography from the recording of an optical wave field by a CCD array, storing it in digital form in computer memory, to the numerical reconstruction of the image field can be interpreted as a coherent optical system yielding a complex wave field which is an image of the wave field originally reflected or refracted by an object. Fourier analysis is a powerful tool to describe coherent and incoherent imaging of optical systems [160]. By

3.5 Wave-Optics Analysis of Digital Holography

its means the *point spread function* (PSF), also called the *impulse response*, of the system is calculated, which characterizes the optical system and specifies the influences of the various parameters.

In this section a Fourier analysis of the digital holography optical system is presented both for the Fresnel reconstruction and for convolution reconstruction [221, 256, 257]. It is shown that the point spread function mainly depends on the CCD aperture. Furthermore the effects of the fill factor of CCD pixels are derived with the help of the point spread function.

3.5.1 Frequency Analysis of Digital Holography with Reconstruction by Fresnel Transform

Digital holography now is interpreted as an optical system imaging an object wave field in the (x, y)-plane to a reconstructed optical field related to the (x', y')-plane, this plane existent only in a computer. To find the impulse response $t(x', y')$ of this imaging system, we follow Goodman [160]. Let the object be a point source characterized by $b(x, y) = \delta(x - x_0, y - y_0)$ in the object plane. Then incident on the CCD-target in the hologram plane, the (ξ, η)-plane, will appear a spherical wave $B(\xi, \eta; x_0, y_0)$ diverging from point (x_0, y_0). In paraxial approximation this wave is written

$$B(\xi, \eta; x_0, y_0) = \frac{1}{i\lambda d} \exp\left\{ \frac{ik}{2d}[(\xi - x_0)^2 + (\eta - y_0)^2] \right\}. \tag{3.122}$$

A reference wave $r(\xi, \eta) = u_r(\xi, \eta) \exp\{i\phi(\xi, \eta)\}$ is superposed in the hologram plane. Both waves interfere, and the resulting intensity distribution is the hologram $h(\xi, \eta; x_0, y_0)$, which is recorded by the CCD, see (3.32):

$$\begin{aligned} h(\xi,\eta; x_0, y_0) &= |B(\xi, \eta) + r(\xi, \eta)|^2 \\ &= \frac{1}{\lambda^2 d^2} + u_r^2(\xi, \eta) + \frac{2u_r(\xi, \eta)}{\lambda d} \exp\left\{ \frac{ik}{2d}[(\xi - x_0)^2 + (\eta - y_0)^2] - \phi(\xi, \eta) \right\} \\ &\quad - \frac{2u_r(\xi, \eta)}{\lambda d} \exp\left\{ -\frac{ik}{2d}[(\xi - x_0)^2 + (\eta - y_0)^2] - \phi(\xi, \eta) \right\} \\ &= \frac{1}{\lambda^2 d^2} + u_r^2(\xi, \eta) + \frac{2u_r(\xi, \eta)}{\lambda d} \sin\left\{ \frac{k}{2d}[(\xi - x_0)^2 + (\eta - y_0)^2] - \phi(\xi, \eta) \right\}. \end{aligned} \tag{3.123}$$

This continuous real valued hologram is recorded by the CCD array and by this transformed into a two-dimensional array of discrete values, which constitute the digital hologram. The resulting digital hologram h_1 is characterized by

$$\begin{aligned} &h_1(\xi, \eta; x_0, y_0) \\ &= h(\xi, \eta; x_0, y_0) \left[\text{rect}\left(\frac{\xi}{\alpha\Delta\xi}, \frac{\eta}{\beta\Delta\eta}\right) \star \text{comb}\left(\frac{\xi}{\Delta\xi}, \frac{\eta}{\Delta\eta}\right) \right] \text{rect}\left(\frac{\xi}{N\Delta\xi}, \frac{\eta}{M\Delta\eta}\right). \end{aligned} \tag{3.124}$$

Here $\text{rect}[\xi/(\alpha\Delta\xi), \eta/(\beta\Delta\eta)]$ represents a single two-dimensional pixel of the CCD. The pixel pitch is $\Delta\xi$ in the ξ-direction and $\Delta\eta$ in the η-direction. The fill-factors in these directions are $\alpha, \beta \in [0, 1]$. The periodic appearance of pixels is expressed by the convolution

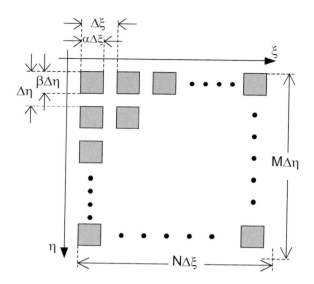

Figure 3.41: Parameters of the CCD-array.

with the comb-function $\text{comb}[\xi/\Delta\xi, \eta/\Delta\eta]$. The whole array has a finite width given by $N\Delta\xi \times M\Delta\eta$ where N and M are the pixel numbers in each direction, Fig. 3.41.

For reconstruction the stored hologram $h_1(\xi, \eta; x_0, y_0)$ is multiplied with the conjugate reference wave r^* and from the resulting product field in the hologram plane the field in the image plane is calculated by the Fresnel diffraction equation which accounts for the propagation over d', see (3.33).

$$b'(x', y'; x_0, y_0) \qquad (3.125)$$
$$= \frac{1}{i\lambda d'} \int_{-\infty}^{\infty} \int_{-\infty}^{\infty} h_1(\xi, \eta; x_0, y_0) r^*(\xi, \eta) \exp\left\{\frac{ik}{2d'}[(x'-\xi)^2 + (y'-\eta)^2]\right\} d\xi\, d\eta.$$

After evaluating the squares in the exponential and the coordinate transform $x'/(\lambda d') \to x'$, $y'/(\lambda d') \to y'$ this represents the Fourier transform

$$b'(x', y'; x_0, y_0) = \mathcal{F}\left\{h_1(\xi, \eta; x_0, y_0) r^*(\xi, \eta) \exp\left[\frac{ik}{2d'}(\xi^2 + \eta^2)\right]\right\}(x', y') \quad (3.126)$$

where phase factors which do not depend on the hologram also are omitted.

For the further discussion of the point spread function $b'(x', y'; x_0, y_0)$ as derived in (3.126), several cases will be treated separately. The first case that will be considered in more detail is the one using the most simple reference wave: the plane wave of unit amplitude impinging normally onto the CCD: $u_r(\xi, \eta) \exp\{i\phi(\xi, \eta)\} = 1$. Since now there are no mixed terms xy, $x'y'$ or $\xi\eta$, the equations can be separated for both directions and it is sufficient to

3.5 Wave-Optics Analysis of Digital Holography

consider only one dimension.

$$b'(x'; x_0) = \mathcal{F}\left\{h_1(\xi; x_0) \exp\left[\frac{ik}{2d'}\xi^2\right]\right\}(x') \qquad (3.127)$$

$$= \mathcal{F}\left\{h(\xi; x_0)\left[\text{rect}\left(\frac{\xi}{\alpha\Delta\xi}\right) \star \text{comb}\left(\frac{\xi}{\Delta\xi}\right)\right]\text{rect}\left(\frac{\xi}{N\Delta\xi}\right)\exp\left(\frac{ik}{2d'}\xi^2\right)\right\}(x').$$

The hologram field, Eq. (3.123), can be expressed in one dimension by

$$h(\xi; x_0) = A + C\exp\left[\frac{ik}{2d}(\xi - x_0)^2\right] + C^*\exp\left[-\frac{ik}{2d}(\xi - x_0)^2\right] \qquad (3.128)$$

where additive and multiplicative terms are summarized in A and C. This gives

$$b'(x'; x_0) \qquad (3.129)$$
$$= \mathcal{F}\left\{\left[\text{rect}\left(\frac{\xi}{\alpha\Delta\xi}\right) \star \text{comb}\left(\frac{\xi}{\Delta\xi}\right)\right]\text{rect}\left(\frac{\xi}{N\Delta\xi}\right)\left[A + \frac{C}{2i}\exp\left(\frac{ik}{2d}(\xi - x_0)^2\right)\right.\right.$$
$$\left.\left. - \frac{C}{2i}\exp\left(-\frac{ik}{2d}(\xi - x_0)^2\right)\right]\exp\left(\frac{ik}{2d'}\xi^2\right)\right\}(x').$$

Performing the multiplication of the last term in square brackets with the exponential yields

$$\left[A + \frac{C}{2i}\exp\left(\frac{ik}{2d}(\xi - x_0)^2\right) - \frac{C}{2i}\exp\left(-\frac{ik}{2d}(\xi - x_0)^2\right)\right]\exp\left(\frac{ik}{2d'}\xi^2\right)$$
$$= A\exp\left(\frac{ik}{2d'}\xi^2\right) + \frac{C}{2i}\exp\left[\frac{ik}{2}\left(\frac{\xi^2}{d} + \frac{\xi^2}{d'} - \frac{2x_0\xi}{d} + \frac{x_0^2}{d}\right)\right]$$
$$- \frac{C}{2i}\exp\left[-\frac{ik}{2}\left(\frac{\xi^2}{d} - \frac{\xi^2}{d'} - \frac{2x_0\xi}{d} + \frac{x_0^2}{d}\right)\right]. \qquad (3.130)$$

The impulse response we are looking for is the Fourier transform of the product resulting when the three terms on the right-hand side of Eq. (3.130) are multiplied with the CCD-term. The presence of a quadratic phase factor in what otherwise would be a Fourier transform relationship will generally have the effect of broadening the impulse response [160]. One has to recognize that only the terms containing the squared variable of integration ξ^2 make trouble, phase factors containing x_0^2 do not affect the Fourier transform. Now the trick is to choose special reconstruction distances d' which make some of these terms identically vanish.

The first term on the right-hand side of Eq. (3.130) corresponds to the d.c.-term. The second term represents the virtual image. It simplifies for $d' = -d$. In the third term, which stands for the real image, the quadratic phase factor is eliminated by the choice $d' = +d$. Let us look at the real image more closely. By choosing $d' = +d$, we obtain a sharp real image without broadening by quadratic exponentials. Nevertheless the d.c.-term remains present and the virtual image is severely broadened so that it appears as a very unsharp cloud in the reconstructed intensity field, see Fig. 3.14b. Now the real image stems from the term $-\frac{C}{2i}\exp\{-\frac{ik}{2d}(-2x_0\xi + x_0^2)\}$ in Eq. (3.130). By application of the convolution theorem

Eq. (3.129) gives

$$b'_{real}(x'; x_0) = \qquad (3.131)$$
$$= \mathcal{F}\left\{\left[\text{rect}\left(\frac{\xi}{\alpha\Delta\xi}\right) \star \text{comb}\left(\frac{\xi}{\Delta\xi}\right)\right] \text{rect}\left(\frac{\xi}{N\Delta\xi}\right) \frac{C}{2\text{i}} \exp\left[-\frac{ik}{2d}(-2x_0\xi + x_0^2)\right]\right\}$$
$$= \alpha\Delta\xi \text{sinc}(\alpha\Delta\xi x')\text{comb}(\Delta\xi x') \star \mathcal{F}\left\{\text{rect}\left(\frac{\xi}{N\Delta\xi}\right) \frac{C}{2\text{i}} \exp\left[-\frac{ik}{2d}(-2x_0\xi + x_0^2)\right]\right\}.$$

Since

$$\mathcal{F}\left\{\text{rect}\left(\frac{\xi}{N\Delta\xi}\right) \exp\left(\frac{ik}{2d}2x_0\xi\right)\right\} = \text{sinc}\left[N\Delta\xi\left(x' - \frac{kx_0}{d}\right)\right] \qquad (3.132)$$

we obtain finally

$$b'_{real}(x'; x_0) = D \text{ sinc}(\alpha\Delta\xi x')\text{comb}(\Delta\xi x') \star \text{sinc}\left[N\Delta\xi\left(x' - \frac{kx_0}{d}\right)\right] \qquad (3.133)$$

where constant intensity and phase factors are contained in D.

Many CCD-arrays have a fill-factor of 100%, meaning $\alpha = \beta = 1$. In this case we use $\text{sinc}(\Delta\xi x')\text{comb}(\Delta\xi x') = \delta(x')$ and Eq. (3.133) simplifies and leads to the compact result

$$b'_{real}(x'; x_0) = D \text{ sinc}\left[N\Delta\xi\left(x' - \frac{kx_0}{d}\right)\right]. \qquad (3.134)$$

Thus the impulse response of the optical system digital holography producing the real image by using the standard plane reference wave is represented by the shifted Fraunhofer diffraction pattern of the aperture defined by the CCD-dimensions. It is a well known fact that if we sample in the spatial domain with a rate $\Delta\xi$ and N samples, then in the related spatial frequency domain the samples calculated by a discrete finite Fourier transform have the distances $\Delta x' = (N\Delta\xi)^{-1}$. Now let the width of the impulse response be defined by the distance between the first zeros of the sinc-function right and left to the origin, then it has the width of two sample distances in the numerically reconstructed image.

Up to now of the three terms on the right-hand side of (3.130) we have considered in more detail only the one related to the real image. The one related to the virtual image can be treated with the same reasoning. On the other hand in Section 3.2.6, we have seen methods for suppressing this image. The d.c.-term of the complex impulse response is determined from (3.130) as

$$b'_{dc}(x'; x_0) = \qquad (3.135)$$
$$= \mathcal{F}\left\{\left[\text{rect}\left(\frac{\xi}{\alpha\Delta\xi}\right) \star \text{comb}\left(\frac{\xi}{\Delta\xi}\right)\right] \text{rect}\left(\frac{\xi}{N\Delta\xi}\right) A \exp\left(\frac{ik}{2d'}\xi^2\right)\right\}$$
$$= A\alpha\Delta\xi \text{sinc}(\alpha\Delta\xi x')\text{comb}(\Delta\xi x') \star \mathcal{F}\left\{\text{rect}\left(\frac{\xi}{N\Delta\xi}\right) \exp\left(\frac{ik}{2d'}\xi^2\right)\right\}.$$

The Fourier transform on the right-hand side of the convolution is calculated in Section A.13. Therefore the d.c.-term manifests as a bright square of width $2\pi N\Delta\xi^2/(d'\lambda)$. A minimization of d' is dictated by the required focusing of the real image. Nevertheless the d.c.-term must have no detrimental effects for there are effective ways to completely eliminate the d.c.-term as introduced in Section 3.2.5.

In the above derivation we have assumed a 100% fill-factor. A large fill-factor results in an averaging of the hologram's intensity distribution in the CCD-plane. A suitable way to analyze the averaging effect of different fill-factors is the sampling of a *chirp function*. Because its frequency increases linearly with the spatial coordinate the influence of the fill-factor to different frequencies can be depicted in a single image. Figure 3.42a displays the real part

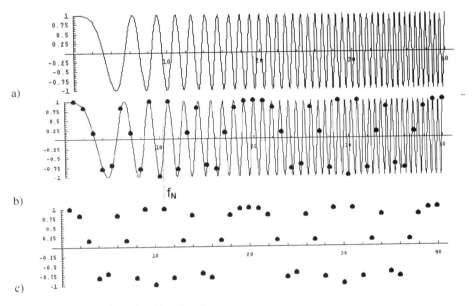

Figure 3.42: Sampling of a chirp function.

of the one-dimensional chirp function $g(\xi) = \exp\{ia\pi\xi^2\}$. The sampling at discrete points, corresponding to a fill-factor approaching zero, is shown in Fig. 3.42b, only the sampled values are shown in Fig. 3.42c. We clearly recognize the Nyquist-limit posted by the sampling theorem at the spatial frequency where one period of the oscillation begins to be sampled by less than two samples. The periodic nature of the samples in Fig. 3.42c stems from the fact that the sampling interval is an integer multiple of a in the formula of $g(\xi)$.

If on the other hand we sample the same chirp function as before with a fill-factor of 100%, we obtain the sampled values of Fig. 3.43a. The continuous chirp function between consecutive sample points is integrated to yield the sampled values. The amplitudes differ from those of Fig. 3.42. While in the foregoing example, Fig. 3.42, the modulus of each sampled complex value was 1, now this modulus varies according to Fig. 3.43b. If we choose a fill-factor of $\alpha = 0.5$ we get the moduli of the sampled complex values as shown in Fig. 3.43c. Figs. 3.43b and 3.43c represent discrete samples of the geometric modulation transfer function (MTF) given in Section 2.8.4 in (2.193). The MTF quantifies the ability of the optical system to transmit spatial frequencies. For fill-factors greater than the zero of pointwise sampling

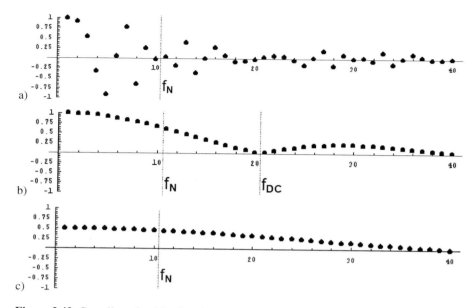

Figure 3.43: Sampling of a chirp function with fill-factors below 100%.

higher frequencies are imaged with less contrast. For a 100% fill-factor, the MTF falls to about 64% of the maximum value at the Nyquist frequency f_N.

For fill-factors less than 100% the factor $\mathrm{sinc}(\alpha\Delta\xi x')\mathrm{comb}(\Delta\xi x')$ in (3.133) will not reduce to a simple δ-impulse at zero. Instead the δ-pulses of the comb-function obtain different amplitudes from the sinc-function which now can take on nonzero values at the points sampled by the comb. The convolution in (3.133) then leads to a periodic replication of the impulse response in (3.134). One such period has the width $1/\Delta\xi$. Since we have N samples, the whole reconstructed image has the width $N\Delta x' = 1/\Delta\xi$ according to (3.36), so in fact we reconstruct only one of the replicated images [258]. Therefore in digital holography with numerical reconstruction this special aspect of the fill-factor can be neglected. On the other hand if we reconstruct digital holograms optically by using a suitable spatial light modulator, e. g. a digital micromirror device (DMD) [227, 259] then all replicated reconstructed images are projected.

The periodic nature of the sampled chirp function as seen in Fig. 3.42c leads to the effect that a finite sized object placed in a way that large angles between object wave and reference wave result and thus high spatial frequencies are produced which are severely undersampled, nevertheless the object surface is faithfully reconstructed. However frequencies now must be restricted to an interval $[n2f_N, (n+1)2f_N]$ in the spatial frequency domain, otherwise ambiguities occur. For fill-factors other than 0% (the pointwise sampling of Fig. 3.42) the intensity of the recorded patterns is reduced. Full theoretical support for this behavior is given by Onural in [132].

Up to now we have only investigated the case of a plane normally impinging reference wave. Next let a spherical reference wave diverge from the point (x_R, y_R, z_R). This wave in

3.5 Wave-Optics Analysis of Digital Holography

the (ξ, η)-plane is

$$r(\xi, \eta) = u_r(\xi, \eta) \exp\left\{\frac{ik}{2z_R}[(\xi - x_R)^2 + (\eta - y_R)^2]\right\}. \quad (3.136)$$

The corresponding hologram distribution in analogy to (3.123) has the form

$$h(\xi, \eta; x_0, y_0) = |u_t(\xi, \eta) + r(\xi, \eta)|^2$$

$$= \frac{1}{\lambda^2 d^2} + u_r^2(\xi, \eta) \quad (3.137)$$

$$+ \frac{2u_r(\xi, \eta)}{\lambda d} \sin\left\{\frac{k}{2}\left[\frac{(\xi - x_0)^2}{d} + \frac{(\eta - y_0)^2}{d} - \frac{(\xi - x_R)^2}{z_R} - \frac{(\eta - y_R)^2}{z_R}\right]\right\}.$$

If we restrict ourselves again to one dimension, we can write

$$h(\xi; x_0) = A + C \sin\left\{\frac{k}{2}\left[\frac{(\xi - x_0)^2}{d} - \frac{(\xi - x_R)^2}{z_R}\right]\right\}. \quad (3.138)$$

In analogy to the way that led to Eq. (3.130) we obtain

$$h(\xi; x_0) \exp\left\{i\frac{k}{2d'}\xi^2\right\} \quad (3.139)$$

$$= A \exp\left\{\frac{ik}{2d'}\xi^2\right\}$$

$$+ \frac{C}{2i} \exp\left\{\frac{ik}{2}\left(\frac{\xi^2}{d'} + \xi^2\left(\frac{1}{d} - \frac{1}{z_R}\right) - 2\xi\left(\frac{x_0}{d} - \frac{x_R}{z_R}\right) + \left(\frac{x_0^2}{d} - \frac{x_R^2}{z_R}\right)\right)\right\}$$

$$- \frac{C}{2i} \exp\left\{-\frac{ik}{2}\left(-\frac{\xi^2}{d'} + \xi^2\left(\frac{1}{d} - \frac{1}{z_R}\right) - 2\xi\left(\frac{x_0}{d} - \frac{x_R}{z_R}\right) + \left(\frac{x_0^2}{d} - \frac{x_R^2}{z_R}\right)\right)\right\}$$

$$= A \exp\left\{\frac{ik}{2d'}\xi^2\right\}$$

$$+ \frac{C}{2i} \exp\left\{\frac{ik}{2}\left(\frac{\xi^2}{d'} + \frac{\xi^2}{\hat{d}} - \frac{2\xi}{\hat{d}}\left(\frac{x_0 z_R - d x_R}{z_R - d}\right) + \frac{1}{\hat{d}}\left(\frac{x_0^2 z_R - d x_R^2}{z_R - d}\right)\right)\right\}$$

$$- \frac{C}{2i} \exp\left\{-\frac{ik}{2}\left(-\frac{\xi^2}{d'} + \frac{\xi^2}{\hat{d}} - \frac{2\xi}{\hat{d}}\left(\frac{x_0 z_R - d x_R}{z_R - d}\right) + \frac{1}{\hat{d}}\left(\frac{x_0^2 z_R - d x_R^2}{z_R - d}\right)\right)\right\}$$

where in the last two lines we have introduced

$$\hat{d} = \frac{d z_R}{z_R - d}. \quad (3.140)$$

Again we only investigate the last part of (3.139) which is related to the real image and also concentrate only on the terms containing ξ^2. One way to perform the numerical reconstruction is to multiply the hologram h with the conjugate of the spherical reference wave (3.136). This eliminates the ξ^2-terms in that part of (3.139) describing the real image and the problem is reduced to the already treated case of a plane reference wave resulting in the same impulse response. Another way to reconstruct is to leave out the multiplication with the conjugate reference wave and to use instead the reconstruction distance $d' = \hat{d}$, which has the same effect of eliminating the ξ^2-terms. This approach is recommended since it avoids some complex multiplications with possible rounding errors, especially in the high spatial frequency regions of the digital holograms.

The important case of lensless Fourier transform holography, see Section 3.2.3, is characterized by $z_R = d = d'$. In this case Eq. (3.138) simplifies to

$$h(\xi; x_0) = A + C \sin\left\{\frac{k}{2}\left[\frac{2\xi(x_R - x_0)}{d} - \frac{x_0^2 - x_R^2}{d}\right]\right\}. \tag{3.141}$$

The multiplication with $\exp\{ik\xi^2/(2d')\}$, e. g. as in Eq. (3.127), is compensated at least in the terms containing ξ^2 by the multiplication with the conjugate of the reference wave. Thus in practice we neither have to multiply with $\exp\{ik\xi^2/(2d')\}$, which is characteristic for the Fresnel transform, nor do we have to carry out the multiplication with the reference wave, we have to perform only the Fourier transform. Equation (3.139) now reduces to

$$h(\xi; x_0) = A + \frac{C}{2i}\left[\exp\left\{\frac{ik}{2d}\left(2\xi(x_R - x_0) + x_0^2 - x_R^2\right)\right\}\right.$$
$$\left. - \exp\left\{\frac{-ik}{2d}\left(2\xi(x_R - x_0) + x_0^2 - x_R^2\right)\right\}\right]. \tag{3.142}$$

The Fourier transform of this expression represents the reconstructed image of lensless Fourier transform holography. It has a d.c.-term that is not broadened but is restrained to a single pixel. Furthermore the choice $z_R = d = d'$ eliminates all ξ^2-terms simultaneously in the real and the virtual image, so both images are reconstructed sharply focused, a behavior we already know.

3.5.2 Frequency Analysis of Digital Holography with Reconstruction by Convolution

Besides reconstruction by the Fresnel transform there exist powerful reconstruction methods based on convolution. This approach is introduced in Section 3.3. As in Section 3.5.1 we perform a frequency analysis to find out the point spread function in the case of reconstruction by convolution. Again we start with an idealized point source at coordinates (x_0, y_0) in the object plane producing the field $B(\xi, \eta; x_0, y_0)$ in the hologram plane as given in (3.122). The resulting hologram $h(\xi, \eta; x_0, y_0)$ as well as the sampled, digitized, recorded, and stored hologram $h_1(\xi, \eta; x_0, y_0)$ are the same as in the aforementioned case of reconstruction by the Fresnel transform and given in (3.123) and (3.124) respectively. From the digital hologram $h_1(\xi, \eta; x_0, y_0)$ multiplied with the conjugate of the reference wave $r^*(\xi, \eta)$ the field in the image plane is now calculated by

$$b'(x', y'; x_0, y_0) = \mathcal{F}^{-1}\left\{\mathcal{F}\left\{h_1(\xi, \eta; x_0, y_0) \cdot r^*(\xi, \eta)\right\} \cdot \mathcal{F}\left\{g(\xi, \eta)\right\}\right\}. \tag{3.143}$$

The equivalence of the image plane and the hologram plane in the case of convolution reconstruction has been discussed in Section 3.4.4, so the coordinates (x', y') are chosen for

3.5 Wave-Optics Analysis of Digital Holography

Figure 3.48: Transfer functions: (a) real part of $G_{1.0m}(n,m)$, (b) real part of $G_{0.5m}(n,m)$, (c) intensity of $G_{1.0m}(n,m)$, (d) intensity of $G_{0.5m}(n,m)$, (e) phase of $G_{1.0m}(n,m)$, (f) phase of $G_{0.5m}(n,m)$.

in Fig. 3.49, but now for the distances 0.33 m and 0.2048 that represents the maximum field for the chosen parameters.

The limited domain in the transfer function now has the effect that not all spatial frequencies eventually contained in the product of hologram and reference wave pass through the multiplication step with the transfer function in the spatial frequency domain. This inherent low-pass filtering eliminates the high frequencies, eventually fine details are averaged out. To avoid this effect a transfer function with positive amplitudes over a rather large frequency band should be chosen for performing the reconstruction. It was already mentioned that the frequency range where the transfer function is non-vanishing increases with decreasing reconstruction distance d. Therefore d should be as small as possible without violating the sampling theorem. This is shown in one dimension in Fig. 3.50. The object consists of bright bars of differing width, Fig. 3.50a. The parameters for the simulation are $\lambda = 0.5$ µm, $\Delta\xi = 5.0$ µm, $N = 1024$. The amplitude of the normally impinging reference wave is 1. The impulse response for $d = 0.4$ m is given in Fig. 3.50b, and Fig. 3.50c shows its Fourier transform amplitude, the amplitude of the transfer function. The reconstructed intensity is shown in Fig. 3.50d with the original intensity inserted for better comparison. Figs. 3.50e to g show the same distributions for $d = 0.2$ m and Figs. 3.50h to j these for $d = 0.1$ m. It is obvious how the steepness of the flanks increases with the increasing frequency bands.

The aim to use transfer functions belonging to small reconstruction distances d leads to the concept of cascaded free space propagation. Let us take the transfer function as defined in (3.77) or (3.79). With the properties of the exponential it can easily be shown that

$$G_d(n,m) = \prod_{i=1}^{k} G_{d_i}(n,m) \qquad \text{with} \qquad \sum_{i=1}^{k} d_i = d \qquad (3.147)$$

where the subscript d in G_d indicates the distance for which the transfer function is defined. As an example

$$G_d(n,m) = G_{d/2}(n,m) G_{d/2}(n,m) \qquad (3.148)$$

which is displayed in Fig. 3.51. Here we see that the diffraction field $b_2(x'', y'')$ can be calculated by free space propagation from the (ξ, η)-plane over the distance d. But alternatively we

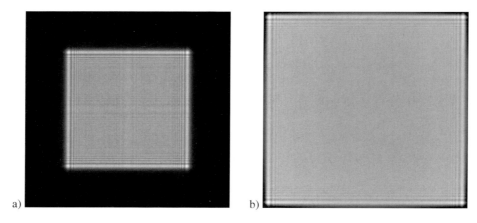

Figure 3.49: Transfer functions: (a) intensity of $G_{0.33m}(n,m)$, (b) intensity of $G_{0.2048m}(n,m)$.

3.5 Wave-Optics Analysis of Digital Holography

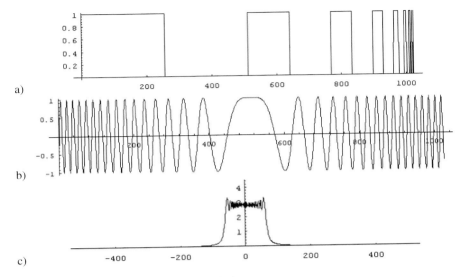

Figure 3.50: Simulated reconstructions of a bar pattern.

can calculate first the field $b_1(x', y')$ by free space propagation over $d/2$ followed by a second propagation over $d/2$ from $b_1(x', y')$ to $b_2(x'', y'')$. The field in the (x', y')-plane is

$$b_1(x', y') = \mathcal{F}^{-1}\{\mathcal{F}\{h \cdot r\} \cdot G_{\frac{d}{s}}\}. \tag{3.149}$$

With this we get

$$\begin{aligned} b_2(x'', y'') &= \mathcal{F}^{-1}\{\mathcal{F}\{b_1(x', y')\} \cdot G_{\frac{d}{s}}\} \\ &= \mathcal{F}^{-1}\{\mathcal{F}\{\mathcal{F}^{-1}\{\mathcal{F}\{h \cdot r\} \cdot G_{\frac{d}{s}}\}\} \cdot G_{\frac{d}{s}}\} \\ &= \mathcal{F}^{-1}\{\mathcal{F}\{h \cdot r\} \cdot G_{\frac{d}{s}} \cdot G_{\frac{d}{s}}\} \end{aligned} \tag{3.150}$$

because $\mathcal{F}\{\mathcal{F}^{-1}\{f\}\} = f$ for any f.

The effect of the cascaded transfer function is shown in an example where the object consists of bright stochastic phase rectangles of different size. These were chosen to demonstrate the fidelity and resolution of the reconstruction, Fig. 3.52a.

The whole object has the size of the d.c.-term, the parameters are $\lambda = 0.5$ µm, $\Delta\xi = \Delta\eta = 10$ µm, $n = m = 1024$, $d = 0.82$ m. The Fresnel-reconstructed intensity is given in Fig. 3.52b. The reconstructions using convolutions and the cascaded transfer functions $G_{0.82m}$, $(G_{0.41m})^2$, $(G_{0.273m})^3$, and $(G_{0.205m})^4$ are shown in Figs. 3.52c, d, e, and f. We recognize large speckles of low frequency in Fig. 3.52c where we reconstructed with $G_{0.82m}$. The fine high-frequency structures of the object are blurred. These fine structures are better recognized in Figs. 3.52d and e. The blurring increases again for $(G_{0.0205m})^4$ but now the reason is that we are at the limit of the sampling theorem for this distance and the chosen parameters.

The results of the theoretical investigations on shifted versions of the chirp function in Section A.13 are valid also for the shifted impulse response functions (3.88) since the chirp

156   3   Digital Recording and Numerical Reconstruction

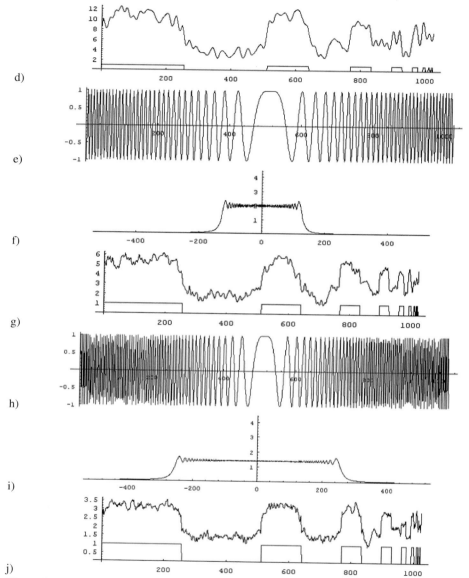

Figure 3.50 (continued): Simulated reconstructions of a bar pattern.

function is an approximation of this impulse response, which is especially evident for the Fresnel approximated version g_F, see (3.80). Shifts (s_k, s_l) move the centers of the ring structures of the impulse responses in the finite rectangular window. It is even possible that the center is far outside the window and only high-frequency partial rings are in the finite field.

This in the spatial frequency domain moves the non-zero part of the transfer function. However, the various reconstructed fields displayed in Fig. 3.27 required different shifts, so the focus that depends on the aperture of the optical system which is given here by the size of the CCD-array, see Section 3.4.4. This limited resolution can be depicted by the so called depth images: In digital holography we cannot only reconstruct in image planes parallel to the hologram plane [90], but also in those orthogonal to that plane, Fig. 3.55. Conventionally

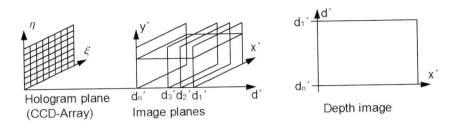

Figure 3.55: Orientation of reconstructed depth image.

we reconstruct in (x', y')-planes for fixed d', but here we calculate the field in a (x', d')-plane for a fixed y'. Figure 3.56a gives the intensity reconstruction in the (x', y')-plane for three simulated particles at different distances to the CCD, each particle having a size of one pixel. The reconstruction depth d' coincides with the distance of the leftmost particle,

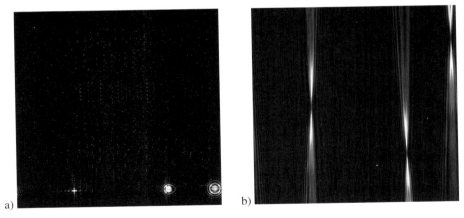

Figure 3.56: Reconstructed particles, image planes oriented parallel (a) and orthogonal (b) to the hologram plane.

therefore the other two particles are unsharp. The corresponding depth image is shown in Fig. 3.56b. The waists of the projected cones roughly indicate the loci of the three particles. Computational time can be saved if not the whole 2D depth image is calculated but only the 1D intensity distribution along a line parallel to the z-axis going through the center of the axisymmetric fringes belonging to a specific particle. The particle is located where this intensity is extremal [96].

A better attempt to determine position and form of the particles in 3D space is by a method related to tomography [253, 265, 266]. Several views of the particle or droplet spray corresponding to different angles of observation have to be recorded simultaneously. This is performed by a multiple pass arrangement. Figure 3.57a shows an arrangement where the collimated laser beam passes the particle stream three times. The deconvoluted light path is depicted in Fig. 3.58. Since in digital holography we can numerically focus to arbitrary distances d', the images reconstructed for different distances correspond to different angular views of the particles.

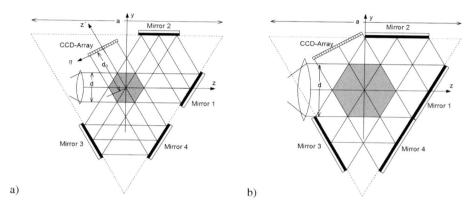

Figure 3.57: Triple pass arrangement for digital holographic particle analysis.

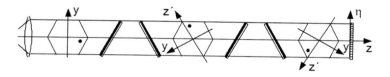

Figure 3.58: Virtual particles in deconvoluted light path.

An analysis of the arrangement of Fig. 3.57a shows that the flat mirrors coincide with the side planes of a prism whose base is an equal-sided triangle. If its sidelength is a then the width d of the collimated laser beam, which also is the width of the measurement volume, has to remain $d \leq a/(2\sqrt{3})$ to prevent shadowing by components. The arrangement with the maximum $d = a/(2\sqrt{3})$ is shown in Fig. 3.57b. For this arrangement the optical path from the origin over two mirrors back to the origin is exactly a, which defines the amount the reconstructed image planes have to be apart.

For the reconstruction the convolution approach instead of the Fresnel transform is now recommended, because all reconstructed images have the same size independent from the individual reconstruction distance. No adaptation of the image coordinates is necessary. An even more refined method for the reconstruction of particle fields in digital holography based on the 2D fractional-order Fourier transformation is presented in [89].

3.6 Non-Interferometric Applications of Digital Holography

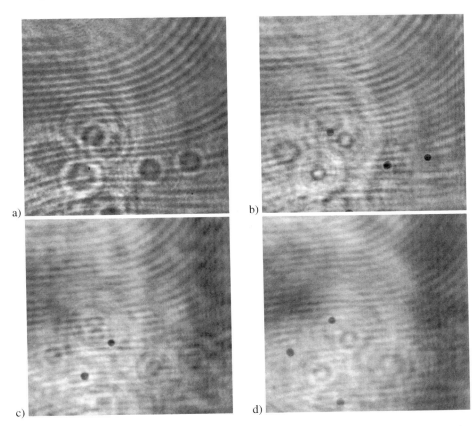

Figure 3.59: Digital hologram of particles (a) and reconstructed intensities for different distances (b), (c), (d).

By numerical focusing onto different distances during reconstruction, the different views of the particle field are gained. Figure 3.59a shows the digital in-line hologram of particles having about 250 µm diameter falling through the collimated beam of a pulsed laser. The CCD-array used in this experiment had 2048 × 2048 pixels each of size 9 µm × 9 µm. The particle fields now are reconstructed from the digital hologram at distances 40 cm, 65.5 cm, and 95.5 cm. The resulting intensity distributions are displayed in Figs. 3.59b, c, and d.

The reconstructed fields now can be evaluated with respect to size and position of the particles. As an intermediate step identical particles in the three reconstructions have to be identified. As long as in a fixed height y_0 there is a single particle image in each reconstruction, these images must stem from the same particle. Ambiguities will occur if there are more than one particle in a certain height y_0. But since we have three views for determining the two-dimensional position (x_0, z_0), this overdetermination can be used for particle identification: The positions of a particle in all three reconstructions must be compatible.

A concept related to the evaluation of different angular views of the same scene is *computerized tomography*. The intensity reconstructions along the line of the same height in all views are used for tomographic processing by e. g. the method of *filtered backprojection*. A schematic sketch of the backprojection principle is given in Fig. 3.60a. A result of this

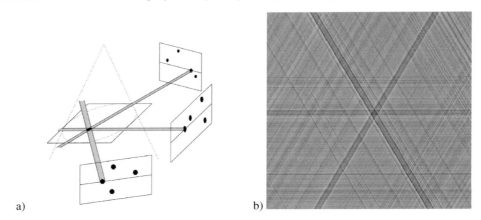

Figure 3.60: Tomographic reconstruction: Principle of filtered backprojection (a), reconstructed plane (b).

approach for the three views of Fig. 3.59 is given in Fig. 3.60b. The particle is located in the cross-section of the three projections visible as the darker streaks. However, the tomographic approach will become feasible if there are much more projections through the particle field than three. Using tomography to detect particle positions from only three views is like taking a sledgehammer to crack a nut. We do not need the ability of tomography to determine continuous transmission values: Our field is binary, the particles are opaque, the space between the particles is transparent.

A straightforward localization of the particles is by measuring their η-coordinate η_i in each view i ($i = 1, 2, 3$) for the investigated intersecting plane and solving the system of equations

$$\begin{aligned} y &= -\sqrt{3}z - 2\eta_3 \\ y &= +\sqrt{3}z - 2\eta_2 \\ y &= \eta_1. \end{aligned} \quad (3.154)$$

The depth coordinate of the particles does not need to be controlled very exactly due to the large depth of focus, nevertheless the lateral η-coordinates are determined very precisely.

The equations (3.154) define three lines whose common intersection gives the position of the particle in the chosen plane through the particle stream, Fig. 3.61a. Since the system is overdetermined, here even two particles in one plane produce no ambiguities but can be located uniquely, Fig. 3.61b.

Another arrangement used in practice also having three angular views is shown in Fig. 3.62. The dashed lines indicate the walls of an octagonal pressure chamber in and out of which the collimated beam is guided through transparent windows. The mirrors, the optics

3.6 Non-Interferometric Applications of Digital Holography

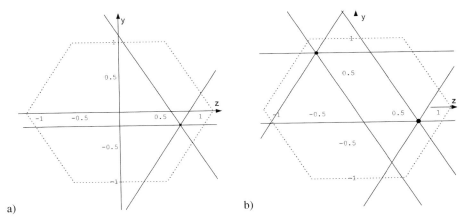

Figure 3.61: Lines indicating particle positions at triple intersection points.

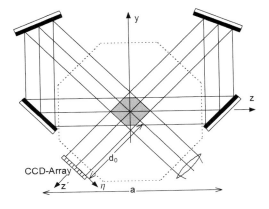

Figure 3.62: Arrangement for particle measurement in an octagonal chamber.

and the CCD-array are outside of the chamber. The related system of equations for this case is

$$\begin{aligned} y &= +z - \sqrt{2}\eta_3 \\ y &= \eta_2 \\ y &= -z - \sqrt{2}\eta_1. \end{aligned} \quad (3.155)$$

More alternatives for multiple pass arrangements are displayed in Fig. 3.63.

The velocity of the particles is obtained by recording digital holograms using double pulsed lasers [261, 265]. Figure 3.64a shows one of the intensity reconstructions of a digital double pulse hologram where the arrangement of Fig. 3.62 was used. Each particle appears twice. From the positions in space and the known pulse delay of 400 µs, a velocity of about 0.6 m s^{-1} was evaluated. The positions of five identified particles determined from several reconstructions are displayed perspectively in Fig. 3.64b. Digital image plane holography as a

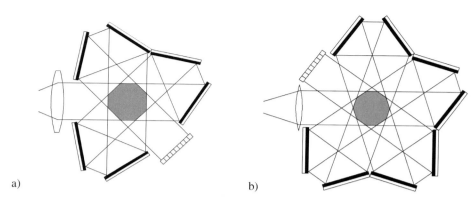

Figure 3.63: Arrangements with four (a) and five (b) passes through the particle stream.

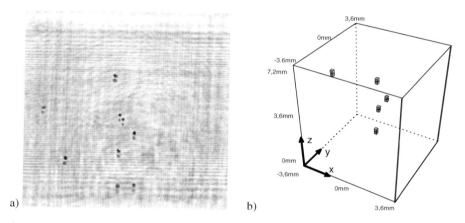

Figure 3.64: Particles reconstructed from digital double exposure hologram (a) and 3D particle distribution (b).

measurement tool for three-dimensional flow velocimetry is presented in [267]. A wavelet approach evaluating digital in-line holograms for multiple plane velocimetry is described in [88].

One may suspect that particles which are positioned one behind the other in the deconvoluted light path would hide each other. This would be so in a purely geometric approach analyzing only the shadow casting. But since the diffracted field of each particle spreads over the whole digital hologram, information from the rear particles is also contained in the hologram. This is demonstrated in the computer simulation shown in Fig. 3.65.

Figure 3.65a gives the digital hologram of particles assumed in three different planes which have the distances 1.0 m, 0.75 m, and 0.5 m from the CCD-array. The positions of the particles in the first plane are in pixel coordinates $\{(600, 300), (500, 800), (100, 900)\}$, in the second plane $\{(600, 300), (500, 800), (800, 400)\}$, and $\{(100, 100), (500, 800), (800, 400)\}$ in the third plane. This means that we have two times two particles one behind the other and once we have three particles lined up. The three intensity reconstructions are displayed in Figs. 3.65b to d, and all particles are uniquely reconstructed.

3.6 Non-Interferometric Applications of Digital Holography 167

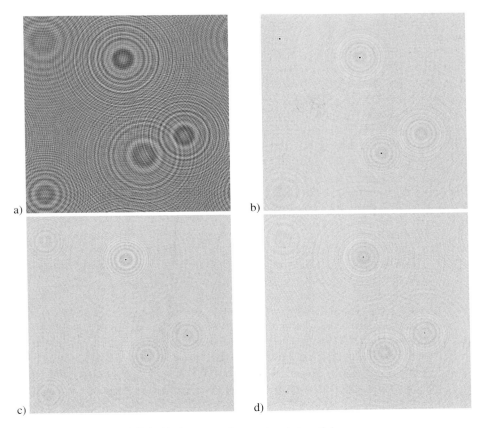

Figure 3.65: Simulated digital hologram and reconstructed particles.

A typical problem in digital in-line holography is the presence of the d.c.-term and the twin image in the reconstructed field, see e. g. Figs. 3.56b, c, and d. While the overall d.c.-term degrades the contrast, the twin images of the particles manifest as unsharp ring structures. Besides the possibilities of d.c.-term suppression of Section 3.2.5 or the employment of phase-shifting digital holography [268], an interesting method for eliminating the twin image in digital in-line holography is proposed by Lai et al. [269]: If not the whole area of the digital hologram is used for numerical reconstruction, but only a non-central part, then for a number of object points in a certain area we have an off-axis geometry with well separated d.c.-term and twin image. By changing the areas of the digital hologram in several reconstruction processes, different areas in the image plane can be reconstructed without a twin image. Finally these undisturbed areas are combined to the total reconstructed field.

A further source for a possible degradation of the reconstructed particle images is the limited size of the CCD array. Consider a single small particle. Its hologram in digital in-line holography consists of infinitely many concentric rings, see e. g. Fig. 3.65a. Due to the limited size of the CCD array only a finite number of such rings fit into the hologram, furthermore the hologram pattern is rectangularly clipped. The consequences from these two facts are now discussed in more detail.

The finite number of rings leads to a blurring of the reconstructed particle image. The radii and thus the number of rings strongly depend on the distance d between particle and hologram: The further away the particle is from the hologram, the larger the ring radii are and the smaller is the number of rings hitting the CCD array. But the rings far away from the center with small distances between are those carrying the high frequency information. If they are cut away the particle image appears low-pass filtered. On the other hand the reconstructed particle is sharper if the distance d is decreased.

The square contour of the CCD array can be modeled by a two-dimensional rectangular function $\text{rect}(ax) \cdot \text{rect}(ay)$, see (A.6). The infinite hologram is multiplied with this function. Therefore the reconstructed particle appears as convolved with the transform of this function, a two-dimensional sinc-function, see Table A.1. Figure 3.66 illustrates this with a simulation of a single particle at different distances.

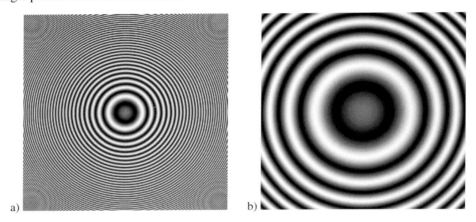

Figure 3.66: Digital holograms of a particle: (a) $d = 1.0$ m, (b) $d = 8.0$ m.

Figure 3.66a shows the hologram of the particle $d = 1.0$ m away from the hologram plane, while Fig. 3.66b displays the hologram for $d = 8.0$ m. The fewer rings in Fig. 3.66b are obvious. Figures 3.66c to f give the magnified central parts of the reconstructed intensity fields. The partial image that has been magnified has the same size in all four displays. The simulations were performed with the distances $d = 1.0$ m (Figs. 3.66a and c), $d = 2.0$ m (Fig. 3.66d), $d = 4.0$ m (Fig. 3.66e), and $d = 8.0$ m (Figs. 3.66b and f). One can recognize the increasing blurring with increasing distance d, also the sinc-function influence. Although the reconstructed patterns here are displayed to the full gray-scale their contrast decreases with increasing d: The contrast $C = (I_{\max} - I_{\min})/(I_{\max} + I_{\min})$ is $C(d = 1.0m) = 0.56$, $C(d = 2.0m) = 0.17$, $C(d = 4.0m) = 0.035$, and $C(d = 8.0m) = 0.0075$. The advice we have to conclude from these facts is that in particle analysis the optical arrangements should be compact to keep the distances d as short as possible.

Applications of digital in-line holography for 3D visualization are numerous. Besides particle tracking, e. g. in the study of erosion processes in sediments [270], there are imaging of

3.6 Non-Interferometric Applications of Digital Holography

This indicates that the reconstruction process can be performed by the following steps:
Step 1: The recorded digital hologram $h(\xi, \eta)$ is multiplied by $4\lambda^2(d^2 + \xi^2 + \eta^2)(1 + d/\sqrt{d^2 + \xi^2 + \eta^2})^{-2}$.
Step 2: The coordinate transformation of (3.165) from the hologram plane to the spatial frequency plane is carried out. This transformation is depicted in Fig. 3.68.
Step 3: From the resulting distribution the inverse Fourier transform is calculated.

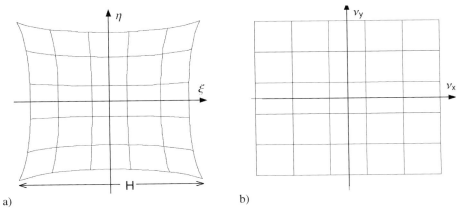

Figure 3.68: Coordinate transformation from the hologram plane (a) to the spatial frequency plane (b).

For the calculation in step 3 the efficient FFT algorithm is employed. Since steps 1 and 2 are pointwise operations, altogether we obtained a fast algorithm for reconstruction. Experimental verifications of this reconstruction method are presented in [100]. The resolvable distance δ obtained by this algorithm can be estimated as follows. Because of $\sqrt{\xi^2 + \eta^2} \leq H/\sqrt{2}$ the highest spatial frequency ν_{max} is given by (3.165) as

$$\nu_{\text{max}} = \frac{1}{\lambda}\sqrt{2 + 4\left(\frac{d}{H}\right)^2}. \tag{3.168}$$

Since the resolvable distance δ obeys the relationship $\delta = \frac{1}{2}\nu_{\text{max}}$ we obtain

$$\delta = \lambda\sqrt{\frac{1}{2} + \left(\frac{d}{H}\right)^2}. \tag{3.169}$$

So the resolution increases, which means that the resolvable distance decreases, when the ratio d/H decreases.

It was assumed that the extent of the investigated object is small. With the help of (3.163) we can determine the allowable object extent. The first approximation in (3.163)

$$kr = k\sqrt{d^2 + \xi^2 + \eta^2 - 2x\xi - 2y\eta}\left[1 + \frac{x^2 + y^2}{2(d^2 + \xi^2 + \eta^2 - 2x\xi - 2y\eta)}\right]$$

$$\approx k\sqrt{d^2 + \xi^2 + \eta^2 - 2x\xi - 2y\eta} \tag{3.170}$$

is valid when the term $(x^2 + y^2)/[2(d^2 + \xi^2 + \eta^2 - 2x\xi - 2y\eta)]$ is small. Let it be smaller than π, then no contrast reversal of fringes would appear

$$x^2 + y^2 < \lambda(d^2 + \xi^2 + \eta^2 - 2x\xi - 2y\eta). \tag{3.171}$$

The right-hand side becomes minimal for $x = \xi$ and $y = \eta$ and we get an allowable extent for the object of

$$x^2 + y^2 < \lambda d^2. \tag{3.172}$$

The approximation in the last line of (3.163) holds if

$$(d^2 + \xi^2 + \eta^2)^{\frac{3}{2}} > \frac{(x\xi + y\eta)^2}{\lambda}. \tag{3.173}$$

With Schwartz's inequality we obtain

$$x^2 + y^2 < \frac{\lambda(d^2 + \xi^2 + \eta^2)^{\frac{3}{2}}}{\xi^2 + \eta^2} \tag{3.174}$$

where the right-hand term is minimal for $\xi^2 + \eta^2 = 2d^2$ yielding

$$x^2 + y^2 < \sqrt{\frac{27}{4}}\lambda d \tag{3.175}$$

which is a less stringent condition than (3.172). Therefore the allowable object width w_{obj} is

$$w_{obj} = \sqrt{2\lambda d}. \tag{3.176}$$

With modified coordinate transformations as presented in [271, 272] the viewing angles can be changed. This allows a variation of the perspective from which the object is viewed. A related approach for reconstruction of particle fields in tilted planes is described in [273].

A slightly different arrangement for digital holographic microscopy is the in-line holography geometry with a divergent wave as shown in Fig. 3.69. The laser beam is focused by a

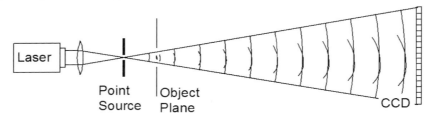

Figure 3.69: Arrangement for point source microscope.

lens with focal length of a few millimeters into a pinhole where the locus of the point source is defined [76]. The object is placed just a few millimeters further along, such that only a

3.6 Non-Interferometric Applications of Digital Holography

fraction of the incident beam is scattered. The scattered wave interferes with the rest of the spherical wave at the detector, which can be a screen that is recorded by a CCD camera, or which can directly be the CCD array.

The numerical reconstruction of these digital holograms of microscopic scenes was developed and investigated by Kreuzer et al. [72, 73, 75]. The first step proposed by them is to calculate the contrast image. Therefore not only the hologram $h(n, m)$ in the detector plane is recorded but also a background image $h_0(n, m)$, where only the illumination is present but the object is removed. Then the contrast image $h_k(n, m)$ is determined pointwise by

$$h_k(n, m) = \frac{h(n, m) - h_0(n, m)}{\sqrt{h_0(n, m)}}. \tag{3.177}$$

This contrast image hereafter is centered around zero intensity by subtracting the average of all $h_k(n, m)$, see Section 3.2.5, so the results will not suffer from a d.c.-term. The modified contrast image $h_k^*(n, m)$ is used to reconstruct the object wave front by the Kirchhoff-Helmholtz transform

$$K(\boldsymbol{r}) = \frac{1}{4\pi} \int_S h_k^*(\boldsymbol{\xi}) \exp\{ik\boldsymbol{\xi} \cdot \boldsymbol{r}/|\boldsymbol{\xi}|\} d\boldsymbol{\xi} \tag{3.178}$$

Here S denotes the screen, which is at a distance D from the source. The screen coordinates are $\boldsymbol{\xi} = (\xi, \eta, D)$, the integration is performed in two dimensions over ξ and η. $K(\boldsymbol{r})$ is significantly different from zero only in the space region occupied by the object. $K(\boldsymbol{r})$ is a complex function and usually the object is represented by its magnitude, although phase information can also be obtained. Normally one reconstructs the wavefront on a number of planes at varying distances from the source in the vicinity of the surface until one obtains the sharpest image of the object.

The general criteria are that the image screen must be large enough so that at least two diffraction orders are recorded and that its spatial resolution must be able to define interference patterns of the order of the wavelength when projected from the screen to the image plane [73].

The arrangement of Fig. 3.69 applies for translucent objects or for objects whose contours are of main interest. However the method is also applicable to investigate opaque surface variations in reflection under glancing incidence [73]. The geometry of this simplest realization of off-axis holography is shown in Fig. 3.70. Here the detector is placed perpendicular to the central reflected beam. The numerical reconstruction is performed analogously to the in-line case.

We now come to the more classic concept of using a microscope objective producing a magnified image of the sample. This image field is used as the object for the hologram generation. In a way this is the inverse approach to that presented in Section 3.1.2 where the wave field reflected from a large object was reduced. The procedure to record and reconstruct the magnified object wave field in digital holographic microscopy is straightforward by using the Fresnel or convolution methods for reconstruction. But digital holography enables us not only to reconstruct intensity contrasts, also we can get phase contrast images. To exploit this additional feature of phase contrast microscopy by digital holography we must recognize the phase aberrations associated with the microscope objective. For an interferometric comparison of two reconstructed wave fields carrying the same aberrations, these eliminate each other and

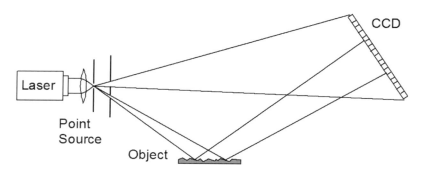

Figure 3.70: Arrangement for point source microscope with glancing incidence.

do not affect the measurement, but in single hologram phase contrast microscopy they matter. However, the phase aberrations can be effectively compensated for by a modification of the digital reconstruction procedure [124].

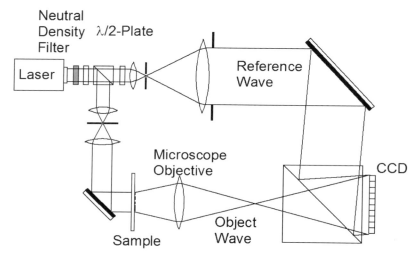

Figure 3.71: Digital holographic microscope, transparent objects.

The arrangement for digital holographic microscopy generally is a Mach-Zehnder interferometer [128]. It can be adapted to measure transparent objects, Fig. 3.71, or can be prepared for reflection imaging, Fig. 3.72. The linearly polarized laser beam passes a neutral density filter and a half-wave plate before it is split into reference and object wave by a polarizing beam splitter. A second half-wave plate is inserted into the reference arm to obtain parallel polarization at the CCD. The reference wave does not hit the CCD in the normal direction, but impinges with a small inclination angle to produce well separated real and virtual images. The transparent sample (or sample holder plus sample) is illuminated by a plane wave, the transmitted light is collected by a microscope objective. For reflection imaging, Fig. 3.72, a

3.6 Non-Interferometric Applications of Digital Holography

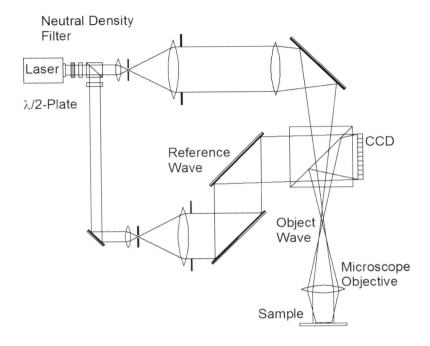

Figure 3.72: Digital holographic microscope, opaque objects.

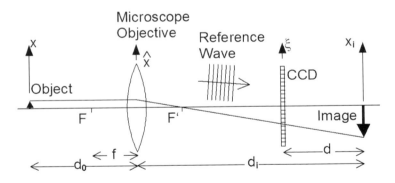

Figure 3.73: Digital holographic microscope, optical configuration.

lens with long focal length is inserted between beam expander and microscope objective. It acts as a condenser, its position is adjusted to illuminate the opaque sample with a collimated beam.

Let us investigate the optical arrangement in the object arm in more detail. It is a single-lens imaging system, Fig. 3.73. The microscope objective produces a magnified image of the object, and the CCD is placed between the objective lens and the image. The distance of the CCD to the image is d. Two special cases are worth mentioning. If the object–hologram distance is zero, $d = 0$, then the image is focused onto the CCD. This would constitute an ESPI or DSPI configuration, which can be used for interference microscopy where phase-shifting interferometry is applied to obtain phase information. The second special case is the sample located in the object focal plane of the microscope objective. Now the distance d_i between objective and image becomes infinite and the Fresnel transform in the reconstruction stage converts to a pure Fourier transform of the digital hologram.

The reconstruction of the image field from the digitally recorded hologram now can be performed by the Fresnel transform, Eq. (3.38). In the numerical model of the reference wave the oblique incidence must be considered, e. g.

$$r(k\Delta\xi, l\Delta\eta) = E_r \exp\left[\frac{\mathrm{i}2\pi}{\lambda}(k_\xi k\Delta\xi + k_\eta k\Delta\eta)\right] \tag{3.179}$$

where k_ξ and k_η are components of the unit vector in the propagation direction of the reference wave [274]. There remains the problem of wavefront deformation by the microscope objective, Fig. 3.74. The resulting phase aberrations are crucial in phase contrast imaging.

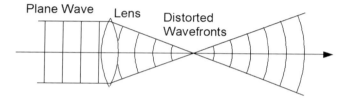

Figure 3.74: Wavefront deformation by the microscope objective.

They can be eliminated by multiplication of the reconstructed field with an appropriate digital phase mask, as derived in [124] or by one of the methods presented in [275].

The relation between the fields $U_0(x, y)$ in the object plane and $U_i(x_i, y_i)$ in the image plane can be described by the superposition integral

$$U_i(x_i, y_i) = \iint h(x_i, y_i; x, y) U_0(x, y) \, dx \, dy \tag{3.180}$$

with the point spread function $h(x_i, y_i; x, y)$. For the moment let us restrict the discussion to two dimensions, say the (x, z)-plane and investigate $h(x_i; x)$. The results are then extended in a natural way to three dimensions. According to Fig. 3.73 we have

$$\frac{1}{d_i} + \frac{1}{d_0} = \frac{1}{f} \tag{3.181}$$

and following [160] we obtain

$$h(x_i; x) = C \exp\left(\frac{\mathrm{i}\pi}{\lambda d_i} x_i^2\right) \exp\left(\frac{\mathrm{i}\pi}{\lambda d_0} x^2\right) \int P(\hat{x}) \exp\left[\frac{-\mathrm{i}2\pi}{\lambda}\left(\frac{x}{d_0} + \frac{x_i}{d_i}\right)\hat{x}\right] d\hat{x} \tag{3.182}$$

where $\hat{x}$ is the coordinate in the plane of the microscope lens, whose pupil function is $P(\hat{x})$. C is a complex constant. We assume perfect imaging with magnification $M = d_i/d_o$ which maps

$$(x, y) \longmapsto (x_i, y_i) = (-Mx, -My). \tag{3.183}$$

Then the integral in (3.182) can be approximated by a Dirac pulse δ. Furthermore we replace $x = -x_i d_o/d_i$ and obtain in three dimensions

$$h(x_i, y_i; x, y) = C' \exp\left[\frac{i\pi}{\lambda d_i}\left(1 + \frac{d_0}{d_i}\right)(x_i^2 + y_i^2)\right] \delta(x_i + Mx, y_i + My). \tag{3.184}$$

This equation states that the image is a magnified replica of the object field multiplied by a paraboloidal phase term. So the phase aberrations are corrected by multiplication of the reconstructed wavefront with the conjugate of this term. In finite discrete coordinates the resulting phase mask is

$$\Psi(n, m) = \exp\left[\frac{-i\pi}{\lambda D}(n^2 \Delta x'^2 + m^2 \Delta y'^2)\right] \tag{3.185}$$

with

$$\frac{1}{D} = \frac{1}{d_i}\left(1 + \frac{d_0}{d_i}\right). \tag{3.186}$$

Altogether we have to calculate $b'(n, m)$ according to (3.38) and multiply it with $\Psi(n, m)$. Then we can get the amplitude contrast image by $I(n, m) = |b'(n, m)\Psi(n, m)|^2$ and the phase contrast image by $\phi(n, m) = \arctan\{\text{Im}[b'(n, m)\Psi(n, m)]/\text{Re}[b'(n, m)\Psi(n, m)]\}$.

The quality of the reconstructed amplitude and phase maps depends strongly upon the exactness of the parameters d, D, k_ξ, and k_η as has been demonstrated in [124]. These parameters can be measured in the optical arrangement, but this can act only as a first rough estimate. A fine tuning of the values of these parameters is obtained by digital variation in the reconstruction stage and checking areas where no phase variations are expected.

An in-line digital microscopic holography system incorporating a long-distance microscope is proposed by Xu et al. [113, 114, 258, 276]. It fulfills the requirements of imaging microstructures with high resolution at sufficient working distances to permit good illumination of the samples.

Digital holographic microscopy is not restricted to visible wavelengths. In [74] in-line holography with a low energy electron point source is considered.

A heterodyne approach to digital microscopic holography is presented in [137, 138]. The reference wave is reflected from a mirror mounted on a saw-tooth voltage driven piezo-translator providing a temporal frequency difference between the two arms of the interferometer. So each scatterer of a three-dimensional object is encoded as a temporally modulated Fresnel pattern, which is recorded by a CCD. Temporal heterodyning of the signal from each pixel results in a single-sideband, on-line holographic record in digital form. Reconstruction of an image focused on a chosen transverse plane in the 3D object volume is performed by digital correlation with a reconstruction function matched to that plane. The method poses relaxed demands on the spatial bandwidth of the detector and does not really need spatially coherent illumination. This spatiotemporal digital holography can be used for imaging through scattering media even when using a white-light source for illumination [137].

The reconstruction process in digital holography involves finite discrete Fourier transforms. If especially in microscopy the digital holograms at opposite border points in the horizontal or vertical direction have discontinuous or non-smooth continuation, the aliasing effects produce spurious diffraction-like patterns. A reduction of these effects can be achieved by border processing [69]. By this the digital hologram is extended to a larger size with the additional part filled by values that minimize, according to a numerical criterion, the erroneous spatial frequencies.

A lensless digital holographic microscope discriminating in depth is reported in [277]. Using a short-coherence-length laser holographic information is only captured if the path lengths of reference and object beam are matched within the coherence length. An aberration compensation for this lensless holography by expansion of the hologram with interpolation of the intensity values and simulation of diffraction using the Rayleigh-Sommerfeld propagation theory is given in [278, 279].

3.6.3 Data Encryption with Digital Holography

We have seen, e. g. Fig. 3.14, that a digital hologram of a 2D or 3D scene looks rather stochastic and seems to give no hint on the content of the original scene. So digital holography on first sight offers a means for *data encryption*. Security of data in the sense of prevention from unauthorized knowledge is a key issue when data from diverse fields are to be stored or transmitted. However, directly recorded digital holograms do not constitute a secure tool for encryption. They are highly redundant, as investigations for effective encoding and data compression [280] have shown. A variation of only a few parameters in a trial and error procedure will quickly lead to the original information.

Nevertheless digital holography with some modifications is a proper means for encrypting two- or three-dimensional optical information. The approach to digital holographic encryption presented here is based on phase-shifting digital holography, Section 3.4.1, and a random phase code that constitutes the key for encrypting and decrypting [134, 135, 281].

The optical system for encrypting 2D or 3D information in this way is schematically shown in Fig. 3.75. It is based on a Mach-Zehnder interferometer architecture. In the displayed configuration data diffusely reflected from the surface of a 3D object are recorded. Instead of the 3D object also a 2D object positioned at the same place can be used. If not the reflective mode but the transmission mode is to be employed, the system of Fig. 3.76 has the object at the same place. Now the object may be a transparency, scattering particles, 3D refractive index distributions etc.

Let the opaque 3D object surface in Fig 3.75 be composed of point sources with complex amplitudes $E_O(x, y, z)$. If we admit a varying depth coordinate z, according to (2.72) the field in the hologram plane is in Fresnel approximation

$$\begin{aligned} E_H(\xi, \eta) &= \frac{1}{i\lambda} \iiint E_O(x, y, z) \frac{1}{z} e^{\frac{i2\pi}{\lambda} z} e^{\frac{i\pi}{\lambda z} \left[(\xi - x)^2 + (\eta - y)^2 \right]} dx\, dy\, dz \\ &= A_H(\xi, \eta) e^{i\phi_H(\xi, \eta)}. \end{aligned} \qquad (3.187)$$

3.6 Non-Interferometric Applications of Digital Holography

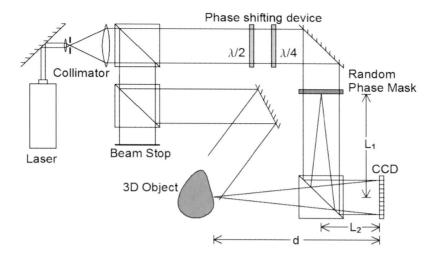

Figure 3.75: Phase shifting digital holographic system for encryption of optical information: Recording 3D object data.

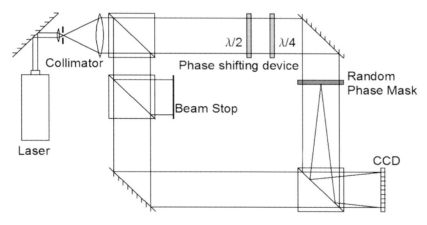

Figure 3.76: Phase shifting digital holographic system for encryption of optical information: Recording the key.

The collimated reference beam travels first through a $\lambda/2$- and a $\lambda/4$-plate, which perform phase retardations of $\Delta\phi_1 = 0$, $\Delta\phi_2 = -\pi/2$, $\Delta\phi_3 = -\pi$, $\Delta\phi_4 = -3\pi/2$, see Section 2.7.4, then through a random phase mask. This phase mask is at distance $L = L_1 + L_2$ from the CCD-array. The reference field in the hologram plane now is

$$\begin{aligned}E_R(\xi,\eta;\Delta\phi_p) &= e^{i\Delta\phi_p} e^{i\frac{\pi}{\lambda L}(\xi^2+\eta^2)} \iint e^{i\phi(x,y)} e^{-i\frac{2\pi}{\lambda L}(\xi x+\eta y)}\,dx\,dy \\ &= A_R(\xi,\eta) e^{i[\phi_H(\xi,\eta)+\Delta\phi_p]}.\end{aligned} \qquad (3.188)$$

Here $\Delta\phi_p$ is the actual phase shift and $\phi(x,y)$ is the random phase introduced by the mask. The phase shifted holograms recorded by the CCD are the intensity distributions given by the coherent superposition of E_H and E_R

$$I_p(\xi,\eta) = |E_H(\xi,\eta) + E_R(\xi,\eta)|^2 \tag{3.189}$$
$$= A_H^2(\xi,\eta) + A_R^2(\xi,\eta) + 2A_H(\xi,\eta)A_R(\xi,\eta)\cos[\phi_H(\xi,\eta) - \phi_R(\xi,\eta) - \Delta\phi_p].$$

A set of four phase shifted digital holograms $I_p(\xi,\eta)$, $p = 1,\ldots,4$ are recorded – three or more than four are also possible, see Section 5.5.2. From the recorded holograms the encrypted amplitude $A_E(\xi,\eta)$ and the encrypted phase $\phi_E(\xi,\eta)$ can be calculated. If phase shifts of $\pi/2$ as mentioned above are employed, these are

$$A_E(\xi,\eta) = A_H(\xi,\eta)A_R(\xi,\eta)$$
$$= \frac{1}{4}\sqrt{[I_1(\xi,\eta) - I_3(\xi,\eta)]^2 + [I_4(\xi,\eta) - I_2(\xi,\eta)]^2} \tag{3.190}$$

and

$$\phi_E(\xi,\eta) = \phi_H(\xi,\eta) - \phi_R(\xi,\eta) = \arctan\frac{I_4(\xi,\eta) - I_2(\xi,\eta)}{I_1(\xi,\eta) - I_3(\xi,\eta)}. \tag{3.191}$$

A reconstruction of the object wave field by inverse Fresnel transform from the encrypted hologram field $A_E(\xi,\eta)\exp[i\phi_E(\xi,\eta)]$ is impossible. We need the knowledge of $A_R(\xi,\eta)$ and $\phi_R(\xi,\eta)$ or at least of $\phi(\xi,\eta)$ and L to recover $A_H(\xi,\eta)$ and $\phi_H(\xi,\eta)$ which then enable the reconstruction of the original object wave field. So $A_R(\xi,\eta)$ and $\phi_R(\xi,\eta)$ or at least $\phi(\xi,\eta)$ and L act as the keys for decryption. We obtain the key functions by using the setup of Fig. 3.76. Now four intensities $I'_p(\xi,\eta)$ are recorded from which the key amplitude $A_K(\xi,\eta)$ is calculated by

$$A_K(\xi,\eta) = A_C A_R(\xi,\eta) = \frac{1}{4}\sqrt{[I'_1(\xi,\eta) - I'_3(\xi,\eta)]^2 + [I'_4(\xi,\eta) - I'_2(\xi,\eta)]^2} \tag{3.192}$$

and the key phase by

$$\phi_K(\xi,\eta) = \phi_C + \phi_R(\xi,\eta) = \arctan\frac{I'_4(\xi,\eta) - I'_2(\xi,\eta)}{I'_1(\xi,\eta) - I'_3(\xi,\eta)}. \tag{3.193}$$

Here A_C and ϕ_C are the constant amplitude and phase of the collimated beam which without loss of generality can be set to $A_C = 1$ and $\phi_C = 0$. Now decryption is performed by

$$A_D(\xi,\eta) = \begin{cases} \frac{A_E(\xi,\eta)}{A_K(\xi,\eta)} & \text{if } A_K(\xi,\eta) \neq 0 \\ 0 & \text{if } A_K(\xi,\eta) = 0 \end{cases} \tag{3.194}$$

and $\phi_D(\xi,\eta) = \phi_E(\xi,\eta) - \phi_K(\xi,\eta)$.

$A_D(\xi,\eta)$ and $\phi_D(\xi,\eta)$ are the amplitude and phase of the decrypted Fresnel hologram, from which by inverse Fresnel transform (3.38) the original object wave field can be reconstructed. Experimental results of this technique and a discussion of the different perspectives belonging to different areas in the hologram plane can be found in [134].

Encryption and decryption by phase shifting digital holography is especially useful for secure storage or transmission of optical 2D or 3D information. The random phase mask or its phase shifted holograms act as the keys. In the presented approach in contrast with other phase encryption methods not only the phase but also the amplitude of the diffraction pattern of the object is modified by the random phase. The Fresnel diffraction offers advantages over Fraunhofer diffraction patterns [135]. One is that phase retrieval algorithms are difficult to apply, which increases security; furthermore the intensity distribution of digital Fresnel holograms is better adapted to the dynamic range of the detectors. An improvement in security can be obtained by a second phase mask in the object arm of the interferometer but for the price of additional computational effort in the decryption stage. Contrary to the digital decryption and reconstruction of the object wave field described above the decrypted digital hologram can be fed into a spatial light modulator for optical reconstruction [98, 142]. If the conditions for the validity of the Fresnel approximation are not fulfilled, meaning the object is too close to the CCD, then the *spectrum manipulating method* (SMM) presented in [91] can be employed for encryption and decryption of multidimensional data.

Another issue in securing digital data is the concept of *watermarking*. In digital watermarking a kind of message is hidden in the digital content in a way that prevents this image from being read by unauthorized persons but lets the message be read as needed. A digital holographic technique that hides a digital watermark as a phase modulation whose Fourier-transformed hologram is superposed on a content image is presented in [282]. A watermarking process using double phase encoding to encode hidden images is described in [283].

4 Holographic Interferometry

In Chapter 2 the interference effect was explained which occurs if two mutually coherent waves are superposed. So we have a means to compare two or more wave fields by checking the resulting interference pattern. Furthermore holography was introduced as a method for recording and reconstructing optical wave fields. These concepts now can be put together in the method of holographic interferometry. Holographic recording and reconstruction of a wave field is precise enough that holographically reconstructed fields can be compared interferometrically either with a wave field scattered directly by the object, or with another holographically reconstructed wave field. Accordingly we define *holographic interferometry* as the interferometric comparison of two or more wave fields, at least one of which is holographically reconstructed [170].

As will become clear in the following only slight differences between the wave fields to be compared by holographic interferometry are allowed. This first concerns the objects, we even have to demand the same microstructure, second the geometry which must be the same for all wave fields to be compared, third the wavelength and coherence requirements for the optical laser radiation used, and fourth the change of the object which is to be measured. All these items – microstructure, geometry, wavelength, and the physical quantities of the object – should be changed in such a small range that only the phase of the scattered wave field is varied, but an alteration of the amplitude can be neglected. If furthermore the change of the wave field is spatially homogeneous enough to vary the phase smoothly from object point to object point, we recognize a macroscopic interference pattern which will be referred to as the *holographic interferogram* or *holographic interference pattern*. To prevent confusion between the stationary interference pattern stored as the hologram and the holographic interferogram, the first mentioned is sometimes called the *microscopic interference pattern* in contrast to the observed and evaluated *macroscopic interference pattern* produced by holographic interferometry. Because holographic interferometry can bring to interference two or more wave fields simultaneously which existed at different times, and since it does not require specular reflecting surfaces, this method has found wide applications as a measurement method. In this book we only deal with laser assisted holographic interferometry, which means we stay in the near infrared, visible or near ultraviolet range of the spectrum. Nevertheless it should be mentioned that holographic interferometry can be performed using microwave or ultrasonic radiation [284, 285].

Handbook of Holographic Interferometry: Optical and Digital Methods. Thomas Kreis
Copyright © 2005 Wiley-VCH Verlag GmbH & Co. KGaA, Weinheim
ISBN: 3-527-40546-1

4.1 Generation of Holographic Interference Patterns

4.1.1 Recording and Reconstruction of a Double Exposure Holographic Interferogram

In the *double exposure method* of holographic interferometry two wavefronts scattered by the same object are recorded consecutively onto the same holographic plate [286]. Here we assume recording onto a photographic plate. But the presented theory also applies for other recording media, especially CCD-arrays, as will be shown in detail in Section 5.8. The two wavefronts correspond to different states of the object, one in an initial condition, Fig. 4.1a, and one after the change of a physical parameter, Fig. 4.1b, e. g. by altering the object loading.

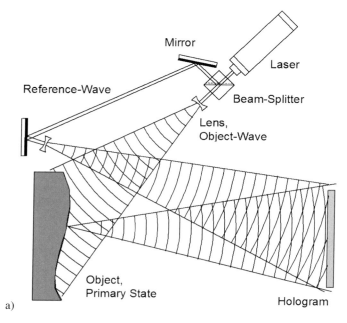

Figure 4.1: Recording (a), (b), and reconstruction (c) of a double exposure holographic interferogram.

Let the complex amplitude of the first wavefront at an object point P be

$$E_1(P) = E_{01}(P)\, e^{i\phi(P)}. \tag{4.1}$$

which is holographically recorded. E_{01} is the real amplitude and $\phi(P)$ is the phase distribution, (2.15). $\phi(P)$ varies spatially in a random manner due to the microstructure of the diffusely reflecting or refracting object. P identifies an object point. For the moment we do not have to discriminate between points of the object and the corresponding points in the image plane.

4.1 Generation of Holographic Interference Patterns

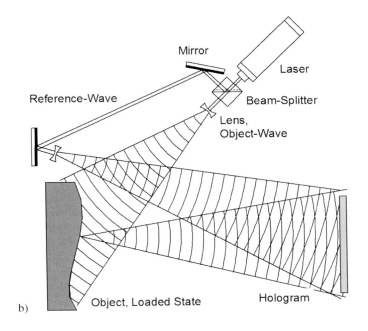

b) Object, Loaded State

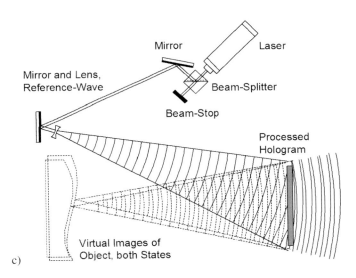

c)

Fig. 4.1: Continued.

The variation of a physical parameter to be measured, e. g. the object shape due to deformation of an opaque object or a change in the refractive index distribution of a transparent object changes the phase distribution at P by $\Delta\phi(P)$. So the complex amplitude of the second wavefront to be recorded holographically onto the same plate is

$$E_2(P) = E_{02}(P)\, e^{i(\phi(P) + \Delta\phi(P))} \qquad (4.2)$$

After development of the holographic plate both wavefronts are reconstructed simultaneously, Fig. 4.1c. They interfere and give rise to a stationary intensity distribution

$$\begin{aligned}
I(P) &= |E_1(P) + E_2(P)|^2 \\
&= (E_{01}(P)e^{i\phi(P)} + E_{02}(P)e^{i(\phi(P)+\Delta\phi(P))})(E_{01}(P)e^{-i\phi(P)} \\
&\quad + E_{02}(P)e^{-i(\phi(P)+\Delta\phi(P))}) \\
&= I_1(P) + I_2(P) + \sqrt{I_1(P)I_2(P)}\left(e^{-i\Delta\phi(P)} + e^{i\Delta\phi(P)}\right) \\
&= I_1(P) + I_2(P) + 2\sqrt{I_1(P)I_2(P)}\cos[\Delta\phi(P)]. \qquad (4.3)
\end{aligned}$$

For identical amplitudes, $E_{01}(P) = E_{02}(P)$, we get

$$I(P) = 2I_1(P)\{1 + \cos[\Delta\phi(P)]\}. \qquad (4.4)$$

The change of the phase $\Delta\phi$ is called the *interference phase difference* or shortly *interference phase*. If the spatial variation of the interference phase over the observed reconstructed surface is low, the intensity distribution (4.3) represents the irradiance of the object, modulated by a cosine-shaped fringe pattern. Bright centers of fringes are the contours, where the interference phase is an even integer multiple of π, dark centers of fringes correspond to odd integer multiples of π. Clearly, if the interference phase changes too rapidly from one observable point to the next, say more than π, so that the sampling theorem is violated, we will recognize only a more or less random intensity distribution, which cannot be evaluated any more.

4.1.2 Recording and Reconstruction of a Real-Time Holographic Interferogram

In *real-time holographic interferometry* only one wavefront, belonging to a reference state of the tested object, is holographically recorded, Fig. 4.2a. After processing, the hologram is replaced exactly in its initial recording position. This replacement has to be performed within sub-wavelength precision. During illumination of the hologram with the original reference wave, the reconstructed virtual image wavefront coincides with the wavefront scattered directly by the object which still is in its original position. A change of the object now changes the scattered wavefront and the superposition with the holographically reconstructed original wavefront produces an interference pattern which can be observed in real time, Fig. 4.2b. Dynamic variations of the object lead to simultaneously observable variations of the interference pattern.

In both real-time and double-exposure holographic interferometry, the intensity variation in the fringe pattern has a cosine shape. However while in the double-exposure technique we have bright fringes where the interference phase is an even integer multiple of π, using amplitude holograms in the real-time method we get bright fringes where $\Delta\phi(P)$ is an odd integer

4.1 Generation of Holographic Interference Patterns

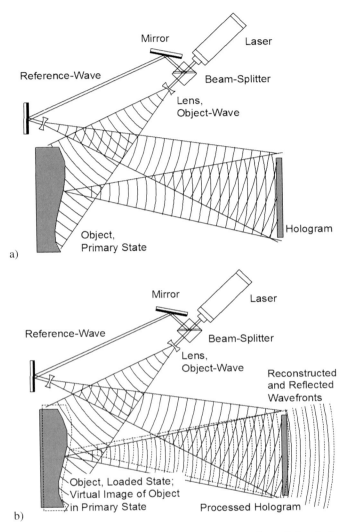

Figure 4.2: Recording (a) and reconstruction (b) of a real-time holographic interferogram.

multiple of π. This is due to the negative sign before the third term of (2.131) which describes the reconstructed virtual image. In the double-exposure technique both reconstructed wavefronts carry the same sign. Using a phase hologram in the real-time technique, we have the same positive sign for the directly scattered and the reconstructed wavefronts, now leading to bright fringes at even integer multiples of π like with the double-exposure technique. As a consequence loci where there is no change, $\Delta\phi(P) = 0$, are the centers of a bright fringe in the double exposure and in the real-time technique with phase holograms, these loci are the centers of a dark fringe for the real-time technique utilizing an amplitude hologram.

4.1.3 Time Average Holography

An in-depth discussion of *holographic vibration analysis* will be given in Section 6.5. Here we only show how fringe characteristics other than the above mentioned cosine shapes arise. One can use the real-time method with a holographically recorded and reconstructed wavefront representing the object in its rest state. Now consider a *harmonic vibration* which gives rise to an interference phase periodically changing in time

$$\Delta\phi(P)\sin(\omega t) \qquad (4.5)$$

where ω is the angular frequency of the vibration and $\Delta\phi(P)$ is related to the maximal amplitude of the vibration at object point P. If we assume illumination and observation in normal direction, and the maximal amplitude as $Z(P)$, we get $\Delta\phi(P) = 4\pi Z(P)/\lambda$, since the light has to travel to and from P along $Z(P)$. During vibration the real-time technique at each instant of time t generates a cosine shaped pattern

$$I(P,t) = 2I_1(P)\{1 - \cos[\Delta\phi(P)\sin(\omega t)]\}. \qquad (4.6)$$

If the frequency ω is higher than the temporal resolution of the eye (having an assumed average response time of $\sim$ 40 ms) or the applied sensor, a time averaged intensity is seen, which according to (C.2) is

$$\begin{aligned} I(P) &= 2I_1(P)\lim_{T\to\infty}\frac{1}{T}\int_0^T\{1 - \cos[\Delta\phi(P)\sin(\omega t)]\}dt \\ &= 2I_1(P)\{1 - \mathrm{J}_0[\Delta\phi(P)]\}. \end{aligned} \qquad (4.7)$$

Here J_0 is the zero-order *Bessel function* of the first kind. These fringes have low contrast, Fig. 4.3. The method can be used for the identification of resonant frequencies by monitoring the fringe pattern while varying the excitation frequency.

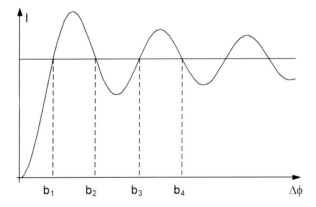

Figure 4.3: Time averaged real-time intensity.

4.1 Generation of Holographic Interference Patterns

The most frequently applied method for holographic vibration analysis is the *time average method*. Here the vibrating surface is recorded holographically with an exposure time which is long compared with the period of the vibration, $T \gg 2\pi/\omega$. Let us consider again the harmonic vibration as given in (4.5). Then we record and reconstruct holographically a continuum of waves each of the form

$$E_{01}(P)e^{i\Delta\phi(P)\sin(\omega t)}. \tag{4.8}$$

This set of wavefronts reconstructed simultaneously will interfere to

$$\begin{aligned}
E_{av}(P) &= \lim_{T\to\infty} \frac{E_{01}(P)}{T} \int_0^T e^{i\Delta\phi(P)\sin(\omega t)} dt \\
&= E_{01}(P)\mathrm{J}_0\left(\Delta\phi(P)\right).
\end{aligned} \tag{4.9}$$

The observable intensity in the reconstructed image then is

$$I(P) = I_1(P)\mathrm{J}_0^2\left(\Delta\phi(P)\right). \tag{4.10}$$

The fringes now are contours of equal vibration amplitudes of the spatial vibration modes. Maximal intensity belongs to $\Delta\phi(P) = 0$, which occurs at the nodes of the vibration mode. Dark centers of the fringes refer to zeros of the Bessel function J_0. The contrast of the fringes decreases with increasing order, Fig. 4.4, see also Fig. C.1.

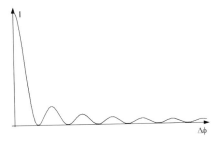

Figure 4.4: Time averaged intensity.

4.1.4 Interference Phase Variation Due to Deformation

In holographic interferometric measurements of the *deformation* of diffusely reflecting opaque object surfaces, the displacement of each surface point P gives rise to an *optical path difference* $\delta(P)$. This is the difference between the paths from the source point S of the illuminating wavefront over the surface point P to the observation point B before and after changing the deformation state [287]. The interference phase $\Delta\phi(P)$ is related to this path difference by

$$\Delta\phi(P) = \frac{2\pi}{\lambda}\delta(P). \tag{4.11}$$

The observed intensity belonging to this interference phase is given in (4.4).

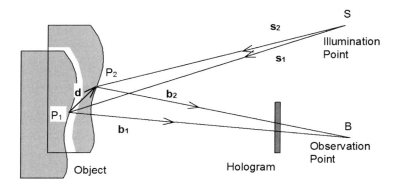

Figure 4.5: Holographic arrangement for measuring the deformation of an opaque surface.

In a holographic arrangement with diverging illumination and converging observation wave fronts, Fig. 4.5, let $S = (x_S, y_S, z_S)$ be the *illumination point* and $B = (x_B, y_B, z_B)$ be the *observation point* given in a Cartesian coordinate system. When the object is deformed, the surface point P moves from $P_1 = (x_{P1}, y_{P1}, z_{P1})$ to the new position $P_2 = (x_{P2}, y_{P2}, z_{P2})$, thus defining the *displacement vector*

$$d(P) = (d_x(P), d_y(P), d_z(P)) = P_2 - P_1. \tag{4.12}$$

Points P_1 and P_2 are different in a microscopic scale, but macroscopically they can be viewed as the same $P_1 = P_2 = P$. The optical path difference $\delta(P)$ now is expressed as

$$\begin{aligned}\delta(P) &= \overline{SP_1} + \overline{P_1B} - (\overline{SP_2} + \overline{P_2B}) \\ &= s_1 \cdot SP_1 + b_1 \cdot P_1B - s_2 \cdot SP_2 - b_2 \cdot P_2B \end{aligned} \tag{4.13}$$

where s_1 and s_2 are unit vectors in the illumination direction, b_1 and b_2 are unit vectors in the observation direction, and SP_i and P_iB are the vectors from S to P_i or P_i to B, respectively. In analogy to Section 2.2.1 let $s(P_1, P_2)$ be the bisector of the unit vectors s_1 and s_2 in illumination direction and $b(P_1, P_2)$ the bisector of the unit vectors in observation direction. $\Delta s(P_1, P_2)$ and $\Delta b(P_1, P_2)$ are the half differences of the unit vectors

$$s(P_1, P_2) = \frac{1}{2}[s_1(P_1) + s_2(P_2)] \qquad \Delta s(P_1, P_2) = \frac{1}{2}[s_1(P_1) - s_2(P_2)]$$

$$b(P_1, P_2) = \frac{1}{2}[b_1(P_1) + b_2(P_2)] \qquad \Delta b(P_1, P_2) = \frac{1}{2}[b_1(P_1) - b_2(P_2)]. \tag{4.14}$$

By definition of the displacement vector $d(P)$ we have

$$\begin{aligned} P_1B - P_2B &= d(P) \\ \text{and} \quad SP_2 - SP_1 &= d(P). \end{aligned} \tag{4.15}$$

Inserting this into (4.13) gives

$$\begin{aligned} \delta &= (s + \Delta s) \cdot SP_1 + (b + \Delta b) \cdot P_1B - (s - \Delta s) \cdot SP_2 - (b - \Delta b) \cdot P_2B \\ &= b \cdot d - s \cdot d + \Delta b \cdot (P_1B + P_2B) + \Delta s \cdot (SP_1 + SP_2) \end{aligned} \tag{4.16}$$

where the arguments have been omitted for clarity.

4.1 Generation of Holographic Interference Patterns

Now the displacements are far smaller than the dimensions of the arrangement geometry – $|d(P_1)|$ is in the micrometer range, the $\overline{SP_i}$ and $\overline{P_iB}$ are in the meter range – so the same relation holds between the lengths of Δs and Δb compared to the lengths of the unit vectors s_i and b_i. Furthermore the vector Δs is nearly orthogonal to $SP_1 + SP_2$ and vector Δb is nearly orthogonal to $P_1B + P_2B$, with the consequence that their scalar products are nearly zero. These scalar products can be neglected and we do not have to distinguish between P_1 and P_2 any more. Altogether the following relation holds for the macroscopic point P

$$\delta(P) = d(P) \cdot [b(P) - s(P)]. \tag{4.17}$$

For divergent illumination and observation the unit vectors $s(P)$ and $b(P)$ at surface point P are computed by

$$s(P) = \begin{pmatrix} s_x(P) \\ s_y(P) \\ s_z(P) \end{pmatrix}$$

$$= \frac{1}{\sqrt{(x_P - x_S)^2 + (y_P - y_S)^2 + (z_P - z_S)^2}} \begin{pmatrix} x_P - x_S \\ y_P - y_S \\ z_P - z_S \end{pmatrix} \tag{4.18}$$

and

$$b(P) = \begin{pmatrix} b_x(P) \\ b_y(P) \\ b_z(P) \end{pmatrix}$$

$$= \frac{1}{\sqrt{(x_B - x_P)^2 + (y_B - y_P)^2 + (z_B - z_P)^2}} \begin{pmatrix} x_B - x_P \\ y_B - y_P \\ z_B - z_P \end{pmatrix}. \tag{4.19}$$

These unit vectors together with the factor $2\pi/\lambda$ (4.11) form the so called *sensitivity vector* $e(P)$

$$e(P) = \frac{2\pi}{\lambda}[b(P) - s(P)] \tag{4.20}$$

so that we get

$$\Delta\phi(P) = d(P) \cdot e(P). \tag{4.21}$$

This means that the interference phase at each point is given by the scalar product of the displacement vector and the sensitivity vector. The sensitivity vector is defined only by the geometry of the holographic arrangement. It gives the direction in which the setup has maximal sensitivity. At each point we measure the projection of the displacement vector onto the sensitivity vector. For displacements orthogonal to the sensitivity vector the resulting interference phase is always zero, independent of the magnitude of the displacement.

The above formula (4.21) is the basis of all quantitative measurements of the deformation of opaque bodies by holographic interferometry and will be used frequently in the following chapters. Throughout this book we assume an illumination point source. For collimated illumination this is infinitely far away, nevertheless there exists a unit vector b defining the direction of the collimated illumination. Holographic interferometry with diffuse illumination from a circular diffuser is investigated in [288].

4.1.5 Interference Phase Variation Due to Refractive Index Variation

Phase objects are *transparent objects* which do not significantly affect the amplitude of an optical wavefront passing through, but only change the phase of this wavefront. Holographic interferometry has widely replaced Mach-Zehnder interferometry for analyzing phase objects, e. g. in applications like flow visualization, plasma diagnostics, or heat transfer analysis, to name just a few [289]. In the analysis of *transparent media* the optical path difference responsible for the interference fringes is generated by a change in the *refractive index distribution* along the optical path. This change may be due to the absence and presence of a phase object or due to a variation of the phase object under test. As discussed before the double exposure or the real time method can be applied to compare the states before and after the variation of the refractive index field.

If the medium to be tested is held in a transparent container, which is not altered between the recording of the two wavefronts, imperfections in the walls of the container, e. g. fabricated from glass, affect both wavefronts in the same way. But since only the differences between the wavefronts generate the holographic interference pattern, the influence of the wall imperfections is canceled out.

A typical off-axis configuration with collimated beams for holographic interferometric measurements at a phase object is shown in Fig. 4.6. Employing the double exposure method,

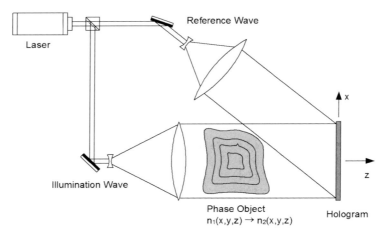

Figure 4.6: Configuration for holographic interferometry at a phase object.

the first recording is done with the refractive index distribution $n_1(x, y, z)$, the second is performed with $n_2(x, y, z)$. Let the rays propagate along straight lines parallel to the z-axis. Then

4.1 Generation of Holographic Interference Patterns

the interference phase distribution $\Delta\phi(x,y)$ in the plane perpendicular to the z-axis is

$$\Delta\phi(x,y) = \frac{2\pi}{\lambda} \int \Delta n(x,y,z) \, dz \tag{4.22}$$

with $\Delta n(x,y,z) = n_2(x,y,z) - n_1(x,y,z)$. The resulting intensity distribution in the holographic interferogram is (4.4)

$$\begin{aligned} I(x,y) &= 2I_1(x,y)\{1 + \cos[\Delta\phi(x,y)]\} \\ &= 2I_1(x,y)\left\{1 + \cos\left[\frac{2\pi}{\lambda}\int \Delta n(x,y,z)\,dz\right]\right\}. \end{aligned} \tag{4.23}$$

Equations (4.22) and (4.23) will play the central role in all quantitative measurements of refractive index distributions at phase objects.

Although the configuration of Fig. 4.6 is conceptually easy, the plane waves have several drawbacks [170]: Any dust particle or scratch on optical elements will diffract light into a nearly spherical wave. These unwanted waves then will interfere with the object wavefront and give annoying concentric ring patterns. Direct observation will exhibit a bright spot at the pinhole of the object beam and the field of view is limited to an area having the size of the iris. Hence the fringes have to be projected onto a screen to be observed. All these disadvantages are avoided by using *diffuse illumination holographic interferometry* [19, 20]. A ground glass plate is used to diffusely scatter the object illumination wave, Fig. 4.7. Now

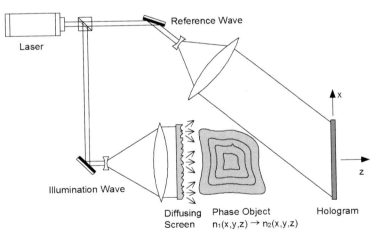

Figure 4.7: Configuration for diffuse illumination holographic interferometry.

the average irradiance is nearly uniform over the hologram, the influence of diffraction rings is minimized, the hologram can be observed with the unaided eye or recorded with a TV camera, and the hologram can be viewed in multiple directions. This last mentioned property will become important for three-dimensional quantitative evaluations, see Section 6.10.

4.1.6 Computer Simulation of Holographic Interference Patterns

Based on the equations (4.3) to (4.23) one can calculate the expected holographic interference pattern for a given setup geometry, including laser wavelength and object shape, as well as the loading applied to the object or the deformation or refractive index distribution [290, 291]. The simulation program basically is a loop over all pixels of the interferogram to be calculated. If we want to calculate an interference pattern of a deformed opaque surface, first each pixel is identified with the Cartesian coordinates of the corresponding point of the surface. With (4.18), (4.19), and (4.20) the sensitivity vectors for these surface points and thus for each pixel are calculated. The scalar multiplication (4.21) with the corresponding displacement vector is performed. The displacement vector has been determined from the loading parameters and the mechanical laws or has been taken from computer memory. While rigid body translations are easiest to handle – the displacement vector remains constant over all surface points – normally we have a combination of a rigid body motion, a deformation, and local deformation variations induced by defects or material inhomogeneities. The cosine (4.4) is applied to the calculated interference phase and gives the intensity distribution. Distortions like Gaussian background or speckle noise also can be taken into account. A Gaussian background may be simulated as an additive term, while the speckles are multiplicative [292]. Their stochastic nature is simulated by a random generator, which mimics the negative exponential probability distribution (2.105).

For the simulation of a real-time or a time average interferogram occuring in vibration analysis, the displacement vector is given by the maximal amplitude at each point, but then the Bessel function (4.7) or the squared Bessel function (4.10) is applied instead of the cosine (4.4). To get a better contrast in the simulated interferogram, the high intensity at the nodal lines, where the interference phase is near at zero, should be truncated, and the remaining intensity is mapped onto the full gray-scale. This results in brighter intensity maxima of higher order. Figure 4.8 shows some simulated holographic interference patterns, where in Fig. 4.8a a rectangular plate was subjected to strain and torsion; the inhomogeneity in the fringe pattern results from an anticipated subsurface void in the plate. Figure 4.8b displays a holographic interferogram of the same kind but with simulated background and speckle noise. Figure 4.8c gives the real-time and Fig. 4.8d the time average interferogram of a vibrating rectangular plate clamped at all four edges.

The simulation of holographic interference patterns is helpful in a number of ways:

- It helps in the planning phase of a holographic interferometric measurement to design optimized holographic arrangements. One can check whether an expected deformation produces fringes of sufficient but not too high a density on the basis of cheap computer experiments instead of expensive practical experiments. The optimal loading amplitudes can be found this way.

- If it is impossible to perform a three-dimensional evaluation with multiple interferograms, one can compare the calculated interferogram of an expected three-dimensional deformation field with the experimentally produced single interference pattern. Although this does not substitute a complete three-dimensional evaluation, it may confirm the results, if also the interferograms of all other possible deformations have been simulated and they differ significantly from the measured one.

4.1 Generation of Holographic Interference Patterns

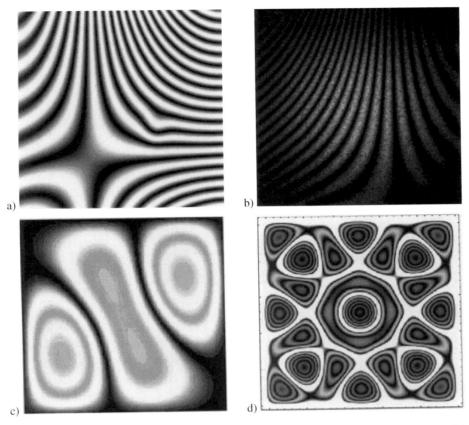

Figure 4.8: Simulated holographic interferograms: (a) double-exposure interferogram, (b) double-exposure interferogram with varying background intensity and speckles, (c) real-time interferogram of a vibrating plate, (d) time average interferogram of a vibrating plate.

- In developing new evaluation methods, e. g. adapted methods for special measurement problems, these methods can be tested first for simulated patterns. Especially the behavior of new methods, when evaluating more or less speckled patterns, can be tested systematically.

- In the application of advanced evaluation methods, e. g. fault detection with neural networks, test data or samples for network training are needed in large numbers and with much diversity [293]. These cannot be all generated experimentally. Based on some experimental examples the vast majority of the training samples are generated by computer simulation.

4.2 Variations of the Sensitivity Vectors

In Section 4.1.4 the foundations for holographic interferometric deformation measurements at diffusely reflecting opaque surfaces have been laid. The displacements of the object surface points together with the fixed sensitivity vectors give rise to the optical path differences which lead to the observable fringe patterns. Since the sensitivity vectors define the directions of the displacement components which are measured with the highest sensitivity, they should be examined in the stage of planning a holographic interferometric measurement experiment. The goal is to find an optimal arrangement that gives maximal accuracy and requires minimum effort for solving a given problem.

Contrary to deformation measurement holographic interferometry can also be applied by leaving the object surface unchanged but altering the sensitivity vectors between the construction of the two wavefronts to be compared interferometrically. This aspect is further addressed in Section 6.6, where it is used for holographic contouring.

4.2.1 Optimization of the Holographic Arrangement

The sensitivity vector is defined by (4.20) where the unit vectors $s(P)$ in illumination direction and $b(P)$ in observation direction are given by (4.18) and (4.19) for the case of divergent illumination and observation beams with a well defined source point S and observation point B. For collimated illumination and observation, Fig. 4.9a, we assume the central points of the collimating lenses as source S and observation point B, and the unit vectors s and b constant over all surface points P. If at least one of the illumination or the observation beam is divergent, then the sensitivity vector varies over the surface, Fig. 4.9b. Only if both the illumination and the observation beam are collimated will the sensitivity vector be the same for all points of the investigated surface, Fig. 4.9a.

The consequences of these two cases can be investigated by looking at the variation of the interference phase (4.21) in e. g. the x-direction

$$\begin{aligned}\frac{\partial}{\partial x}\Delta\phi &= \frac{\partial}{\partial x}[d_x e_x + d_y e_y + d_z e_z] \\ &= \frac{\partial d_x}{\partial x}e_x + d_x\frac{\partial e_x}{\partial x} + \frac{\partial d_y}{\partial x}e_y + d_y\frac{\partial e_y}{\partial x} + \frac{\partial d_z}{\partial x}e_z + d_z\frac{\partial e_z}{\partial x}.\end{aligned} \quad (4.24)$$

The arguments P have been omitted for clarity. We see that a change in the interference phase may stem from a variation of the displacement vector as well as from a variation of the sensitivity vector. For a constant sensitivity vector the interference phase depends solely on the displacement vector variation.

From a theoretical standpoint, only in the case of rigid body translations to be measured does it seem desirable to have varying sensitivity vectors to cover the object with fringes; in all other cases of deformations and rotations the evaluation will become easier with constant sensitivity vectors. But practically this is only possible for objects or illuminated areas on the objects of less than or equal to the collimating lens dimensions. A detailed analysis of sensitivity errors in interferometric deformation metrology, especially for divergent illumination and non-planar objects, is given in [294]. Nevertheless if the illumination and observation points are far away from the examined surface, compared with the surface dimensions, we can

4.2 Variations of the Sensitivity Vectors

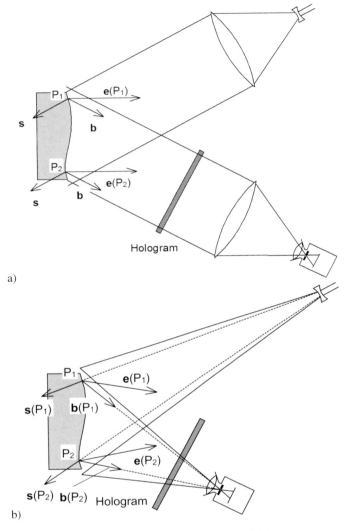

Figure 4.9: Constant (a) and not constant (b) sensitivity vectors.

assume nearly constant sensitivity vectors. The remaining errors may be estimated with the help of (4.24) by taking the maximum anticipated values for all variables and summing the absolute values.

In this way the problem of measuring only the out-of-plane displacements is tackled as follows. The Cartesian coordinate system is defined so that for a central surface point P we have $P = (0,0,0)$ and furthermore $y_S = y_B = 0$, $x_S = -x_B$, and $z_S = z_B$. Then the sensitivity vector at least at P reduces to $\boldsymbol{e}(P) = (0, 0, e_z(P))^T$ with

$$e_z(P) = \frac{2\pi}{\lambda}[z_B(P) - z_S(P)] = \frac{4\pi \cos \Theta}{\lambda} \qquad (4.25)$$

where Θ is the angle between the z-axis and the illumination direction which due to the assumptions agrees with the angle between the z-axis and the observation direction. The condition $x_S = -x_B$ and $z_S = z_B$ can be relaxed to only identical angles Θ for both directions. For constant or for nearly constant sensitivity vectors we now can determine the z-component of the displacement with good accuracy by

$$d_z(P) = \frac{\lambda \Delta \phi(P)}{4\pi \cos \Theta}. \qquad (4.26)$$

The basic ideas contained in this reduction of complexity can be worked out to a concept of optimizing the holographic setup [295]. In nearly all metrologic problems, we have a lot of previous information, e. g. about the directions of the deformations we expect. Then it is good practice to configure a holographic setup with maximum sensitivity in the expected direction and minimum sensitivity in directions of eventual distortions. As a help for this aim Jüptner et al. [296] have introduced the sensitivity functions and Abramson [297–301] designed the so called *holo-diagram*. A quantification of possible sensitivity errors in interferometric deformation measurements is given by Farrant and Petzing [294].

In [296] a Cartesian coordinate system based on the positions of the illumination source point and the observation point is defined. All lengths are normalized by the distance between these two points. The *sensitivity functions* indicate the sensitivity with respect to the different displacement components. Now if the direction of the object deformation can be anticipated, one has to search for regions where the sensitivity function for this direction varies linearly and the sensitivity functions for the other directions remain as close as possible to zero. It is worth mentioning that in this approach sensitivity is a property varying in space, only depending on the positions of the illumination and observation points, but there is no object already in the holographic arrangement.

The other aid for optimizing the holographic arrangement is the *holo-diagram* [297–301]. This consists of the two-dimensional projection of confocal ellipsoids with the focal points being the point source of illumination S and the point of observation B. By definition the path length from S to B is constant when travelling via any point along one ellipsoid. Adjacent ellipsoids are given path lengths from S to B via this ellipsoid which differ by λ. The distance between adjacent ellipses, the projections of the ellipsoids, varies with the directions to S and B, respectively. But it is constant along arcs of circles, or toroids in the three-dimensional case, known as k-circles. A displacement of an object point P from one ellipse to the next corresponds to one wavelength and thus to one fringe in the interference pattern. Displacements along the ellipses will cause no fringes. On the other hand the required displacement of P for inducing one fringe is minimum when its motion is along the normal to the ellipse. This normal corresponds to the sensitivity vector $\boldsymbol{e}$. As before the holo-diagram depicts the sensitivity of the holographic setup in space without the presence of an object.

For optimization of the holographic configuration the object has to be positioned in the holo-diagram in an orientation that the main direction of the anticipated displacement becomes

4.3 Fringe Localization

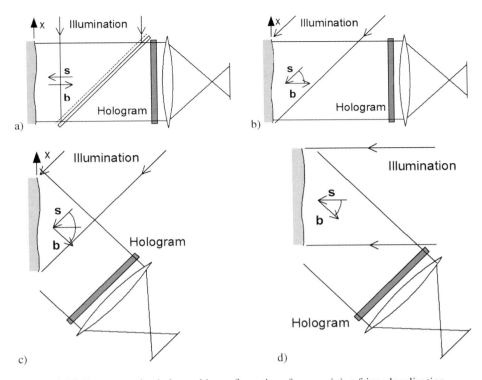

Figure 4.16: Representative holographic configurations for examining fringe localization.

If the object is rotated by an angle θ about an axis parallel to the surface, but not lying in it, the fringes localize off the surface. Let the rotation axis be parallel to the y-axis and lying a distance r behind the surface which is in the x-z-plane, then

$$\boldsymbol{d}(x,y,0) \approx (r\theta, 0, -\theta x). \tag{4.40}$$

Let us furthermore for simplicity only consider illumination, observation, and object points in the x-z-plane, then $b_y = s_y = 0$ and from (4.34) we get

$$z = -\frac{b_z[(b_y^2 + b_z^2)r + b_x b_z x]}{(b_z - s_z)} \tag{4.41}$$

which again is a plane like in (4.37). If viewed normally, $b_z = 1$, $b_x = b_y = 0$, Figs. 4.16a and b, the fringes are in a plane parallel to the object surface but behind it, namely by $-r(1 - s_z)$.

A rotation about the z-axis by an angle θ gives

$$\boldsymbol{d}(x,y,0) \approx (-\theta y, \theta x, 0) \tag{4.42}$$

and the localization conditions (4.34) and (4.35) are

$$z_1 = -\frac{b_z\left[(b_y^2+b_z^2)y+b_xb_yx\right]}{b_y-s_y}$$
$$z_2 = -\frac{b_z\left[(b_x^2+b_z^2)x+b_xb_yy\right]}{b_x-s_x}. \tag{4.43}$$

If now the object is illuminated and viewed in the normal direction, Fig. 4.16a, the displacement vector is orthogonal to the sensitivity vector, and no fringes are formed at all. For illumination at an angle of 45° and normal observation, Fig. 4.16b, we have $s_x = -s_z = \sqrt{2}/2$, $b_x = b_y = 0$, and $b_z = 1$, now the fringes are localized in the line

$$z = \sqrt{2}x, \qquad y = 0 \tag{4.44}$$

The configurations of Figs. 4.16c and d give $z = x/2$ and $z = x$, respectively.

When deformations are involved, the displacement vector d may be a nonlinear function of the object surface coordinates and the fringes appear to localize along a curve in space, or in a curved surface. An elongation of e. g. a tensile test specimen is described by

$$\boldsymbol{d}(x,y,0) \approx (\varepsilon x, 0, 0) \tag{4.45}$$

where ε is the linear strain. For the configurations of Figs. 4.16a and c we will get no fringes, since there is orthogonality between displacement and sensitivity vectors. For the configuration of Fig. 4.16b the localization is at $z = -\sqrt{2}x$
and for Fig. 4.16d atn $z = x/2$.

The bending of a cantilever beam of length L fixed at one end is described by

$$\boldsymbol{d}(x,y,0) \approx \left(0,0,\frac{d}{2}\left[3(\frac{x}{L})^2-(\frac{x}{L})^3\right]\right) \tag{4.46}$$

with d the deflection at the loose end. The fringes here are localized in $z = 0$ for arrangements of Figs. 4.16a and b , in

$$z = \frac{L}{6\sqrt{2}}\frac{3(\frac{x}{L})-(\frac{x}{L})^2}{2-(\frac{x}{L})} \tag{4.47}$$

for Fig. 4.16c and in

$$z = \frac{-L}{6(1+1/\sqrt{2})}\frac{3(\frac{x}{L})-(\frac{x}{L})^2}{2-(\frac{x}{L})} \tag{4.48}$$

for Fig. 4.16d.

Generally it is possible to bring a fringe pattern that is localized far from the object surface to it by tilting the reference beam adequately [311].

4.3.3 Fringe Localization with Spherical Wave Illumination

If the curvature of the illuminating wavefront at the object surface cannot be neglected, differentiation of the sensitivity vectors in (4.29) leads to additional $\partial s_i(P)/\partial x$- and $\partial s_i(P)/\partial y$-terms in (4.31). The derivatives of $\mathbf{s}$ are calculated analogously to (4.32) and with the abbreviation $R = \sqrt{(x_P - x_S)^2 + (y_P - y_S)^2 + (z_P - z_S)^2}$ as

$$\frac{\partial}{\partial x}s_x(P) = -\frac{s_y^2(P) + s_z^2(P)}{R} \qquad \frac{\partial}{\partial y}s_x(P) = \frac{s_x(P)s_y(P)}{R}$$

$$\frac{\partial}{\partial x}s_y(P) = \frac{s_x(P)s_y(P)}{R} \qquad \frac{\partial}{\partial y}s_y(P) = -\frac{s_x^2(P) + s_z^2(P)}{R} \qquad (4.49)$$

$$\frac{\partial}{\partial x}s_z(P) = \frac{s_x(P)s_z(P)}{R} \qquad \frac{\partial}{\partial y}s_z(P) = \frac{s_y(P)s_z(P)}{R}$$

and instead of (4.33) now we get

$$\left(\left[\frac{b_z}{z_B}(b_y^2+b_z^2) - \frac{1}{R}(s_y^2+s_z^2)\right]d_x - \left[\frac{b_z}{z_B}b_x b_y - \frac{1}{R}s_x s_y\right]d_y - \left[\frac{b_z}{z_B}b_x b_z - \frac{1}{R}s_x s_z\right]d_z\right.$$
$$\left. - (b_x - s_x)\frac{\partial d_x}{\partial x} - (b_y - s_y)\frac{\partial d_y}{\partial x} - (b_z - s_z)\frac{\partial d_z}{\partial x}\right)dx$$
$$+ \left(\left[\frac{b_z}{z_B}(b_x^2+b_z^2) - \frac{1}{R}(s_x^2+s_z^2)\right]d_y - \left[\frac{b_z}{z_B}b_x b_y - \frac{1}{R}s_x s_y\right]d_x - \left[\frac{b_z}{z_B}b_y b_z - \frac{1}{R}s_y s_z\right]d_z\right.$$
$$\left. - (b_x - s_x)\frac{\partial d_x}{\partial y} - (b_y - s_y)\frac{\partial d_y}{\partial y} - (b_z - s_z)\frac{\partial d_z}{\partial y}\right)dy = 0 \quad (4.50)$$

This is the most general condition for fringe localization and can be applied to the different rigid body motions and deformations as well as to the varied holographic configurations in the same manner as (4.33) did. Then one recognizes that the variation in the sensitivity vector introduces curvature into the surface or curve of localization and can lead to localization at finite distances even for rigid body translations. A detailed analysis of this case shows that the fringes now localize in a curved surface near $z = R$.

4.3.4 Fringe Localization with Phase Objects

Using the same approach as in the sections before now we investigate the localization of the holographic interference fringes arising from refractive index changes in phase objects [170]. Here only diffuse illumination holographic interferometry is considered, Fig. 4.7, for it is the practically relevant case. The analysis in the following will employ straight rays, ray bending due to refractive index gradients will be neglected. Figure 4.17 shows the geometry on which the following analysis is based. In front of the illuminated diffuser is a phase object introducing the refractive index variation

$$f(\mathbf{r}) = f(x,y,z) = \frac{2\pi}{\lambda}\Delta n(x,y,z) \qquad (4.51)$$

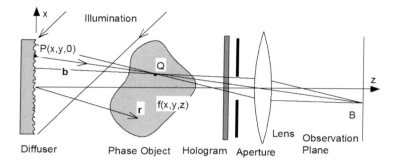

Figure 4.17: Fringe localization of phase objects.

where $\Delta n(x, y, z)$ was already defined in (4.22). Without loss of generality a Cartesian coordinate system is fixed to the diffuser with the x-y-plane tangential and the z-axis normal to the diffuser surface. The viewing system should be focused on a plane containing the point Q. As already seen, the holographic interference fringes are localized at points Q where the phase difference $\Delta\phi$ is nearly constant over the cone of ray pairs passing through Q and imaged onto the observation plane.

In analogy to (4.22) the optical path difference is

$$\Delta\phi = \int_S f(x, y, z) ds \qquad (4.52)$$

where s is along the ray of direction $\boldsymbol{b}$ through the phase object. It is assumed that $f(x, y, z) = 0$ outside a finite region S – the intersection of the observation ray with the phase object – and that f exhibits no discontinuities.

The condition for fringe localization is given analogously to (4.29) by

$$d(\Delta\phi) = \frac{\partial \Delta\phi}{\partial x} dx + \frac{\partial \Delta\phi}{\partial y} dy = 0. \qquad (4.53)$$

Since the integration limits are fixed, we can interchange the order of integration and differentiation, and with the chain rule we get

$$\frac{\partial \Delta\phi}{\partial x} = \int_S \nabla f \cdot \frac{\partial \boldsymbol{r}}{\partial x} ds \qquad (4.54)$$

with $\nabla f = (\partial f/\partial x, \partial f/\partial y, \partial f/\partial z)^T$. Since

$$\boldsymbol{r} = \begin{pmatrix} x \\ y \\ 0 \end{pmatrix} + \begin{pmatrix} b_x \\ b_y \\ b_z \end{pmatrix} s \qquad (4.55)$$

4.3 Fringe Localization

gives the derivative

$$\frac{\partial \mathbf{r}}{\partial x} = \begin{pmatrix} 1 + \frac{\partial b_x}{\partial x} s \\ \frac{\partial b_y}{\partial x} s \\ \frac{\partial b_z}{\partial x} s \end{pmatrix}, \tag{4.56}$$

(4.54) is written as

$$\frac{\partial \Delta \phi}{\partial x} = \int_S \left[\left(1 + \frac{\partial b_x}{\partial x} s\right) \frac{\partial f}{\partial x} + \frac{\partial b_y}{\partial x} \frac{\partial f}{\partial y} s + \frac{\partial b_z}{\partial x} \frac{\partial f}{\partial z} s \right] ds. \tag{4.57}$$

In the same way the derivative with respect to y is

$$\frac{\partial \Delta \phi}{\partial y} = \int_S \left[\left(1 + \frac{\partial b_y}{\partial y} s\right) \frac{\partial f}{\partial y} + \frac{\partial b_x}{\partial y} \frac{\partial f}{\partial x} s + \frac{\partial b_z}{\partial y} \frac{\partial f}{\partial z} s \right] ds. \tag{4.58}$$

Again Δx and Δy can be varied independently, so each of (4.57) and (4.58) must vanish. With the derivatives of the components of $\mathbf{b}$ given in (4.32) we get from (4.57)

$$\int_S \left[\left(1 - \frac{b_z}{z}(b_y^2 + b_z^2)s\right) \frac{\partial f}{\partial x} + \frac{b_z}{z} b_x b_y \frac{\partial f}{\partial y} s + \frac{b_z}{z} b_x b_z \frac{\partial f}{\partial z} s \right] ds = 0 \tag{4.59}$$

and from (4.58)

$$\int_S \left[\left(1 - \frac{b_z}{z}(b_x^2 + b_z^2)s\right) \frac{\partial f}{\partial y} + \frac{b_z}{z} b_x b_y \frac{\partial f}{\partial x} s + \frac{b_z}{z} b_y b_z \frac{\partial f}{\partial z} s \right] ds = 0. \tag{4.60}$$

For the distance s measured from the diffuser we can write $s = z/b_z$. Now splitting the integrals and solving for the z_l where the fringes localize gives the two conditions

$$z_{l1} = \frac{\int_S \left[(b_y^2 + b_z^2) \frac{\partial f}{\partial x} - b_x b_y \frac{\partial f}{\partial y} - b_x b_z \frac{\partial f}{\partial z} \right] z \, dz}{\int_S \frac{\partial f}{\partial x} dz}, \tag{4.61}$$

$$z_{l2} = \frac{\int_S \left[(b_x^2 + b_z^2) \frac{\partial f}{\partial y} - b_x b_y \frac{\partial f}{\partial x} - b_y b_z \frac{\partial f}{\partial z} \right] z \, dz}{\int_S \frac{\partial f}{\partial y} dz}. \tag{4.62}$$

Equations (4.61) and (4.62) generally define two surfaces, the interference fringes localize along the intersection of these two surfaces.

For the special case of viewing along the z-axis we have $b_x = b_y = 0$ and $b_z = 1$, then the conditions (4.61) and (4.62) reduce to

$$z_{l1} = \frac{\int_S \frac{\partial f}{\partial x} z\, dz}{\int_S \frac{\partial f}{\partial x} dz}, \tag{4.63}$$

$$z_l l2 = \frac{\int_S \frac{\partial f}{\partial y} z\, dz}{\int_S \frac{\partial f}{\partial y} dz}. \tag{4.64}$$

Numerical and experimental examples depicting the localization of fringes due to refractive index variations are given in [170]. For single radially symmetric fields the analysis of (4.63) and (4.64) shows that the fringes localize in the center plane of the object. The plane is normal to the line of sight and contains the axis of symmetry. Looking through two identical radially symmetric objects, one placed behind the other, gives a localization just in the plane midway between the two objects. For more general phase objects the curves of fringe localization become quite convoluted.

4.3.5 Observer Projection Theorem

A concept for considering the effect of variations of the observation vector $\mathbf{b}(P)$ is given by the so called *observer projection theorem*. This theorem is implicit in the geometric optics approach to fringe localization which is followed in this chapter, but can also be derived by wavefront analysis in wave optics [35]. This theorem states that, if the holographic interferometric fringes are localized off the object surface, they can be observed as if projected onto the object surface radially from the center of the aperture of the viewing system.

The theorem is useful when optically reconstructed fringes to be recorded by a CCD-camera are localized well off the object surface, especially if their spatial frequency is high. In order to record the fringes and the object surface simultaneously, the aperture of the camera must be small. But a small aperture will cause unacceptable noise in the form of speckles, as pointed out in Section 2.5. Following the observer projection theorem, however, one may record the fringes while focused with a large aperture in their localization surface. Then the object surface is recorded separately from the same camera position. After appropriate relative magnification these two stored images can be superimposed digitally. Nevertheless this last step can be omitted if the correspondence between the pixel coordinates and the object surface is known and considered in the further processing.

The treatment of localization of holographic interferometric fringes based on geometric optics given here followed closely that of Vest [170]. Its aim was not to give a detailed description of all localization aspects, but to introduce briefly into the problems and possibilities of fringe localization, to present the basic mathematical relations, and to give some easy examples. A more thorough description can be found in [170] in the context of a global description of the fringe patterns. An introduction to localization may further be found in [21, 22]. An exhaustive study of localization was performed by Stetson, presented in a series of papers [31–35, 312–314]. His treatment is based on wavefront analysis. More facts about localization can be found in [307, 315–319].

4.4 Holographic Interferometric Measurements

4.4.1 Qualitative Evaluation of Holographic Interferograms

Holographic interferometry produces unique two-dimensional patterns of fringes, whose density and form depend on the loading of a tested structure, the structure itself and the geometry of the holographic arrangement. Since even minute loading amplitudes may cause deformation amplitudes that generate observable interference fringes, holographic interferometry is an extraordinary well suited tool for *nondestructive testing* (*NDT*): faults and flaws in a technical component like cracks, voids, thin areas, debonds, etc. under proper load induce a characteristic local deformation, which can be detected in the holographic interferogram [320–322]. A defect in a structure may be critical with regard to one loading type or loading direction but not to another, Fig. 4.18. Therefore it is recommended to test a component with the type and

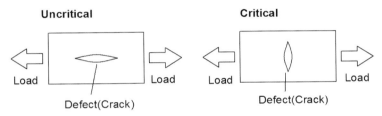

Figure 4.18: Critical and uncritical defects with regard to applied tensile stress.

direction of the intended operational load, but with low loading amplitude just sufficient to produce enough interference fringes to enable a reliable evaluation. This way we get a validation of the tested structure relative to the applied load. The defects are not visualized on their own but their response to the load is registered. The defects may reach the surface, e. g. surface cracks, but also subsurface defects like voids or material inhomogeneities are detectable. All these potentialities and advantages lead to the field of *holographic nondestructive testing* (*HNDT*).

Development and utilization of an HNDT scheme or device for a particular application involves three interrelated tasks [323]: Selecting a loading – type, direction, and amplitude – which causes a detectable anomaly in the fringe pattern, if a defect is present; designing the optical system – sensitivity vectors in the proper direction –; and interpreting the results. Still for most applications in HNDT the interferograms are observed and judged by skilled personnel. But there are attempts to perform an automatic computer controlled evaluation in HNDT, see Section 6.11. Since the interpretation of the interferograms consists only in the detection of typical patterns indicating the existence of a defect, but no quantitative interference phase determination is carried out, one speaks about *qualitative evaluation*.

Application areas for HNDT are manifold: components like tires or other automotive parts, turbine blades, honeycomb panels, pressure transducers, pressure vessels, aircraft parts, satellite tanks, musical instruments, artwork, composites, and laminates are tested holographically, to name only a few [324–335]. A 100 percent holographic inspection is performed e. g. on tires for special purposes, like retreaded aircraft tires, or on the nitrogen tanks of European satellites.

4.4.2 Holographically Measurable Physical Quantities

Holographic interferometry measures optical pathlength differences, which affect the interference phase, e. g. according to (4.11) or (4.22), which in turn gives rise to the observable interference pattern, (4.4) or (4.10). There are a number of physical processes where the optical path length is modified by various physical quantities. If the induced modification of the path length can be controlled to change only the phase of the light field and the lateral displacement to stay in the range less than the average speckle diameter, then these physical quantities are measurable by holographic interferometry.

The expression 'measurable' here can be understood in a very broad sense. It ranges from the quantitative derivation of the precise values of the physical quantity to a qualitative judgment by trained persons, where by looking at the interferogram areas of inhomogeneous variations of the underlying physical quantity can easily be detected, giving insight into the integrity of the object under test.

In this section only a short glimpse of the many holographically measurable physical quantities are given, the detailed physical laws governing the path length variations and the strategies of how to evaluate them are presented in Chapters 5 and 6.

One class of holographic interferometric measurements concerns the displacements and deformations of diffusely reflecting opaque surfaces, as introduced in Section 4.1.4. Some measurable quantities based on the displacement of surface elements are:

- *One-dimensional displacements* of the surface points in the direction of the sensitivity vector can be measured. If it is known that the displacement of all surface points is in the same direction, as is often the case, an optimized holographic setup as described in Section 4.2.1 should be used. The evaluation then is directed by (4.26).

- The *three-dimensional displacement* vector field of the surface points can be measured. This vector field may be defined by *rigid body translations*, *rigid body rotations*, general *deformations*, or a combination of these. For determining the three-dimensional vectors, a system of at least three linearly independent equations of the form (4.21) has to be solved for each surface point.

- Since holographic interferometry measures with high spatial resolution, numerical differentiation for the derivation of *strains* and *stress* becomes possible. The *in-plane strains* in the case of *plane stress* can be determined [336]. Stresses are then calculated by the related proportionalities. *Bending moments*, *thermal expansion coefficients*, or the *Poisson ratio* have been successfully evaluated. Likewise *surface tensions* of fluids have been measured by holographic interferometry [337].

- Contrary to these *static displacements* also *dynamic displacements* are measured holographically. The motions can be frozen by defined triggering of pulsed illumination for

5 Quantitative Determination of the Interference Phase

Quantitative evaluation of holographic interference patterns for measurement purposes consists in the pointwise determination of the numerical value of the physical quantity which has produced the optical path length change at each point and thus has given rise to the intensity distribution. The way how the physical quantity to be measured is contained in the observed intensity distribution was shown generally in Chaps. 2 and 4. There the predominant role of the interference phase was discussed.

In the early days of holographic interferometric metrology the fringes in the interferograms, or in photographs of the interferograms, were manually counted to obtain an approximation to the interference phase distribution. Later on the macroscopic interference patterns were recorded by video-cameras – nowadays CCD- or CMOS-cameras – and were stored after digitization and quantization in a computer for further evaluation. In the actual employment of digital holography for interferometric measurement purposes the wave fields are digitally reconstructed from the hologram data – not optically – and the interferograms principally might be generated digitally in computer. But this step normally can be skipped and the interference phase is determined directly from the reconstructed complex wave fields.

A computer aided quantitative evaluation thus is composed of two principal steps: First the interference phase distribution is determined from the recorded holographic interferogram or from the numerically reconstructed phases, see Section 5.8, and second the interference phase is combined with the sensitivity vectors to achieve the spatial distribution of the physical quantity to be measured [325, 362–375]. In this chapter the first of these two steps is addressed and the different methods to perform this task are discussed. The determination of the physical quantities out of the interference phase distribution is discussed in Chapter 6.

Quantitative evaluation stands in contrast to a qualitative evaluation, where not the precise values of the interference phase are interesting, but the global and local shape of the interference pattern. By a qualitative interpretation of the pattern one decides on the existence of areas of extraordinary local deformation in experimental stress analysis or holographic non-destructive testing (HNDT). These issues are discussed in Section 6.11.

5.1 Role of Interference Phase

In (4.4) it was shown how in the double-exposure or in the real-time techniques the intensity distribution $I(x,y)$ of the holographic interference pattern depends on the interference phase distribution $\Delta\phi$ by the cosine-function. There identical amplitudes for all surface points have been assumed. If we want to recognize the more general case of differing amplitudes, Section 2.2.3, and modifications of the intensity by environmental effects, noise, or other distortions, we can write

$$I(x,y) = a(x,y) + b(x,y)\cos\left[\Delta\phi(x,y)\right]. \tag{5.1}$$

This is the intensity distribution $I(x,y)$ that is recorded by e. g. a CCD-camera, digitized into an array of $N \times M$ pixels, quantized into L discrete gray-values and in this form stored digitally in the computer memory. The (x,y) denote the pixel coordinates. The mapping of object points to the individual pixel coordinates also is discussed in Chapter 6. The $a(x,y)$ and $b(x,y)$ contain the intensities of the interfering wave fields (2.40) and the various disturbances. Generally one can say that $a(x,y)$ contains all additive contributions and $b(x,y)$ comprises all multiplicative influences.

The objective is to extract the interference phase distribution $\Delta\phi(x,y)$ from the more or less disturbed intensity distribution $I(x,y)$ of (5.1). Constant values of $\Delta\phi$ define fringe loci on the object's surface, therefore $\Delta\phi(x,y)$ is called the *fringe-locus function* by some authors [35, 376].

5.1.1 Sign Ambiguity

When one tries to extract the interference phase $\Delta\phi(x,y)$ out of the intensity distribution $I(x,y)$ by a kind of inversion of (5.1), the problem arises that the cosine is not a one-to-one function, but is even and periodic

$$\cos\Delta\phi = \cos(s\Delta\phi + 2\pi n) \qquad s \in \{-1, 1\}, \quad n \in \mathbb{Z}. \tag{5.2}$$

An interference phase distribution determined from a single intensity distribution remains indefinite to an additive integer multiple of 2π and to the sign s.

Each inversion of expressions like (5.1) contains an inverse trigonometric function. All inverse trigonometric functions are expressed by the *arctan-function* like $\arccos(x) = \arctan(\sqrt{1-x^2}/x)$. The arctan-function of a single variable has its *principal value* in the interval $]-\pi/2, +\pi/2]$. But in most algorithms for quantitative evaluation of holographic interferograms the argument of the arctan-function is accomplished by a quotient, where the numerator characterizes the sine of the argument and the denominator corresponds to the cosine of the same argument. Then it is good practice to consider the signs of the numerator and the denominator separately, as is done by the FORTRAN- or C-function ATAN2(X,Y), for in this case the principal value is determined consistently in the interval $]-\pi, +\pi]$. The four situations of the sine- and cosine signs are shown in Fig. 5.1.

But there still remains a modulo 2π uncertainty as well as the sign ambiguity. Figure 5.2 displays a part of a graph which extends to infinity upward and downward. Each path from the left to the right through this graph represents a one-dimensional interference phase distribution belonging to the intensity distribution given at the bottom of the figure [377, 378].

A practical way to get rid of the *sign ambiguity* is to use side-informations about the experimental conditions leading to the measured optical path length changes and so to the interference phase distribution. In many applications one can assume that the measured interference phase distribution is not only continuous but differentiable as well, meaning a smoothly varying function. Depending upon the known direction in which the load acts on the object, e. g.

5.1 Role of Interference Phase

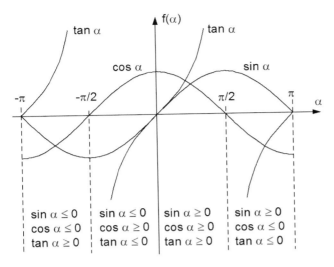

Figure 5.1: Signs of the trigonometric functions in $[-\pi, +\pi]$.

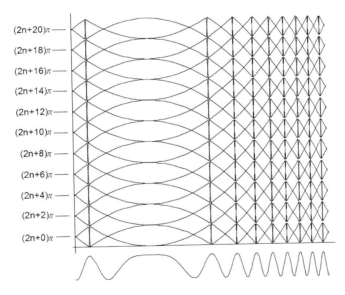

Figure 5.2: Ambiguity of the interference phase.

compressive or tensile, a proper sign distribution assignment norm can be fixed based on the known increase or decrease of the interference phase.

Reliable ways to eliminate the sign ambiguity without using prior knowledge about the underlying experiment consist of either recording multiple phase stepped interferograms or introducing experimentally a linear phase carrier with a positive slope higher than the steepest descent of the interference phase, thus producing only increasing phase maps [379, 380]. The concepts also are known as *infinite fringes* and *finite fringes*. If for $\Delta\phi(x, y) = 0$ a field of uniform irradiance, that is, an infinitely wide fringe, results, one speaks of an *infinite fringe interferogram*. Since $-\Delta\phi(x, y) = 0$ and $+\Delta\phi(x, y) = 0$ yield the same fringe pattern, there remains a *sign ambiguity*. If reference fringes are introduced by an additional carrier phase gradient of known sign, the resulting interferograms are called *finite fringe interferograms* [170]. It must be mentioned that in digital holography we obtain the interference phase $\Delta\phi(x, y)$ by a subtraction of the calculated phases, as is presented in more detail in Section 5.8. Then we know which is the first and which is the second state – this information is lost in double exposure holography – and no sign ambiguity remains.

5.1.2 Absolute Phase Problem

The 2π *ambiguity* manifests in the evaluated interference phase distributions by wrapping the phase *modulo* 2π. Since only principal values of the arctan-function in the interval $]-\pi, +\pi]$ – or equivalently in $]0, 2\pi]$ – are determined, as soon as an extreme value of the interval is reached, the phase jumps to the other extreme value, although the correct phase proceeds smoothly increasing or decreasing.

The modulo 2π effects are corrected by the processing step called *demodulation, continuation,* or *phase unwrapping,* see Section 5.9. By addition or subtraction of integer multiples of 2π the phase jumps are eliminated. The correct additive term in some applications can be determined if there is a point P in the pattern where the exact value of the displacement or equivalently the interference phase is known. Preferably this value is $\Delta\phi(P) = 0$. If a continuous variation of the interference phase can be assumed, the 2π-multiples at each point can be determined by counting the 2π-jumps from P to this point along an uninterrupted path. Sometimes an elastic ribbon is tied from one point P on the tested surface to a point in the holographic arrangement which has undergone no displacement but is lying in the observable interference pattern. Then the fringes along this ribbon are counted starting from zero [381].

If one is only interested in the deformation relative to an arbitrary point of the investigated surface but not in any additional rigid body translation, then one might define the interference phase as zero at this starting point. But this is only admissible for constant sensitivity vectors. This approach can be envisaged as evaluating the variation of the displacement from the variation of the interference phase. Now (4.24) shows that the variation of the interference phase in the case of varying sensitivity vectors not only depends on the variation of the displacement vectors but also on the direct values of the displacements. Consequently, if the sensitivity vector varies over the surface, the constant additive term has to be taken into account, meaning that the *absolute phase* including the correct multiples of 2π must be evaluated. Sometimes it may be sufficient to estimate the maximum errors by assuming minimal and maximal multiples of 2π and put them into (4.21) and (4.24) using the extreme sensitivity vectors of the actual holographic setup.

The error of not recognizing the absolute phase is demonstrated in Fig. 5.3 for the example of a cantilever beam clamped at the left end and bent by a point load at the right end. Curve 1 in Fig. 5.3a gives the z-displacement $d_z(x)$ along the x-axis of the cantilever beam of length 100 mm. The z-component of the sensitivity vector is shown in Fig. 5.3b, when the coordinates of the illumination point are $S = (-200, 0, 250)$ mm, and those of the observa-

5.3 Fringe Skeletonizing

internal pressure. Figure 5.4a shows the holographic interferogram together with an intensity profile along the dotted horizontal line. The enhanced intensity distribution, after averaging and shading correction, is given in a gray-scale display in Fig. 5.4b. The enhanced intensity is segmented, Fig. 5.4c, where the ridges are white, the valleys are gray, and unidentified pixels are black. This pattern is enhanced by region-growing and binary filtering; the result is given in Fig. 5.4d. These regions are then thinned to a skeleton as shown in Fig. 5.4e. Numbering and interpolation lead to a continuous phase distribution as the one given in Fig. 5.4f.

5.3.3 Skeletonizing by Fringe Tracking

In methods based on fringe tracking, the algorithm looks for neighboring pixels which correspond to local maxima or minima in the gray-value distribution [391, 409–413]. The first step in recognizing the fringes is to locate a starting point on each fringe. Only for a restricted number of interference patterns is it sufficient to traverse a line which is known to cut all fringes, and then to take the maxima on this line as the starting points [391]. Due to the manifold of possible holographic interference patterns usually the starting points are manually defined by the user. If there are locally parallel fringes in the pattern, straight lines of finite length perpendicular to these fringes can be constructed. Along these lines the starting points and the direction of search can be defined [404].

The tracking then either follows the curved ridges characterized by local intensity maxima or traces the boundaries between adjacent fringes by using derivatives of gray levels. Since the starting point is not necessarily at either end of the fringe, the program has to trace in one direction from the starting point and afterwards has to go back to the starting point and trace in the other direction. A search for the next point in the range from $-90°$ to $+90°$ relative to the forward direction or even a smaller range prevents the program from getting caught in small loops around insufficiently filtered speckles [391]. This results in an automatic check of five of the eight nearest neighbors of the current pixel for the maximum intensity.

The tracking procedure stops if a pixel is encountered that is already marked as belonging to a skeleton line. Reaching the starting point either means to have correctly traced a closed fringe, or the process is erroneously caught in a small loop. Another reason for reaching an already marked skeleton point is the attempt to cross another fringe which normally is forbidden. In this case it has to be checked whether a bifurcation of the fringes or hyperbolically formed fringes due to a saddle point in the interference phase distribution have occurred. Most programs at this point require interaction by the user. Also the operator may link data points belonging together, continue if obstacles are met, correct wrong decisions, and finally check that no fringes have been overlooked.

5.3.4 Other Fringe Skeletonizing Methods

The *phase lock method* uses a sinusoidal phase modulation obtained by, for example, a piezo-electrically excited axially oscillating mirror [414]. The resulting intensity is written as

$$I(x,y,t) = a(x,y) + b(x,y) \cos[\Delta\phi(x,y) + L \sin \omega t]. \tag{5.10}$$

$L < \lambda/2$ is the amplitude and $\nu = \omega/2\pi$ the frequency of this oscillation. A bandpass filter, centered on $\sin \omega t$, determines the amplitude $U_\omega = 2b(x,y)J_1(L) \sin \Delta\phi(x,y)$, which is zero at the points (x, y), where $\Delta\phi(x, y) = N\pi$. These points can thus be detected and give a skeleton whose lines correspond to the interference phase differences of $\pi/2$.

A special fringe contour detection scheme is proposed in [415]. It only works with interferograms fulfilling the prerequisites: (1) the presence of a dominant *spatial frequency* associated with the fringe pattern, (2) the near invariance of this frequency with position. Then, along lines normal to the fringes, the one-dimensional *Fourier spectrum* is calculated by the *FFT algorithm*. The phase of the dominant spatial frequency computed for each image line is a quantitative measure of fringe displacement at each line.

Skeletonizing methods nowadays are used if there is no way to produce multiple phase shifted interferograms. An alternative in this case is the Fourier transform evaluation, Section 5.6, which in its general form also requires only a single interferogram. From the interference phase distribution determined by the Fourier transform algorithm, (5.51) to (5.53), a skeleton may be produced by taking only those pixels where the 2π-jumps occur. Even further intermediate skeleton lines are possible if the pixels nearest to certain phase values are selected. The evaluation of a single interferogram generally is lacking in the sign, but taking the absolute value circumvents this problem by producing two skeleton lines. As an example let the interference phase distribution $\Delta\phi(x, y)$ modulo 2π be in the interval $]-\pi, +\pi]$. Selecting all points (x, y) with $|\Delta\phi(x, y)| \approx \pi/2$ yields the two skeleton lines belonging to $-\pi/2$ and to $+\pi/2$, but no information as to whether a part of a skeleton line belongs to $-\pi/2$ or to $+\pi/2$.

A rather new approach to skeletonizing uses concepts of *artificial neural networks* [416]. Points of the fringe skeleton are found by Kohonen's *self organizing feature map*. At the beginning a number of points, the *neurons*, are spread randomly over the interference pattern. The processing step consists of a random choice of an interferogram point, searching for the nearest neuron to this point, and moving this neuron towards the selected interferogram point. The amount of this motion is proportional to the distance between the point and the neuron, to the intensity of the interferogram point, and to an actual learning rate. This forces a higher probability for motion towards high intensity fringe centers than towards dark fringe areas. If this step is repeated sufficiently often, all neurons will concentrate at the bright fringe centers. If the number of neurons is high enough, the neurons in each fringe can be automatically connected by a nearest neighbor criterion without ambiguity, thus yielding the skeleton.

5.3.5 Fringe Numbering and Integration

After the skeleton lines are found we have to perform the *fringe numbering*, meaning to define a *fringe order* to each line. The integer fringe orders $n(x, y)$ correspond to the interference phase values $\Delta\phi(x, y)$ via $\Delta\phi(x, y) = 2\pi n(x, y)$. Even if we are not obliged to map the absolute fringe orders to the skeleton lines, see Section 5.1.2, the relations between the fringe orders must be fulfilled. So in particular local fringe order maxima and minima have to be uniquely detected. Generally, if a continuous interference phase distribution can be assumed, neighboring skeleton lines can differ in order only by -1, 0, or $+1$, lines of different order must not intersect or merge, lines do not end inside the field of view, and the skeleton line number differences, integrated along any closed line through the interferogram, always yield zero [403]. Automatic fringe numbering algorithms based on these constraints still may re-

5.4 Temporal Heterodyning

quire manual interaction by the user [391, 403, 405, 406, 417]. A graph-theoretic approach to automatic fringe numbering is given in [418].

The problem is made easier if a substantial degree of tilt can be added to the object deformation, leading to essentially parallel fringes with the measurement parameter being encoded as the deviation from straightness of the fringes [393]. The most important fact is that now the interference order behaves monotonically if moving roughly perpendicular to the skeleton lines, enabling an easy fringe numbering.

After assignment of the interference order to each skeleton line, the interference phase values are known along these lines, which represents a rather irregular distribution of points. To determine the values at all points of a regular grid, the interference phase values have to be interpolated for the grid points based on the phase values at the skeleton lines [419]. Three *interpolation* methods are presented in [420]:

- Interpolation based on *one-dimensional splines* fits cubic polynomials to the skeleton points in horizontal and vertical directions through the grid point and gets the phase value for each grid point from the two spline values at this point.

- In another approach at each grid point the four next neighboring skeleton points right and left as well as up and down are taken and from their phase values the interference phase at the grid point is calculated by *bilinear interpolation*.

- Closely related to this scheme is the *interpolation by triangulation*, where the whole skeleton is covered with small triangles. Two vertices of each triangle lie on one skeleton line, the third vertex is on an adjacent skeleton line. The phase at each grid point is found by linear interpolation from the phases at the vertices of the triangle in which it is contained.

A comparison of the resulting accuracy has shown that of these methods the interpolation by triangulation yields the highest accuracy [420].

5.4 Temporal Heterodyning

5.4.1 Principle of Temporal Heterodyning

The basis of *temporal heterodyning* is the interference of two optical waves of different frequencies. As shown in Section 2.2.2, two mutually coherent harmonic waves, differing in frequency by $2\Delta f$, produce an intensity oscillating with the *beat frequency*, which equals the *frequency difference* $2\Delta f$.

To translate this principle into *heterodyne holographic interferometry*, the two interfering wave fields are both holographically reconstructed with different optical frequencies f_1 and f_2, or the reconstructed wave field has another frequency than the reflected or refracted one [303, 305, 421–424]. Due to the context with the interference phase, more often the angular frequencies $\omega_i = 2\pi f_i$, $i = 1, 2$ are employed in the theoretical description.

The frequency of the holographically reconstructed wave field is defined by the frequency of the reconstructing reference wave. To achieve a frequency shift between the two reconstructed wave fields in the *double exposure method*, they have to be recorded and reconstructed with a *two reference beam holography* arrangement as described in Section 4.2.2.

Now the optical frequency of one of the two reference beams is shifted with respect to the other. This in most cases is done by a pair of *acoustooptical modulators* as is indicated in Fig. 5.5, which shows the typical holographic arrangement with two reference waves for performing the temporal heterodyne method.

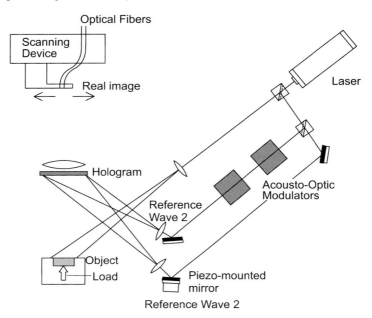

Figure 5.5: Two reference beam arrangement for double exposure temporal heterodyning.

The two reference beam method in conjunction with double exposure holography is the standard for temporal heterodyning. *Real-time holographic interferometry* with a wave field reconstructed by a reference wave having a frequency shift relative to the illuminating wave and being reflected by the object is possible in principle but not feasible, due to the extreme stability requirements and the high sensitivity to distortions from outside.

The holographic interference pattern resulting from reconstruction with two reference waves having a mutual frequency shift of $\Delta\omega = \omega_2 - \omega_1$ is

$$I(x,y,t) = a(x,y) + b(x,y)\cos[\Delta\phi(x,y) + t\Delta\omega] \qquad (5.11)$$

where $a(x,y)$ and $b(x,y)$ are the additive and multiplicative distortions and $\Delta\phi(x,y)$ is the interference phase distribution to be determined. If the frequency offset $\Delta\omega/2\pi$ is adjusted low enough to be resolved by opto-electronic sensors, say < 100 MHz, the interference phase can be measured with high accuracy by an electronic phasemeter. One would ideally want a two-dimensional sensor which should simultaneously detect oscillations of some MHz at a high number of pixels. Unfortunately such a sensor does not exist, therefore point-sensors fulfilling the demanded temporal resolution have to be scanned mechanically over the reconstructed real image.

5.4 Temporal Heterodyning

Figure 5.11: Test object.

istered by the ends of two of the five optical fibers, Fig. 5.6, and transmitted to photodiodes. Their signals go via an amplifier and filter to the phasemeter, the evaluated phase difference is transmitted via an IEEE-interface to the computer, which not only collects and evaluates the phase data, but also controls the stepper motors of the scanning device.

In this way 36 interference phase differences of point pairs have been recorded. This number is determined by the length of the object in the real image and the fixed distance of the fiber ends. Integration yields the continuous interference phase distribution which is proportional to the deflection curve of the cantilever beam shown in Fig. 5.12. The small deviations

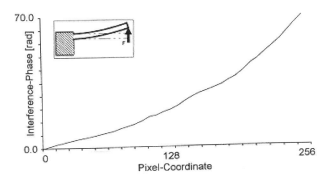

Figure 5.12: Evaluated interference phase by temporal heterodyning.

from the theoretical deflection line are caused by air turbulences during the measuring time. After each scanning step some seconds elapse until all mechanics come to rest. The errors accumulate to about 8% of the whole deflection [426].

5.5 Phase Sampling Evaluation

Although the heterodyne method is the widely used state of the art in interferometric length measurement, in holographic interferometry it has only found limited application. This is mainly due to the necessity of sensors with high temporal bandwidth, which is only accomplished by point detectors. On the other hand the two-dimensional holographic interferograms appeal for image detectors like TV tubes or CCD arrays which do not reach the required bandwidth. A way out of this dilemma is offered by the *phase step* or the *phase shift methods* [202].

The frequency shift in one of the interfering light waves of the heterodyne method can be envisaged as a continuous shift of the mutual phase between the light waves. Now this phase may be varied very slowly, in the extreme case stepwise, so that the intensity can be sampled corresponding to different values of this reference phase. The intensity distributions $I_n(x, y)$ recorded in this way are expressed by the so called *phase sampling equation*

$$I_n(x,y) = a(x,y) + b(x,y)\cos[\Delta\phi(x,y) + \phi_{Rn}] \qquad n = 1, \ldots, m \qquad m \geq 3 \quad (5.15)$$

where $a(x, y)$ and $b(x, y)$ are the additive and multiplicative distortions, $\Delta\phi(x, y)$ is the interference phase distribution to be determined, and ϕ_{Rn} is the shifted reference phase belonging to the n-th intensity distribution $I_n(x, y)$.

The phase step and phase shift methods which record and evaluate a set of intensity distributions (5.15) represent the widely accepted state of the art in the automatic evaluation of interference patterns especially of optically reconstructed holographic interference patterns. Due to the redundant information contained in the I_n's the interference phase is calculated with high accuracy at all pixels of the interference pattern and without sign ambiguity. The discrete sampling of several phase-shifted digital holograms from the dynamically varying heterodyne signal by a CCD is described in [84]. Due to the above mentioned conceptual relation to heterodyning these methods sometimes are called *quasi heterodyne methods* [305, 427–431].

The different components one can use to perform the phase shifts are described together with the other components of the holographic setup in Section 2.7. A frequently used option is a mirror, which reflects the reference wave, mounted on a piezo-crystal to shift the mirror by fractions of the used wavelength. This is depicted in Figs. 5.13 and 5.14 to indicate any phase shifting component.

Since the holographically reconstructed wave field has an optical phase distribution defined by the phase of the reconstructing reference wave, phase shifting the reference wave during reconstruction of a double exposure hologram would shift both reconstructed images in phase and so no effect on the interference fringes will be seen. Therefore in the *double exposure method* of holographic interferometry the two states have to be recorded with different reference waves and reconstructed by both of them simultaneously, but this approach brings all the problems of *two reference beam holography*, see Section 4.2.2. The phase shift here is introduced by shifting only one of the two reference waves, Fig. 5.13.

The other way is to employ the real-time method, Fig. 5.14. The wave field coming directly from the object is not affected by the phase shift in the reference wave which only modifies the optical phase of the reconstructed wave field. Thus the reference phase ϕ_R in (5.15) can be varied.

In most applications the phase shifted interferograms are recorded subsequently. The simultaneous recording of phase shifted interferograms, e. g. when measuring *transient events*,

5.5 Phase Sampling Evaluation

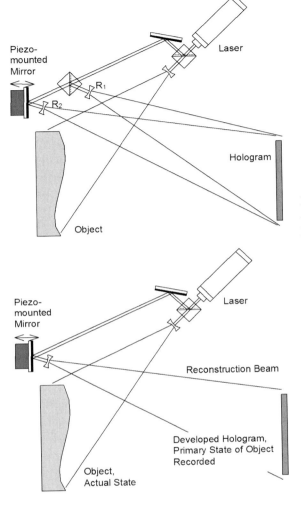

Figure 5.13: Two reference beam arrangement for double exposure phase sampling evaluation.

Figure 5.14: Real-time method for phase sampling evaluation.

is enabled by the introduction of a *diffraction grating* in the object beam [205,432]. The fringe patterns are obtained from direct interference of wavefronts propagating in the n-th diffraction order directions behind the grating and the hologram. For phase shifting the grating is shifted transversely between the two exposures of double exposure holography or between recording and reconstruction when employing real-time holographic interferometry, Section 2.7.4.

5.5.1 Phase Shifting and Phase Stepping

Although it is possible to perform arbitrary phase shifts and recognize these in the evaluation, it is recommended to use constant phase steps $\Delta\phi_R = \phi_{Rn+1} - \phi_{Rn}$ to keep the analysis easy. The phase can be shifted linearly in time, Figs. 5.15a and b, or in discrete steps, Figs. 5.15c

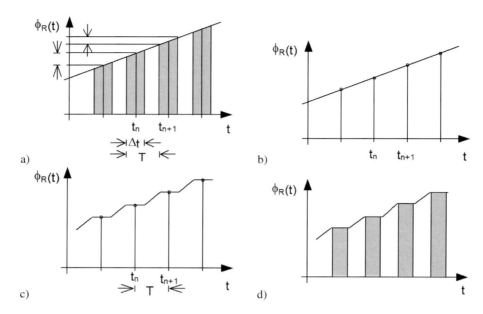

Figure 5.15: Phase shift vs. time: (a) phase shifting, (b)–(d) phase stepping; (a) and (b) linear phase shift, (c) and (d) stepwise phase shift.

and d. In practical applications, the time delay T between the recording of the phase shifted intensities should be as short as possible, e. g. only limited by the video frequency. For short T spurious vibrations induced by the environment have a minimal influence on the interference phase, in a rather worse case they can be modeled as an additional constant phase shift.

Theoretically one may discriminate between *phase shifting*, Fig. 5.15a, where each intensity is integrated over the time interval Δt during which the phase varies linearly, and *phase stepping*, Figs. 5.15b to d, where we have fixed phase values as assumed in (5.15). The result of integration in phase shifting can be calculated:

$$I_n(x,y) = \frac{1}{\Delta t} \int_{t_n - \Delta t/2}^{t_n + \Delta t/2} a(x,y) + b(x,y) \, \cos[\Delta\phi(x,y) + \phi_R(t)] \, dt. \tag{5.16}$$

Assuming a linear variation of $\phi_R(t)$ with t, this integral can be evaluated applying a substitution and the formula $\sin(\alpha + \beta) - \sin(\alpha - \beta) = 2\cos\alpha\sin\beta$ to yield

$$I_n(x,y) = a(x,y) + \mathrm{sinc}\left(\frac{\Delta\phi_R}{2}\right) b(x,y) \, \cos[\Delta\phi(x,y) + \phi_{Rn}]. \tag{5.17}$$

This expression corresponds to (5.15), only the contrast term $b(x,y)$ is modified by the constant factor $\mathrm{sinc}(\Delta\phi_R/2)$. Here $\Delta\phi_R$ denotes the phase shift during the time interval Δt. In this sense phase shifting is equivalent to phase stepping, both names in the following are used synonymously. One can see that for $\Delta\phi_R = 0$, the phase stepping of Figs. 5.15b to d, the

5.5 Phase Sampling Evaluation

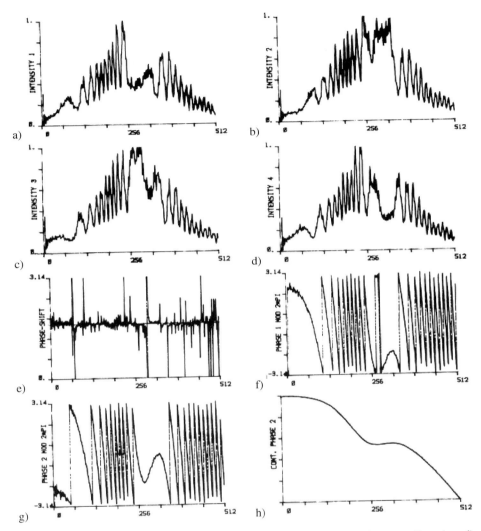

Figure 5.18: Evaluation of phase stepped holographic interferograms along one line: (a – d) intensity distributions; (e) phase shift, (f) interference phase 1, (g) interference phase 2, (h) demodulated interference phase

calculated interference phase distribution is clean and smooth. The changes in slope, decreasing to increasing and vice versa, are uniquely detected.

A two-dimensional evaluation by the same procedure of phase sampling is demonstrated for the example of a thermally loaded panel, consisting of an internal aluminum honeycomb structure with surface layers of carbon fiber reinforced plastic. Figures 5.19a to d exhibit the four phase stepped interferograms arising from a temperature difference of 2° C. The size of the panel was about 80×80 cm. The resulting interference phase distribution, which can be interpreted as proportional to the normal displacement field due to the optimized sensitivity vectors, is presented in Fig. 5.20. Several debonds of the surface layer from the internal structure are clearly detectable.

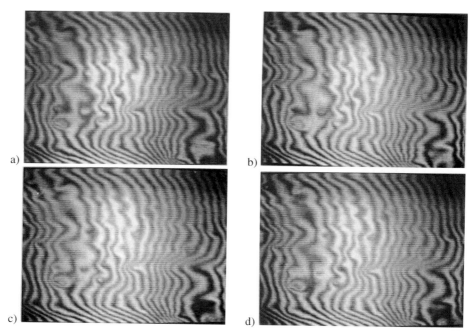

Figure 5.19: Four phase stepped holographic interferograms of a thermally loaded panel.

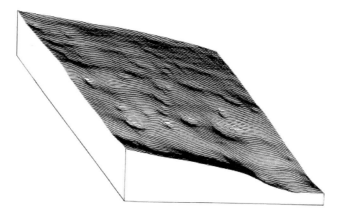

Figure 5.20: Evaluated interference phase distribution.

5.5.5 Discussion of Phase Shift Evaluation Methods

The phase shift methods for determining the interference phase distributions in holographic interferometry offer a number of advantages. The price one has to pay for it is the additional technical effort to perform the phase shifts, the increased requirements necessary for stability as for the *two reference beam holography* with double exposure or for the *real-time method*, as well as the additional storage capacity of the computer, all compared to the recording of a single interference pattern. But the benefits are manifold [440, 441]:

- The evaluation procedure can be fully automated.

- The interference phase is calculated at all pixels, not only at the fringe centers. Thus we get the best *spatial resolution* possible with the available electronics. No interpolation between skeleton lines is necessary.

- Due to the multiple recorded interferograms and the knowledge about their sequential order the *sign ambiguity* is resolved automatically. The evaluated interference phase increases and decreases in the same manner as the original one.

- The additive and multiplicative noise components are inherently recognized and compensated for in the automatic evaluation process.

- There are many different algorithms which fit to different circumstances, e. g. environmental distortions, phase shifter miscalibration, unknown phase shifts, etc.

- The intermediate results of the evaluation procedure allow a detection, whether the whole measurement was erroneous, e. g. zero phase shifts, whether there are points with contrast too low to guarantee a reliable evaluation, or whether there are points where no object surface was existent at all, e. g. at holes or outside the object margins.

- A phase shift which is varying over the pattern can be detected by some algorithms and sometimes can be fitted by a low degree polynomial, especially when the phase of the reference wave is shifted by a piezo-mounted mirror which undergoes an additional tilt.

- The resolution and accuracy of the determined interference phase is better than $1/20$ of 2π and with some experimental skill reaches $1/100$ of 2π.

Although the evaluation scheme, (5.21) and (5.24), applies for all sets of reference phases ϕ_{Rn}, $n = 1, \ldots, m$ which do not produce a singular matrix in (5.21), the best choice for the phase shifts between successive interferograms is between $30°$ and $150°$. This recommendation concerns to the methods with known as well as to these with unknown phase shifts. For the methods with known phase shifts it is not necessary that all phase shifts are constant, but this simplifies the evaluation procedure.

The influence of different error sources on the phase shift methods has been investigated by a large number of researchers [387–389, 442–444]. The results can be summarized by the statement that all these error sources like insufficient quantization, spurious diffraction or reflection patterns, aberrations of the optics, vibrations, air turbulence, inhomogeneity of the reference beam wavefront etc., which can be modeled as additive or multiplicative noise degrade the precision and accuracy independently of the choice of the evaluation algorithm.

Exceptions are detector nonlinearities and false phase shifts, e. g. by phase shifter miscalibration, whose consequences depend on the specific evaluation algorithm. Their influence has been tested in a comparison of different algorithms [445].

These tests have shown that a linear phase shift error degrades the calculated interference phase for all algorithms which assume a known constant phase shift. The degradation is reduced with the increase of the number of interferograms, especially for $m = 5$ or $m = 7$, [446–448]. The algorithms which calculate the unknown phase shift from the recorded interference patterns compensate for a linear phase shift error inherently and remain exact.

The same trend holds for nonlinear, quadratic phase shift errors. Although their influence remains present when applying the algorithms with unknown phase shifts, it is least then, compared to the algorithms assuming a known constant phase shift and employing the same number of frames.

A higher number of frames as well as making use of the algorithms with unknown phase shifts also is favorable if detector nonlinearities have to be considered.

5.6 Fourier Transform Evaluation

5.6.1 Principle of the Fourier Transform Evaluation Method

In *Fourier transform evaluation* [43,44,449–455] essentially a linear combination of *harmonic spatial functions* is fitted to the recorded and stored intensity distribution $I(x, y)$, given by (5.1). The admissible spatial frequencies of these harmonic functions are defined by the user via the *cutoff frequencies* of a *bandpass filter* in the spatial frequency domain.

To do this the intensity function is expressed with the help of the complex exponential. Introducing

$$c(x, y) = \frac{1}{2} b(x, y) \, e^{i \Delta \phi(x, y)} \tag{5.49}$$

the intensity $I(x, y)$ of (5.1) becomes

$$I(x, y) = a(x, y) + c(x, y) + c^*(x, y) \tag{5.50}$$

with $a(x, y)$, $b(x, y)$ as described in Section 5.1, i being the imaginary unit and * denoting complex conjugation. The discrete two-dimensional Fourier transform via the *FFT algorithm* applied to $I(x, y)$ yields

$$\mathcal{I}(u, v) = \mathcal{A}(u, v) + \mathcal{C}(u, v) + \mathcal{C}^*(u, v) \tag{5.51}$$

with (u, v) being the spatial frequency coordinates. Since $I(x, y)$ is a real distribution in the spatial domain, $\mathcal{I}(u, v)$ is a *Hermitean* distribution in the spatial frequency domain, which means

$$\mathcal{I}(u, v) = \mathcal{I}^*(-u, -v). \tag{5.52}$$

The real part of $\mathcal{I}(u, v)$ is even and the imaginary part is odd. The *amplitude spectrum* $|\mathcal{I}(u, v)|$ thus looks point-symmetric with respect to the d.c.-term $\mathcal{I}(0, 0)$. $\mathcal{A}(u, v)$ contains the

5.6 Fourier Transform Evaluation

zero-peak $\mathcal{I}(0,0)$ and the low frequency variations of the background. $\mathcal{C}(u,v)$ and $\mathcal{C}^*(u,v)$ carry the same information as evident from (5.52).

By *bandpass filtering* in the *spatial frequency domain*, $\mathcal{A}(u,v)$ and one of the terms $\mathcal{C}(u,v)$ or $\mathcal{C}^*(u,v)$ are eliminated. The remaining spectrum, $\mathcal{C}^*(u,v)$ or $\mathcal{C}(u,v)$, is no longer Hermitean, so the *inverse Fourier transform* applied to, e.g. $\mathcal{C}(u,v)$, gives a complex $c(x,y)$ with non-vanishing real and imaginary parts. The interference phase can be calculated by

$$\Delta\phi(x,y) = \arctan \frac{\operatorname{Im} c(x,y)}{\operatorname{Re} c(x,y)}. \tag{5.53}$$

The inverse transform of $\mathcal{C}^*(u,v)$ instead of $\mathcal{C}(u,v)$ would result in $-\Delta\phi(x,y)$. The uncertainty about which of the symmetric parts of the spectrum belongs to $\mathcal{C}(u,v)$ and which to $\mathcal{C}^*(u,v)$ is a manifestation of the *sign ambiguity*, (5.2).

A one-dimensional example of the Fourier transform evaluation is given in Fig. 5.21. The

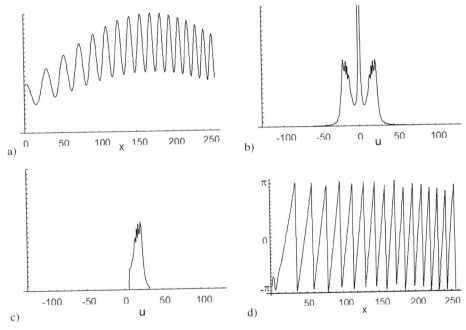

Figure 5.21: Fourier transform evaluation.

intensity distribution with varying contrast and varying background of Fig. 5.21a is Fourier transformed, the amplitude spectrum is shown in Fig. 5.21b. After filtering, the spectrum of which the amplitude is given in Fig. 5.21c remains. The application of the inverse transform and (5.53) result in the phase modulo 2π shown in Fig. 5.21d.

Although the Fourier transform calculated by the FFT algorithm normally has its zero frequency at the left and the two-dimensional Fourier transform has its zero frequency in the upper left corner when displayed, in the examples demonstrated here and in the following, we use the reordered display of the amplitude spectra, as determined by (A.77).

5.6.2 Noise Reduction by Spatial Filtering

The bandpass filtering in the spatial frequency domain not only makes the spectrum non-Hermitean, but also enables a reasonable *image enhancement* [451]. Low frequency *background variations*, e. g. a Gaussian illumination, lead to spectral components centered around the zero-component. Their influence is minimized by a bandpass filter which eliminates all spectral components up to a certain lower *cutoff frequency*. High frequency components, like *speckle noise*, are suppressed to a reasonable amount if the filter stops all frequencies higher than an upper cutoff frequency.

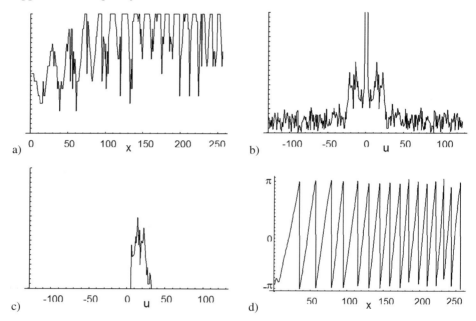

Figure 5.22: Image enhancement by Fourier transform evaluation.

This capability is shown in Fig. 5.22. The intensity, Fig. 5.22a, is degraded by a Gaussian background, varying contrast, speckle noise, reduced *quantization* into 16 gray-levels, and nonlinear response with *saturation*.

We recognize the presence of higher spatial frequencies in the amplitude spectrum, Fig. 5.22b.

Most of them are cut away in the filtered spectrum, Fig. 5.22c, so that the finally evaluated interference phase distribution, Fig. 5.22d, is clean and fully modulated.

In eliminating high frequencies, one has to consider the fact that the *finite discrete Fourier transform* assumes a periodic input signal, whereas, in practice, data consist of one non-periodic stretch of finite length. The discontinuities from the right to the left or from the lower to the upper edge of the image lead to high frequency components in the spectrum. If these are filtered away, the resulting phase distribution suffers from a *wrap-around pollution* by a forced smooth continuation at the edges of the frame. Thus the marginal pixels at the edges of the frame get no reliable phase values; the number of these pixels depend on the choice of the upper cutoff frequency. The run-out at the left edge in Fig. 5.22d is caused by this effect.

Making the spectrum non-Hermitean means setting to zero one of the spectral values at each spatial frequency or at its symmetric counterpart. Since there is no general way to decide whether $\mathcal{I}(u,v)$ at a certain (u,v) belongs to $\mathcal{C}(u,v)$ or to $\mathcal{C}^*(u,v)$, or is a combination of contributions belonging to both of them, the easiest way is to eliminate one halfplane of the spatial frequency plane. Some of these halfplanes are displayed in Fig. 5.23. In Fig. 5.23a

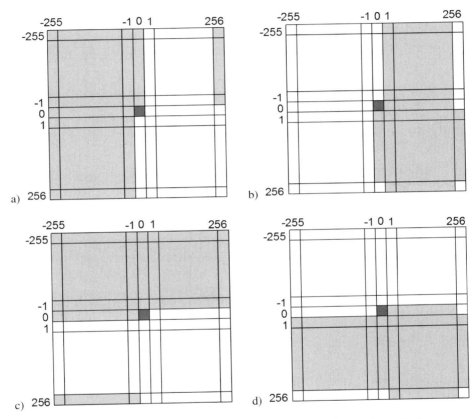

Figure 5.23: Halfplanes for filtering in the spatial frequency domain, reordered display.

the passband is the $+u$-halfplane, displayed in white, and the shaded region is the stopband. In Fig. 5.23b the passband is the $-u$-halfplane, in Fig. 5.23c the passband is the $+v$-halfplane and in Fig. 5.23d we have the passband in the $-v$-halfplane. For Fourier transform evaluation the component at spatial frequency $(0,0)$ is always set to zero. Filters which destroy the Hermitean property and eliminate low-frequency background as well as high-frequency speckle noise have the form shown in Fig. 5.24, with a passband in one halfplane and with defined lower and upper cutoff frequencies.

260   5 Quantitative Determination of the Interference Phase

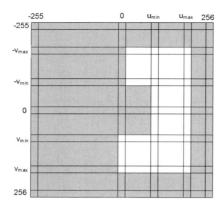

Figure 5.24: Bandpass filter, reordered display.

In many holographic applications it is possible to record and store the illuminated surface before the interference pattern is produced [450, 456]. So in *real-time holographic interferometry* one may:

- Record the object surface illuminated only by the object wave with the reference wave blocked. The hologram plate may be present or absent.

- Record a holographic reconstruction of the object with the object wave blocked. Only the reference wave illuminates the hologram plate.

- Record the zero interference pattern, still without fringes, before starting the loading.

Let this recorded background be $a'(x, y)$ and its Fourier transform be $\mathcal{A}'(u, v)$. A normalized version of $\mathcal{A}'(u, v)$ is then subtracted from $\mathcal{I}(u, v)$

$$\mathcal{I}'(u, v) = \mathcal{I}(u, v) - \frac{\text{Re } \mathcal{I}(0, 0)}{\mathcal{A}'(0, 0)} \mathcal{A}'(u, v). \tag{5.54}$$

Since the imaginary part of the d.c.-term of a Fourier transform is always zero, we have zeroed the d.c.-term and eliminated the background. Now from $\mathcal{I}'(u, v)$ one halfplane is eliminated and after the inverse transform has been obtained the interference phase is calculated as described above. This procedure is recommended if the interference pattern or the object does not cover the whole frame or if we have complicated background variations in the frequency range of the interference pattern.

5.6.3 Spatial Filtering and Sign Ambiguity

The filter of Fig. 5.23a, where only positive *spatial frequencies* in the horizontal, the u-direction, and both positive and negative frequencies in the vertical, the v-direction, can pass, gives an interference phase distribution with an increasing interference phase in the horizontal direction, but increasing and decreasing interference phase in the vertical direction. For a passband in the $-u$-halfplane the phases are decreasing in the horizontal direction, Fig. 5.23b. In the cases of Figs. 5.23c and d, the roles of the directions interchange with respect to Figs. 5.23a and b.

5.6 Fourier Transform Evaluation

Fourier transform evaluation now using (5.49) this intensity is written

$$I(x,y) = a(x,y) + c(x,y)\exp(2\pi i f_0 x) + c^*(x,y)\exp(-2\pi i f_0 x). \tag{5.61}$$

The Fourier transform of the intensity with respect to x then yields [479]

$$\mathcal{I}(u,y) = \mathcal{A}(u,y) + \mathcal{C}(u - f_0, y) + \mathcal{C}^*(u + f_0, y). \tag{5.62}$$

Since the chosen spatial carrier frequency f_0 is higher than the spatial variations of $a(x,y)$, $b(x,y)$, and $\Delta\phi(x,y)$, the partial spectra $\mathcal{A}$, $\mathcal{C}$, and $\mathcal{C}^*$ are well separated. $\mathcal{A}$ is concentrated around the d.c.-term at $u = 0$ and carries the low frequency background illumination. $\mathcal{C}$ and $\mathcal{C}^*$ are placed symmetrically to the d.c.-term and are centered around $u = f_0$ and $u = -f_0$. If by an adequate bandpass filter first $\mathcal{A}$ and $\mathcal{C}^*$ are eliminated, and then $\mathcal{C}(u - f_0, y)$ is shifted by f_0 toward the origin, the carrier is removed and we obtain $\mathcal{C}(u, y)$. Taking the inverse Fourier transform of $\mathcal{C}(u, y)$ with respect to u yields $c(x,y)$ defined by (5.49). From this $c(x,y)$ the interference phase is calculated by (5.53) with phase values lying between $-\pi$ and $+\pi$. An alternative to the Fourier transform evaluation is the fitting of modified sinusoids to the carrier fringes [480].

A combination of spatial and temporal heterodyning is used for the simultaneous recording of multiple phase objects on a single *space-time interferogram* [207]. Although the method was proposed originally for Michelson- and Mach-Zehnder-type interferometers, it may be used in holographic interferometry as well. An interferogram of the form

$$I(x,y,t) = a(x,y,t) + b(x,y,t)\cos[\Delta\phi(x,y,t) + 2\pi(f_{0X}x + f_{0Y}y + f_{0T}t)] \tag{5.63}$$

with the spatial carrier frequencies (f_{0X}, f_{0Y}) and the temporal carrier frequency f_{0T} can be evaluated by the three-dimensional form of the above evaluation scheme. Now phase variations with wider spatio-temporal bandwidths can be determined than would be possible by using only a single carrier frequency. Partial spectra $c_n(u, v, \omega)$ not separated by using either one of the carrier frequencies are separable when using both spatial and temporal carrier frequencies. Furthermore the spatio-temporal frequency bandwidth available for the system can be effectively utilized. Several images having bandwidths less than the image detection system can be recorded multiplexed on a single interferogram.

5.6.6 Spatial Synchronous Detection

Spatial heterodyning consists of filtering and shifting of frequency components in the spatial frequency domain. The analogue in the spatial domain is the *spatial synchronous detection* method of [481, 482]. This method assumes a holographic interference pattern with a spatial carrier as described by (5.60). The spatial carrier may be artificially introduced by tilting one of the interfering wavefronts or is a constituent of the interference pattern.

In spatial synchronous detection the recorded and stored intensity, (5.60), is multiplied by R_1 and R_2 given by

$$\begin{aligned} R_1(x,y) &= \cos 2\pi f_0 x \\ R_2(x,y) &= \sin 2\pi f_0 x. \end{aligned} \tag{5.64}$$

The multiplication results in

$$\begin{aligned}
I(x,y)R_1(x,y) &= [a(x,y) + b(x,y)\cos(\Delta\phi(x,y) + 2\pi f_0 x)][\cos 2\pi f_0 x] \quad (5.65)\\
&= a(x,y)\cos 2\pi f_0 x + \frac{b(x,y)}{2}\cos(\Delta\phi(x,y) + 4\pi f_0 x) + \frac{b(x,y)}{2}\cos(\Delta\phi(x,y)).
\end{aligned}$$

and

$$\begin{aligned}
I(x,y)R_2(x,y) &= [a(x,y) + b(x,y)\cos(\Delta\phi(x,y) + 2\pi f_0 x)][\sin 2\pi f_0 x] \quad (5.66)\\
&= a(x,y)\sin 2\pi f_0 x + \frac{b(x,y)}{2}\sin(\Delta\phi(x,y) + 4\pi f_0 x) - \frac{b(x,y)}{2}\sin(\Delta\phi(x,y))
\end{aligned}$$

Of the three terms in (5.65) and (5.66) the first is of high frequency, just the carrier frequency. The second term has an even higher frequency, but the third term represents a low spatial frequency component which can be separated from the remaining terms by low-pass filtering. Afterwards the interference phase is calculated from the isolated cosine- and sine-terms by (5.53).

The filtering in the spatial domain, intended to extract the third terms, is done by convolving the products IR_1 and IR_2 with a window function, which must be several fringe periods wide. This window function may be a rectangular window, or with better results, a Hanning window or one of the numerous others.

5.7 Dynamic Evaluation

Up to now in this chapter we have considered methods evaluating the interference fringes which appear superimposed over the reconstructed virtual image of the object. The fringes are observed from a fixed point of observation. These approaches to the evaluation of holographic interference patterns can be summarized as the *static evaluation methods* [483].

On the other hand the holographic interference fringes are generally localized in space, Section 4.3, so different observation points generate different interference patterns. A continuous variation of the observation point thus will induce a continuous change of the observed interferogram during reconstruction. This is the basic idea behind the so called *dynamic evaluation methods*, which were proposed very early after the invention of holographic interferometry [37, 287, 484]. But despite some conceptual advantages the dynamic methods have not found the as widespread applications as like the static methods.

5.7.1 Principles of Dynamic Evaluation

In the dynamic evaluation methods the interference orders are counted which are moving over a single observed object point while the observation point is continuously changed. Let us consider the case of a deformation of an opaque surface. An arbitrary point P of the surface may be displaced by the vector $\mathbf{d}(P)$ between the two exposures of a *double exposure hologram*, Fig. 5.27. Let this point be observed from an observation point which continuously moves from B_1 to B_2, then according to (4.20) and (4.21) the interference phase changes from

5.7 Dynamic Evaluation

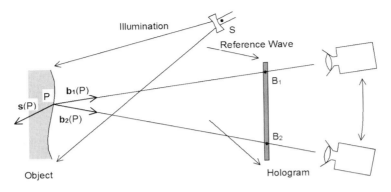

Figure 5.27: Geometry for dynamic evaluation.

$\Delta\phi_1(P) = \frac{2\pi}{\lambda} d(P) \cdot [b_1(P) - s(P)]$ to $\Delta\phi_2(P) = \frac{2\pi}{\lambda} d(P) \cdot [b_2(P) - s(P)]$. What has to be measured is the interference phase difference $\Delta\Delta\phi_{1,2}(P)$ between the two observation points B_1 and B_2, which is

$$\Delta\Delta\phi_{1,2}(P) = \Delta\phi_2(P) - \Delta\phi_1(P)$$
$$= \frac{2\pi}{\lambda} d(P) \cdot [b_2(P) - b_1(P)]. \tag{5.67}$$

This is the central equation for the dynamic evaluation methods.

Two advantages of the dynamic methods now become obvious: (1) the interference phase difference does not depend on the illumination vector $s(P)$, and (2) there is no absolute phase problem, since the additive phase term which is identical in $\Delta\phi_1(P)$ and $\Delta\phi_2(P)$, is canceled in the subtraction leading to $\Delta\Delta\phi_{1,2}(P)$. What remains is the necessity to determine $\Delta\Delta\phi_{1,2}(P)$ with the proper sign distribution; especially sign changes which may occur for complicated curves of localization must be recognized exactly.

The main sensitivity of the dynamic evaluation methods is in direction $b_2(P) - b_1(P)$, which is roughly parallel to the hologram plane and thus in most applications nearly tangential to the object surface. This complements the static methods, where the main sensitivity is for displacements in direction $b(P) - s(P)$, which is nearly normal to the surface.

If a one-dimensional displacement parallel to the hologram plane has to be evaluated, a single measurement of $\Delta\Delta\phi_{1,2}(P)$ suffices and a scalar form of (5.67) has to be solved for $d(P)$. For a two-dimensional displacement two measurements $\Delta\Delta\phi_{1,2}(P)$ and $\Delta\Delta\phi_{3,4}(P)$ are required, which may be evaluated from a single hologram if both displacement components are parallel to the hologram. In the general three-dimensional case three observations are necessary and the three-dimensional system of equations (5.67) has to be solved. The three observations can be taken through two holograms whose normals should form a significantly large angle, $> 30°$, to the best a $90°$ angle. An alternative to the second hologram can be a mirror close to the object, so that the investigated object point can be observed in two directions employing a single hologram [37].

5.7.2 Dynamic Evaluation by a Scanning Reference Beam

A straightforward approach to dynamic evaluation is to record many static reconstructions during a shift of the camera whose optical axis is kept intersecting a fixed object surface point in the virtual image. The angular separation between the frames must be small enough that the fringe count cannot be lost during data analysis. In [483] this was performed with 39 data frames, which were recorded and evaluated with the image processing facilities of those days. The experimental error could be restricted to about 5 %.

A more direct fringe counting is possible if the reconstruction of the holographic interferogram is performed by a thin reconstruction beam conjugate to the original reference beam [485, 486]. In this way, a real object image is obtained at the position of the original object, Fig. 5.28. The position of the reconstruction beam on the hologram plate defines the actual observation point. When the position of the reconstruction beam on the hologram is changed, the projected fringes move across the real image of the object surface. The varying intensity is recorded by a photodetector placed at the object point whose displacement is to be evaluated. To get the displacements at many points the detector position must be scanned over the real image, the same way as in temporal heterodyning, Section 5.4.

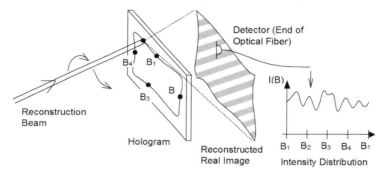

Figure 5.28: Dynamic evaluation by a scanning reconstruction beam.

Due to the small beam diameter, the aperture at the hologram is low and thus fulfills the requirements of the *observer projection theorem*, Section 4.3.5. That means the fringes are projected onto the real image plane independently of the localization in space.

The hologram may be scanned along several lines by a conjugate reference beam, which is produced by a spherical mirror or a converging lens corrected for spherical aberrations. Since more than three measurements now are taken, (5.67) is evaluated by the least squares method [486].

A thin conjugate reference beam can scan along closed loops. A repeated closed loop scan may filter temporal fluctuations by averaging. Arbitrary pairs of observation points B_i now can be chosen from the loop, Fig. 5.28 [485, 487–489].

Nothing is found in the literature about how the fringes, especially fractions of the fringes, are counted, or about the translation of the recorded intensity to interference phase. A change from increasing to decreasing interference order, or vice versa, must be detected. In this context the closed loop scanning offers some advantages: The phase difference along one whole

5.8 Digital Holographic Interferometry

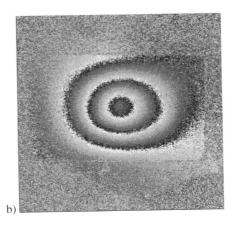

Figure 5.32: Phase difference without d.c.-term suppression (a) and after reconstruction by convolution (b).

The reconstructions leading to the results in Fig. 5.31 were performed by the Fresnel transform. A reconstruction by the convolution approach and the same phase subtraction as before gives the interference phase distribution of Fig. 5.32b. Despite the different pixel sizes and thus the other format of the object in the reconstructed image we have the same resolution of the measured interference phase values. The sharper appearance of the smaller object in Fig. 5.31c is only due to the pixel size. If we average the phase values over several pixels in the distribution of Fig. 5.32b to obtain pixels of the same size as in Fig. 5.31c we will get an identical image.

5.8.2 Enhancement of Interference Phase Images by Digital Filtering

The interference phase distributions generated by digital holographic interferometry as introduced in 5.8.1 exhibit already low noise. The signal-to-noise ratio is typically in the same range as that of phase differences determined by one of the phase shift methods. However, in the sawtooth-like display of the modulo 2π distributions even low-amplitude distortions appear visually annoying, since any single correct value below π that is slightly shifted above π occurs as an isolated black point in a neighborhood of bright points. In the same way a value that is erroneously below $-\pi$ although it correctly must be slightly above $-\pi$, will establish a bright pixel surrounded by dark ones. Therefore to produce a visually attractive interference phase image and to enable a reliable phase unwrapping, a smoothing of the interference phase by digital filtering should be performed.

A simple low-pass filtering, e. g. by replacing each pixel value by the average value of all neighboring pixel values, only is feasible when applied to already demodulated interference phase distributions. If applied to phase differences modulo 2π it will grind down the sharp edges at the 2π-jumps, thus introducing additional errors in the subsequent demodulation process.

It is well known that median filters behave similarly to low-pass filters. They effectively eliminate erroneous isolated points, but only to a certain degree they are edge-preserving. *Median filters* replace the value of each pixel by the median of the values of all pixels in a neighborhood of this pixel. The median of a finite set of values is that value for which there exist the same number of higher values as lower values in the set. The main parameter is the cardinality of the set, here the size of the neighborhoods. Frequently used square neighborhoods have 3×3, 5×5, or 7×7 pixels. The higher the pixel number of the used neighborhoods, the smoother the filtered phase distribution will look, but simultaneously the trend to round out the 2π-jumps also increases. This is demonstrated by an example in Fig. 5.33 where an

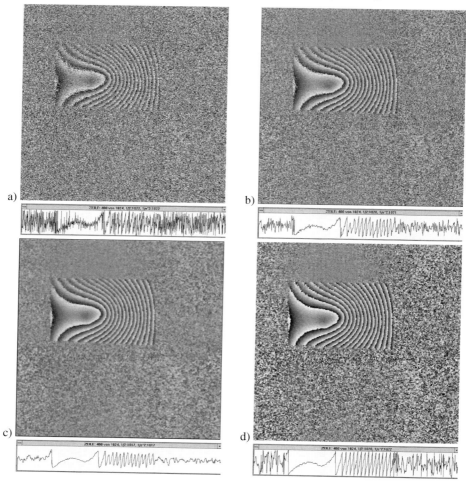

Figure 5.33: Filtering of interference phase distributions.

aluminum plate clamped at the edges was shifted horizontally and centrally pressed from the back side. A negative lens formed a small virtual image of the large object as described in

5.8 Digital Holographic Interferometry

Section 3.1.2. A normally impinging plane reference wave was used. The interference phase distribution modulo 2π obtained by reconstruction with elimination of the d.c.-term and by subtracting the phases is shown in Fig. 5.33a. Figure 5.33b gives the result after application of a median filter that uses 3×3-neighborhoods, Fig. 5.33c displays the same for a median filter considering 7×7-pixel neighborhoods. The effects of the smoother phase variation, but also the amplitude decrease at the 2π-jumps, are obvious.

However, there is a filter that performs even better in smoothing the phase while keeping the 2π-jumps at full amplitude. For this filter first the distributions $s(n,m)$ and $c(n,m)$ are calculated by

$$\begin{aligned} s(n,m) &= \sin[\Delta\phi(n,m)] \\ c(n,m) &= \cos[\Delta\phi(n,m)]. \end{aligned} \qquad (5.72)$$

Although $\Delta\phi(n,m)$ is modulo 2π, $s(n,m)$ and $c(n,m)$ show no 2π-discontinuities. They are both smoothed by a conventional low-pass filter, e. g. one replacing each value by the average value over square $k \times k$-pixel neighborhoods. After this filtering of $s(n,m)$ and $c(n,m)$ the filtered version of $\Delta\phi(n,m)$ is determined by

$$\Delta\phi_f(n,m) = \arctan \frac{s_f(n,m)}{c_f(n,m)}. \qquad (5.73)$$

The subscript f denotes the low-pass filtered versions of the corresponding distributions. The result of such a filter applied to the interference phase distribution of Fig. 5.33a is given in Fig. 5.33d. Here 5×5-pixel neighborhoods have been used.

A low-pass filtering of the numerator and denominator of (5.71) individually will not succeed, since their amplitudes vary from pixel to pixel. The procedure via (5.71) and (5.72) leads to normalized amplitudes.

Normalized $s(n,m)$ and $c(n,m)$ can be determined by

$$\begin{aligned} s(n,m) &= \frac{b_{r1}'(n,m)b_{i2}'(n,m) - b_{r2}'(n,m)b_{i1}'(n,m)}{\sqrt{[b_{r1}'^2(n,m) + b_{i1}'^2(n,m)][b_{r2}'^2(n,m) + b_{i2}'^2(n,m)]}} \\ c(n,m) &= \frac{b_{r1}'(n,m)b_{r2}'(n,m) + b_{i1}'(n,m)b_{i2}'(n,m)}{\sqrt{[b_{r1}'^2(n,m) + b_{i1}'^2(n,m)][b_{r2}'^2(n,m) + b_{i2}'^2(n,m)]}}. \end{aligned} \qquad (5.74)$$

These relations also result if $\sin(\arctan x) = x/\sqrt{1+x^2}$ and $\cos(\arctan x) = 1/\sqrt{1+x^2}$ are applied to (5.71).

5.8.3 Evaluation of Series of Holograms

The availability of the individual reconstructed phase distributions in digital holographic interferometry offers flexible metrologic applications. Thus one may record a series of digital holograms with gradually increased load amplitude. In the evaluation stage the convenient states can be interferometrically compared. The best combination of reconstructed phase distributions from all holograms is selected. A criterion for this selection can be the detectability of the interference orders in the resulting phase difference images. An analogous approach in optical holographic interferometry is known as *sandwich holography*: The states to be combined are recorded onto individual hologram plates, which are combined to a sandwich in the evaluation stage. This sandwich is reconstructed like a double-exposure hologram.

In the following a typical experiment of this kind is presented. The object is an aluminum plate clamped at its edges and positioned on a micro-stage. The reduced virtual image of the plate's surface is at a distance $d = 0.254$ m from the CCD array. This has $N \times M = 1024 \times 1024$ pixels of dimensions $\Delta\xi = \Delta\eta = 6.8$ µm. The laser wavelength is $\lambda = 0.6328$ µm, a normally impinging plane reference wave is guided to the CCD over a beam splitter cube. The refractive index of this cube is already recognized in the distance d. Two different displacements of the surface are possible:

- a deformation by a central pointwise pressure load from the back side of the aluminum plate, and

- a horizontal in-plane rigid body shift of the plate.

Eight digital holograms have been recorded with the displacements applied in accordance to the scheme shown in Fig. 5.34.

Figure 5.34: Scheme of the displacements and recordings of holograms.

The single capitals denote the recorded digital holograms while the terms between the boxes characterize the motions: DF denotes the deformations and TR stands for a translation. TR+ indicates a translation of higher amplitude than TR.

After reconstruction and calculation of the phase distributions of all digital holograms A to H combinations of phase distributions have been used for subtraction. Figure 5.35 shows some of the resulting interference phase distributions modulo 2π. The combination of holograms A and B displays the typical pattern of a bump produced by a pointwise pressure load. The comparison of holograms B and C exhibit the vertical fringes characteristic for the horizontal in-plane translation. The combination of these two motions leads to the pattern resulting from subtraction of the phases of holograms A and C. The high amplitude translation between holograms G and H leads to high density fringes which are just detectable, on the other hand the even higher amplitude translation between F and G lets the fringes disappear in noise.

In quantitative holographic interferometry it is generally not possible to determine multidimensional displacement vectors from a single interference pattern. One needs several interferograms, recorded with various sensitivity vectors, to obtain the data to define a multidimensional system of equations, which has to be solved. However, if the displacement is composed of elementary motions, e. g. rigid body translations, and if we have the ability to perform these motions separately with recording of the digital holograms between the single motions, then we can separate the motions numerically as is demonstrated for holograms A, B, and C of the aforementioned example.

On the object surface we fix three evaluation points P_l, P_c, and P_r, where the subscripts stand for left, center, and right. The angles between the optical axis and all lines between object points and CCD array points remain small, so the unit vector in the observation direction

5.8 Digital Holographic Interferometry

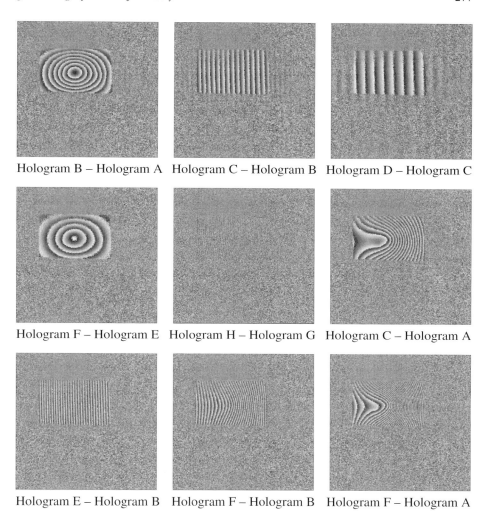

Figure 5.35: Evaluation of a series of digital holograms: interference phase distributions of various combinations.

can be set $\boldsymbol{b} = (0, 0, 1)$ for all three points. The illumination poses no special conditions on the angles. From the geometry of the holographic arrangement we get the unit vectors in the illumination direction $\boldsymbol{s}_l = (0.38, 0, -0.93)$, $\boldsymbol{s}_c = (0.49, 0, -0.87)$, and $\boldsymbol{s}_r = (0.58, 0, -0.82)$. The differences of the unit vectors are $\boldsymbol{e}_l = (-.38, 0, 1.93)$, $\boldsymbol{e}_c = (-0.49, 0, 1.87)$, and $\boldsymbol{e}_r = (-0.58, 0, 1.82)$. The interference pattern resulting from comparison of holograms A and B describes a continuous deformation that has only a component in the z-direction. Due to the clamping the interference orders at the left and the right edges are $n = 0.$, to the center we count $n = 6.5$. According to $n\lambda = \boldsymbol{d} \cdot \boldsymbol{e} = d_z \cdot e_z$ we obtain the z-displacement at the central point caused by the load from the back $d_{zc} = 6.5 \times 0.6328 \ \mu\text{m}/1.87 = 2.2 \ \mu\text{m}$. From the interferogram stemming from holograms B and C we detect an interference order at the right edge that is 15 periods less than that at the left edge, i. e. $n_r = n_l - 15$. The absolute fringe order is unknown, but along the whole surface we must have a constant translation d_x. From the two equations $n_l = d_x \cdot e_{xl}/0.6328$ µm and $n_l - 15 = d_x \cdot e_{xr}/0.6328$ µm we obtain the translation $d_x = 47.5$ µm as a solution. Now we can conclude on the interference orders in the interference pattern of the combined holograms A and C. The displacement vectors to the left, in the center, and to the right are $\boldsymbol{d}_l = (47.5, 0, 0)$, $\boldsymbol{d}_c = (47.5, 0, 2.2)$ and $\boldsymbol{d}_r = (47.5, 0, 0)$. From this the absolute interference orders $\boldsymbol{d}_l \cdot \boldsymbol{e}_l/\lambda = -28.52$, $\boldsymbol{d}_m \cdot \boldsymbol{e}_m/\lambda = -30.28$, and $\boldsymbol{d}_r \cdot \boldsymbol{e}_r/\lambda = -43.54$ result. Between the left edge and the center we have a difference of the interference orders of 1.76, between the center and the right edge this difference is 13.26. This can be controlled visually at the corresponding interferogram in Fig. 5.35. It is not possible to extract the absolute interference orders from only this single interferogram.

5.8.4 Compensation of Motion Components

In digital holographic interferometry the individual phase distributions are available numerically, therefore one may attempt to compensate motion components which lead to high fringe densities.

To make things not too complicated only rigid body translations in the plane of a plane surface should be investigated and we restrict the theoretical treatment to one dimension. A first attempt tries to shift one of the two holograms before the reconstruction process. But this does not lead to success, as can be proven theoretically with the help of the Fresnel approximation.

We have seen that if we shift a distribution $b(x)$ by a to $b(x-a)$ then the Fresnel transform $\mathcal{FR}\{b(x-a)\}$ is the Fresnel transform of $b(x)$ that is shifted by α and multiplied by a phase-factor β

$$\mathcal{FR}\{b(x-a)\} = e^{-i\pi\beta} B(\nu - \alpha) \tag{5.75}$$

where $B(\nu) = \mathcal{FR}\{b(x)\}$, $\alpha = 2\pi a/(d\lambda)$, and $\beta = 2a\nu - a^2/(d\lambda)$.

Let $R(\nu)$ be a reference wave for recording the hologram. Now we will show that even if α and β are known we cannot determine $h_1 = |R(\nu)+B(\nu)|^2$ from $H_2 = |R(\nu)+e^{-i\pi\beta}B(\nu-\alpha)|^2$. It is

$$\begin{aligned} h_1 &= (R_r + iR_i + B_r + iB_i)(R_r - iR_i + B_r - iB_i) \\ &= R_r^2(\nu) + R_i^2(\nu) + B_r^2(\nu) + B_i^2(\nu) + 2R_r B_r(\nu) + 2R_i B_i(\nu). \end{aligned} \tag{5.76}$$

On the other hand

$$\begin{aligned} h_2 &= (R_r + iR_i + \cos\pi\beta B_r + \sin\pi\beta B_i + i\cos\pi\beta B_i - i\sin\pi\beta B_r) \\ &\quad \times (R_r + iR_i + \cos\pi\beta B_r + \sin\pi\beta B_i - i\cos\pi\beta B_i + i\sin\pi\beta B_r) \\ &= R_r^2(\nu) + R_i^2(\nu) + B_r^2(\nu-\alpha) + B_i^2(\nu-\alpha) + 2\cos\pi\beta R_r(\nu)B_r(\nu-\alpha) \\ &\quad + 2\cos\pi\beta R_i(\nu)B_i(\nu-\alpha) + 2\sin\pi\beta R_r(\nu)B_i(\nu-\alpha) - 2\sin\pi\beta R_i(\nu)B_r(\nu-\alpha). \end{aligned} \tag{5.77}$$

In (5.77) we have dependencies on the arguments ν as well as on $\nu - \alpha$, which shows that in this general case a shift is not possible.

The ν-dependency in (5.77) only concerns the reference wave, therefore the special case of the normally impinging plane reference wave $R_r(\nu) = 1.$, $R_i(\nu) = 0$ is considered because this reference wave does not depend on ν. We get

$$h_1 = 1 + B_r^2(\nu) + B_i^2(\nu) + 2B_r(\nu) \tag{5.78}$$

and

$$h_2 = 1 + B_r^2(\nu - \alpha) + B_i^2(\nu - \alpha) + 2\cos\pi\beta B_r(\nu - \alpha) + 2\sin\pi\beta B_i(\nu - \alpha). \tag{5.79}$$

A transformation of h_2 into h_1 for all ν by a shift would imply $\cos\pi\beta = 1$ and $\sin\pi\beta = 0$. By definition of β this is only possible for $a = 0$, which is a contradiction to the assumed non-vanishing shift. Altogether it is proven that by a mere shift of the digital hologram we cannot compensate any motion components.

However, a feasible approach is first to reconstruct the phase distributions from the holograms and to add or subtract a certain phase to one of the two or equivalently to the phase difference [491, 492]. This additional phase can be a linearly varying phase distribution. Although we will get no additional information by this procedure, sometimes fine details in the display of the interference phase distribution are better recognized.

This is shown in three examples in Fig. 5.36. Some phase distributions of the holograms leading to Fig. 5.35 have been used for this purpose. The reconstructed phases of holograms B and C have been compared with a phase subtracted that increases linearly from the left to the right. At the left edge of the reconstructed pattern this phase is 0 rad, it increases to the right edge by 150 rad $\approx 24 \times 2\pi$. The second example is with holograms B and E, the

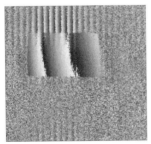

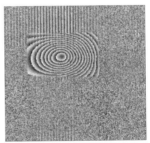

Hologram C – Hologram B   Hologram E – Hologram B   Hologram F – Hologram A

Figure 5.36: Compensation of motion components by subtraction of linearly varying phases.

third one uses the holograms A and F. In both these cases the linear phase increase is 300 rad $\approx 48 \times 2\pi$. The density of the fringes on the reconstructed surface is significantly reduced. In the remaining low-density fringes now we see a tilt, which is not recognized before the compensation, Fig. 5.35. The reason for this inclination of the fringes is a spatial variation of the sensitivity vector. In the last of the three examples we had a combination of a rigid body translation and a bump. Of these the translation has been nearly completely eliminated.

One may notice interference fringes in the area of the suppressed d.c.-term. This pattern corresponds to the phase distribution of the subtracted linearly increasing phase. A similar approach as presented in this section is used by Gombköto et al. [493].

5.8.5 Multiplexed Holograms Discriminated in Depth

In most practical applications of holographic interferometric deformation measurement one is only interested in the one-dimensional displacement distribution in the normal direction. This measurement is performed by using a sensitivity vector (4.20) normally oriented to the object's surface and varying as little as possible. Ideally one uses a collimated expanded object illumination beam. However, in various applications the whole three-dimensional displacement vector field has to be determined. Then several holographic interferograms belonging to different sensitivity vectors have to be produced and evaluated. The result of the evaluation of three interferograms is a set of three interference phase distributions $\Delta\phi_i(P)$, $i = 1, 2, 3$, which allow a solution of the system of equations (4.21)

$$\begin{pmatrix} \Delta\phi_1(P) \\ \Delta\phi_2(P) \\ \Delta\phi_3(P) \end{pmatrix} = \begin{pmatrix} d_x(P) \\ d_y(P) \\ d_z(P) \end{pmatrix} \begin{pmatrix} e_{1x}(P) & e_{1y}(P) & e_{1z}(P) \\ e_{2x}(P) & e_{2y}(P) & e_{2z}(P) \\ e_{3x}(P) & e_{3y}(P) & e_{3z}(P) \end{pmatrix}. \tag{5.80}$$

Different sensitivity vectors can be obtained by manifold ways: one can vary the illumination direction, one can vary the observation direction, or both simultaneously. A cheap, robust and reliable method is the recording of different views of the object's surface, corresponding to different observation directions, onto one hologram. The different views come from appropriately placed mirrors, a procedure already suggested by Schönebeck [494, 495].

This approach now should be transfered to digital holography. One has to be careful not to obtain too large angles between object and reference waves, see Section 3.1.1, therefore an arrangement as shown in Fig. 5.37 is suggested. It allows the simultaneous recording of the

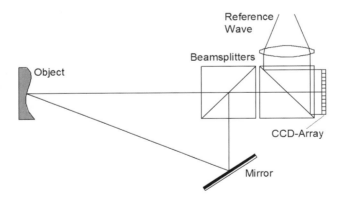

Figure 5.37: Arrangement for multiplexed digital holography.

wave field coming directly from the object and the one coming over the mirror. A third observation direction can be designed with a third beamsplitter and a further mirror, both oriented 90° to the drawing plane of Fig. 5.37. The hologram now contains the information about several angular views of the object, hence we can call it a *multiplexed hologram*. There is

5.8 Digital Holographic Interferometry

no extra problem with the object's stability since all recordings from different angles are performed simultaneously. The individual interference patterns can be separated by their different distances from the hologram plane. For the geometry of Fig. 5.37 with the two observation directions separated by angle β, the distances d_1 and d_2 as indicated in Fig. 5.38 are given by

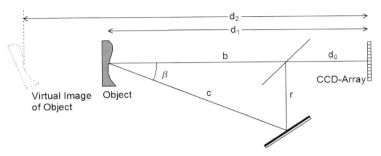

Figure 5.38: Geometry of multiplexed digital holography.

$$\begin{aligned} d_1 &= d_0 + b \\ d_2 &= d_0 + r + c \\ &= d_0 + b\left(\tan\beta + \frac{1}{\cos\beta}\right). \end{aligned} \qquad (5.81)$$

Experiments have shown that it is advantageous to have large $|d_2 - d_1|$. If the reconstructed objects are too close together they will disturb each other. This is shown in Fig. 5.39. The two

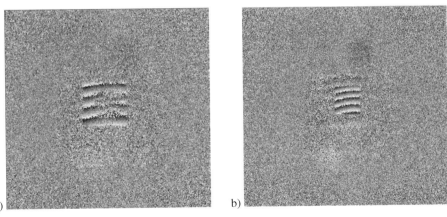

Figure 5.39: Interference phase distributions reconstructed from multiplexed digital hologram at 50 cm (a) and 75 cm (b) distance.

interference phase distributions are reconstructed from a single multiplexed digital hologram. The different sizes of the reconstructed surface are caused by the Fresnel transform (3.43). The distances in this experiment were $d_1 = 50$ cm and $d_2 = 75$ cm. We can recognize in

Fig. 5.39a the disturbance in the center of the interference phase distribution reconstructed for the distance $d_1 = 50$ cm that was caused by the other reconstruction. An alternative way instead of large $|d_2 - d_1|$ is a slight tilt of the mirror so that the reconstructed sharp and unsharp object wave fields do not overlap. The interference phase distributions resulting from such an experiment are given in Fig. 5.40.

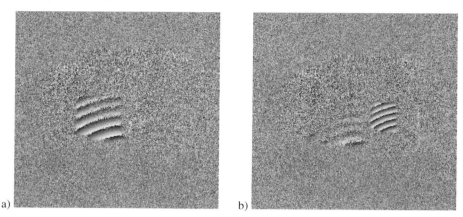

Figure 5.40: Interference phase distributions reconstructed from multiplexed digital hologram at 50 cm (a) and 75 cm (b) distance, tilted mirror.

5.8.6 Multiplexed Holograms with Discrimination by Partial Spectra

We have seen in Sections 3.2.5 and 3.2.6 that in the computer we have access to the frequency spectrum of the digital hologram, which can be manipulated to influence the reconstructed image in a desired way. There the suppression of the d.c.-term as well as the twin image by filtering in the spatial frequency plane was demonstrated. This approach also is feasible for recording simultaneously multiple holograms onto the CCD which can then be discriminated by filtering in the spatial frequency domain. Therefore the multiplexed holograms must exhibit different spatial frequencies. This can be obtained by using different reference waves for the generation of the single holograms. To guarantee that only the corresponding waves interfere and the resulting holograms superpose incoherently avoiding any unwanted cross-interferences, we have basically two possibilities:

- Two single holograms are separated by polarization: one pair of corresponding reference and object waves is linearly polarized in a certain direction, the other wave-pair is linearly polarized in the orthogonal direction [496, 497].

- The light paths are of different lengths, which can be achieved by delay lines, eventually by using optical fibers. The length differences then have to be larger than the coherence length of the used laser light [62, 306]. This option allows even more than two incoherently overlayed holograms.

If we arrange the source points of the divergent spherical reference waves in such a manner that the resulting partial spectra do not overlap, then we can extract the partial spectra by simple band-pass filtering in the spatial frequency domain and reconstruct from the individually filtered digital hologram [498]. So we only have to find the source points of the reference waves enabling this.

The first fact we have to recognize in doing this is the nature of the digital hologram as an intensity distribution described by real numbers. For this reason the spectrum of the hologram as well as the spectrum of each partial hologram is Hermitean, which means it consists of two parts placed symmetrically with respect to the zero frequency. In order to put a certain number of spectra non-overlapping into the spectral range, the size of each partial spectrum is restricted. Therefore first we have to choose an object-to-CCD distance which on the one hand fulfills the sampling theorem, and on the other hand produces holographic structures which have a spectrum smaller than the predetermined partial spectrum size. For an initial test hologram we employ an arbitrary divergent reference wave, which only has its source point in the plane where the source points of the desired reference waves are positioned. This plane should be parallel to the CCD-array, so that for the reference wave source points $(x_R^{(i)}, y_R^{(i)}, z_R^{(i)})$ we have $z_R^{(i)} = z_R$ for all i. The i term numbers the partial holograms and thus the different reference waves. To find proper $(x_R^{(i)}, y_R^{(i)})$, we shift the individual partial spectra in the spatial frequency domain, which means we investigate and manipulate the Fourier transform of the digital hologram. Now we remember that the Fresnel transform reconstruction in the case of lensless Fourier-transform holography reduces to a mere numerical Fourier transform, and the reference wave is a divergent one with its source point in the plane of the object surface. We can assume $(x'_R, y'_R, z'_R) = (0., 0., d)$ as such a reference wave with d the object-to-CCD distance. The choice $x'_R = y'_R = 0$. can be done without restriction of generality, other values lead to phase factors which do not shift geometrically the reconstructed image.

In the wave field reconstructed with $R' = (0., 0., d)$, which is an unsharp replica of the object wave field recorded with the help of $(x_R^{(i)}, y_R^{(i)}, z_R)$, we now define a point $P = (x_P, y_P, z_P)$ which should be shifted to $P' = (x'_P, y'_P, z_P)$. These data are put into the holographic imaging equations (2.154), (2.155), and (2.156), which describe the shift of points in the reconstructed image if the reference wave source is shifted. The solution of the imaging equations is

$$\begin{aligned} x_R &= x_P \frac{z_R}{z_P} + x'_P \left(1 - \frac{z_R}{z_P} - \frac{z_R}{z'_R}\right) \\ y_R &= y_P \frac{z_R}{z_P} + y'_P \left(1 - \frac{z_R}{z_P} - \frac{z_R}{z'_R}\right) \\ z_R &= \text{predetermined.} \end{aligned} \quad (5.82)$$

This procedure is carried out for all desired spectra and we obtain the reference wave source points. The x_P, y_P, x'_P, y'_P can be expressed in pixel coordinates n, m; $n = 1, \ldots, N$; $m = 1, \ldots, M$ or in real lengths $n\Delta x', m\Delta y'$.

The whole process now is demonstrated in a simulated example of black and white rectangles as object. The object has size 256×256 pixels in a 1024×1024-pixel field. The other parameters are $d = 2.0$ m, $\lambda = 0.6328$ μm, $\Delta \xi = \Delta \eta = 6.8$ μm. The hologram is recorded with the help of a spherical reference wave emitted from $R = (0.01, 0.02, 0.5)$ m. The hologram is displayed in Fig. 5.41a, the reconstructed intensity image is shown in Fig. 5.41b. The spectrum of this hologram is given in Fig. 5.41c. It is an image reconstructed using the reference wave $\hat{R} = (0., 0., 2.)$ m. In the plane $d = -1.0$ m this reference wave reconstructs sharply, the intensity display in this plane we see in Fig. 5.41d.

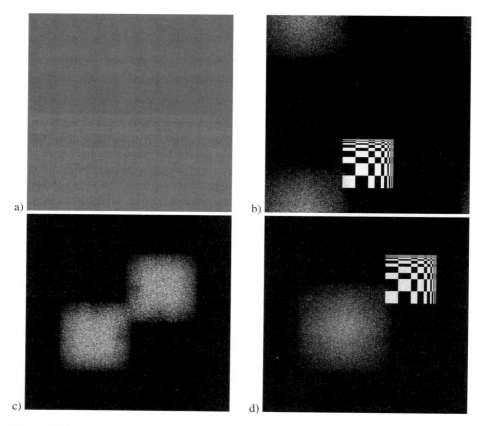

Figure 5.41: Simulated digital hologram (a), reconstructed image (b), spectrum of the hologram (c), reconstructed image in the plane $d = -1.0$ m (d).

Now the two modes of the spectrum of Fig. 5.41c should be taken apart along the diagonal, furthermore a second spectrum along the other diagonal should be generated from the same object. To solve the first of these two tasks we note that the upper right edge of the object is at the pixel coordinates $(860, 199)$, which corresponds to $P = (78.155, 18.085, -1000.)$ mm in real world dimensions, since the pixel size in this plane is $\Delta x = \Delta y = 1$ m $\cdot$ 0.6328 μm$/(1024 \cdot 6.8$ μm$) = 90.88$ μm. This P should be shifted to $P' = (93.053, 0.091, -1000.)$ mm which corresponds to $(1024, 1)$ in pixel coordinates. But P itself is a shifted point, because the hologram was generated using $R = (0.01, 0.02, 0.5)$ m,

5.8 Digital Holographic Interferometry

and P was reconstructed by $\hat{R} = (0., 0., 2.)$ m. We have

$$0.078155\,\text{m} = x_P = \frac{\hat{x_P} z_R \hat{z}_R - x_R z_P \hat{z}_R}{z_R \hat{z}_R - z_P \hat{z}_R + z_P z_R} = \frac{\hat{x_P} - 0.01 \cdot 2. \cdot 2.}{1. - 4. + 1.} \frac{\text{m}^3}{\text{m}^2}$$

$$= \left(\frac{0.04}{2} - \frac{\hat{x_P}}{2}\right)\text{m} \qquad (5.83)$$

$$\hat{x_P} = 0.04\,\text{m} - 2. \cdot 0.078155\,\text{m} = -0.116\,\text{m}.$$

In the same way we calculate

$$0.018085\,\text{m} = y_P = \frac{\hat{y_P} z_R \hat{z}_R - y_R z_P \hat{z}_R}{z_R \hat{z}_R - z_P \hat{z}_R + z_P z_R} = \left(\frac{\hat{y_P}}{2} - \frac{0.02 \cdot 4}{2}\right)\text{m}$$

$$\hat{y_P} = 0.08\,\text{m} + 2. \cdot 0.018085\,\text{m} = +0.116\,\text{m}. \qquad (5.84)$$

This point $\hat{P} = (-0.116, 0.116, -1.0)$ m now is shifted to P' by the reference wave $R' = (x'_R, y'_R, 0.5\,\text{m})$. The expression $0.093\,\text{m} = (-0.116\,\text{m} - 4x'_R)/-2$ is solved by $x'_R = 0.0175$ m and $0.000091\,\text{m} = (0.116\,\text{m} - 4y'_R)/-2$ is solved by $y'_R = 0.029$ m. Thus the desired first reference wave is $R' = (0.0175, 0.029, 0.5)$ m.

The second reference wave R'' shifts $\hat{P}$ to $P'' = (0.093, 0.070, -1.0)$ m. As in the preceeding case we get $x''_R = 0.0175$ m, but now we obtain $0.07\,\text{m} = (0.116\,\text{m} - 4y''_R)/-2$, which is solved by $y''_R = -0.006$ m. So we have $R'' = (0.0175, -0.006, 0.5)$ m. The hologram recorded with R' and reconstructed with $\hat{R}$ gives the sharp image shown in Fig. 5.42a, the hologram recorded with R'' and reconstructed likewise with $\hat{R}$ leads to the image given in Fig. 5.42b.

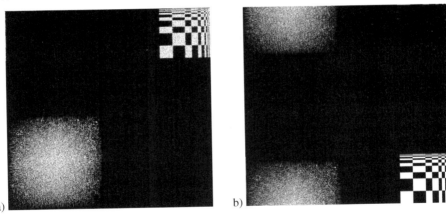

Figure 5.42: Images in the plane $d = -1.0$ m, both reconstructed using $\hat{R}$, but (a) recorded with R' and (b) with R''.

Now we record two holograms of the object, one with the reference wave $R' = (0.0175, 0.029, 0.5)$ m, one by using $R'' = (0.0175, -0.006, 0.5)$ m. The holograms are summed incoherently. The spectrum of this sum-hologram is displayed in Fig. 5.43a. If we reconstruct from the unfiltered sum-hologram using reference wave $R' = (0.0175, 0.029, 0.5)$ m, we obtain an image which contains the object twice, Fig. 5.43b. But if we filter the sum-

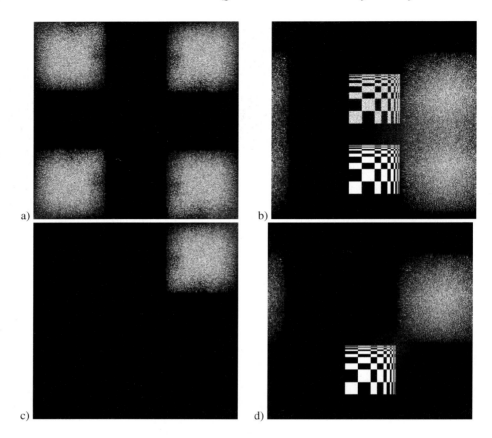

Figure 5.43: Spectrum of sum-hologram (a), reconstructed intensity from sum-hologram (b), filtered spectrum (c), reconstructed intensity from filtered hologram (d).

hologram in such a way that only the spectrum of Fig. 5.43c remains, after reconstruction with $R' = (0.0175, 0.029, 0.5)$ m we get the image of Fig. 5.43d. After the application of a band-pass filter leaving only the spectrum of Fig. 5.43e the reconstruction using the corresponding reference wave $R'' = (0.0175, -0.006, 0.5)$ m gives the field shown in Fig. 5.43f. Because in both cases we used the corresponding reference waves, the image of the object is at the same place in each reconstructed field.

Spectral multiplexing of digital holograms is a way to record fast events on a CCD target. Liu et al. [144] report on the recording of three frames of a fast event in a single CCD frame. The resolution of a single shot was 5.9 ns, defined by the pulsed laser used, the frame interval was 12 ns, generated by a specially designed multiple pulse-generation cavity. The holograms of the three states of the event are well separated in the spatial frequency domain.

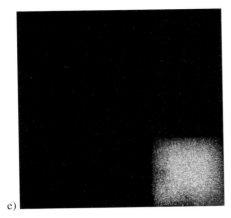 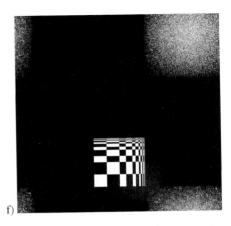

Figure 5.43 (continued): (e) filtered spectrum, (f) reconstructed intensity from filtered hologram.

5.9 Interference Phase Demodulation

As outlined in Section 5.1.1, the evaluated interference phase distributions are ambiguous in having only values between $-\pi$ and $+\pi$. They are said to be wrapped modulo 2π [499, 500]. Although in most practical applications a continuous interference phase distribution is expected, we get a sawtooth-like phase as the one shown in Fig. 5.44a. The process of resolving the 2π-discontinuities by adding a step function consisting only of 2π-steps, Fig. 5.44b, is called *continuation*, *phase unwrapping* or *demodulation*.

5.9.1 Prerequisites for Interference Phase Demodulation

There are a number of requirements to be fulfilled in order for a reliable demodulation. The first is that continuous phase data are adequately sampled. Each demodulation procedure checks interference phase differences of neighboring pixels. If due to a violation of the sampling theorem this difference exceeds π, the demodulation must fail because of the introduction of unnecessary phase jumps. On the other hand undersampled high frequency fringe data may result in small phase differences between two pixels but in reality a phase jump between these pixels is missed. Nevertheless an approach to phase recovery from undersampled interferograms is suggested by Munoz et al. [501, 502].

Statistical noise like speckle noise is a common cause for false identification of phase jumps. As soon as the amplitude of the noise approaches π, the actual phase jumps become obscured. A low pass filter for smoothing the interference phase data is not recommendable, since it will wash out the sharp 2π-steps. Instead median filtering is a good choice, since it is 2π-step preserving [394], and a still better choice is the sin-cos-filter of Section 5.8.2.

Each demodulation procedure assumes a continuous interference phase distribution which is wrapped into an interval of width 2π. So discontinuities resulting from large height steps or holes or edges of objects not filling the full frame have to be avoided. The region in which the demodulation should be performed has to be defined by masking. If the region for de-

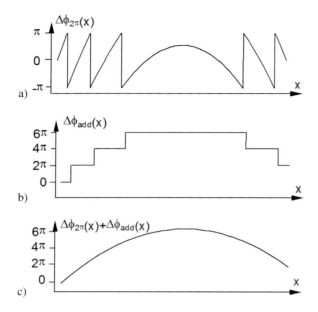

Figure 5.44: Demodulation: (a) interference phase modulo 2π, (b) step function to be added, (c) unwrapped interference phase distribution.

modulation is not topologically connected, then no correct relations of the interference phase in the unconnected parts can be derived without additional side information, e. g. particular points in each part where the exact displacement is known. An approach to automatic phase unwrapping in the presence of surface discontinuities is described in [503].

Especially if the interference phase distribution has been determined by the Fourier transform method without additionally introduced carrier fringes, Section 5.6, then sign-errors may occur. The sign-error must be corrected, e. g. as described in Section 5.6.3, before demodulation. Each demodulation requires sign-correct interference phase distributions modulo 2π [504], only a global sign-change may be tolerable, leading to a global sign-change in the unwrapped continuous phase.

The demodulation procedures can roughly be categorized into *path dependent demodulation* techniques, where the order in which the pixels are investigated and unwrapped is predetermined by the process, and into *path independent demodulation* techniques. In this latter case the order of investigated pixels is determined by the phase values at the pixels.

5.9.2 Path-Dependent Interference Phase Demodulation

A straightforward attempt to demodulate a one-dimensional interference phase distribution $\Delta\phi(x)$ is done by checking the phase differences of adjacent pixels $\Delta\phi(x+1) - \Delta\phi(x)$. If this difference is less than $-\pi$, an additional 2π is added to $\Delta\phi$ from $x+1$ onwards; if the difference is greater than $+\pi$, one more 2π is subtracted from $\Delta\phi$ starting at pixel $x+1$.

5.9 Interference Phase Demodulation

After a number of local iterations – a number, that mainly depends on the maximum distance of 2π-steps in the original phase-map – oscillation between successive patterns occurs. When this oscillatory state is reached, the arithmetic average of the two states is computed at each pixel and the process starts anew. The averaging is called *global iteration*. If a steady state is reached after global iteration, the process stops.

The process of cellular automata demodulation is shown in Fig. 5.46 for the example of a simulated 32×32-pixel interference phase modulo 2π, Fig. 5.46a. The first oscillation takes place after 6 local iterations, Fig. 5.46d. After 3 global iterations, the phase is successfully demodulated, Fig. 5.46i.

Figure 5.46: Cellular automata demodulation: (a) interference phase modulo 2π, (b) phase after 1 local iteration, (c) phase after 2 local iterations, (d) phase after 6 local iterations, (e) phase after 1 global iteration, (f) phase after 1 global and 1 local iterations, (g) phase after 2 global iterations, (h) phase after 2 global and 1 local iterations, (i) phase after 3 global iterations.

The algorithm is robust against distortions and noise, but there are possible points or regions that would lead to new 2π-discontinuities after each global iteration, so that the algorithm would never terminate. Therefore a *check for consistency* has to be performed. This consists of summing the four phase differences along the closed path connecting the four pixels of a 2×2-neighborhood. Here each phase difference is taken modulo 2π. If the sum is zero, the neighborhood is said to be consistent, otherwise it is interpreted as being inconsistent. Figure 5.47a gives a consistent 2×2-array, since $[-2.77 - 2.67] + [-2.04 +$

Figure 5.47: Consistent (a) and inconsistent (b) 2×2-arrays of interference phase values modulo 2π.

$2.77] + [1.94 + 2.04] + [2.67 - 1.94] = 0.84 + 0.73 - 2.30 + 0.73 = 0$. Here $[.]$ denotes the value modulo 2π in $]-\pi, \pi]$. Figure 5.47b is an example of an inconsistent array, for $[-2.53 - 2.45] + [-1.15 + 2.53] + [1.08 + 1.15] + [2.45 - 1.08] = 1.30 + 1.38 + 2.23 + 1.37 = 6.28$. The inconsistent regions are masked and discarded from being tested in the local iterations. The inconsistency test may be performed once before the first local iteration [527] or additionally after each global iteration [528]. Nevertheless there remain the problems that the inconsistency check does not guarantee convergence to the steady state for all possible phase-maps, especially those with phase dislocations induced by aliasing [527], or that the inconsistency regions will increase in size [528].

5.9.5 Further Approaches to Interference Phase Demodulation

The demodulated interference phase distribution is assumed to have no sharp discontinuities which would give rise to high frequency components in the Fourier spectrum. A check of the bandwidth of a one-dimensional phase distribution by the Fourier transform is the base for the *bandlimit demodulation* [529, 530]. Here all possible step functions with one $+2\pi$- or -2π-step are tested, the step function that minimizes the bandwidth is selected as one that unwraps at one position. This process is repeated until no further reduction of the bandwidth is obtained. The method has been shown to be robust in the presence of signal-independent random additive noise, provided that the correct interference phase function is slowly varying. The method will fail if the number of phase steps increases. The extension of the bandlimit approach to two-dimensional phase-maps on a more effective basis than mere repeated row or column processing has still to be done.

An actual approach to demodulation is the application of a *neural network* of the Hopfield type [531, 532]. In *Hopfield networks* all possible connections between all neurons are existent. The synaptic weights only take on the inhibitory value 0 or the excitatory value 1. Hopfield networks are used to minimize a so-called Hopfield energy function. For demodulation purposes with each pixel a number of neurons is associated. A merit function $E(\phi_{add})$ is defined, which measures the deviation from smoothness of the phase distribution unwrapped by ϕ_{add}. The $n(x)$ in $\phi_{add}(x) = 2\pi n(x)$ is represented by the number of the excited neu-

5.9 Interference Phase Demodulation

rones of all those associated with pixel x. This merit function is related to the Hopfield energy function in a discrete-time Hopfield network. According to the equations of the network dynamics, the neurons change their states in such a manner that the value of the energy function decreases with the temporal evolution of the network until a stable state is reached.

As with the cellular automata approach in defining the merit function here we may use 4- or 8-pixel neighborhoods. Furthermore inconsistencies have to be excluded. This is done as described for the cellular automata demodulation, Section 5.9.4. But now no cut-lines as in [519] are defined, instead the synaptic connections across the cut-lines are prohibited.

All demodulation techniques presented so far spatially unwrap a single interference phase distribution. In *temporal phase unwrapping* as proposed in [533, 534] a number of measurements with increasing load is recorded, where the time increments are chosen small enough to fulfill the sampling theorem requirements. Now unwrapping is performed along the time axis for each pixel independently from its neighboring pixels. So objects not filling the full frame, unconnected regions, or regions with poor signal-to-noise ratios will not corrupt the demodulation of good data points. The requirement of sign-correct phase data is fulfilled by using the phase shift method with four interferograms and a mutual phase shift of $90°$, see (5.24). By a proper combination of the equation for phase shift evaluation (5.24) and the calculation of the phase difference between successive time instants, this temporal phase difference can be calculated with values in $]-\pi, +\pi]$, as long as the sampling theorem is fulfilled. The overall interference phase without any 2π-discontinuity then for each time instant is determined by summation of the temporal phase differences. Temporal phase unwrapping can be used not only for interference phase distributions but also for the phase maps of complex wave fields in digital image plane holography [535].

A demodulation by two-dimensional fitting of unwrapped phase maps in a least squares sense is proposed in [513]. The problem of two-dimensional *least squares phase unwrapping* is shown to be equivalent to the solution of Poisson's equation on a discrete rectangular grid with Neumann boundary conditions. It is solved by application of a fast discrete cosine transform [536]. If less reliable pixels or regions are known, resulting from regionally varying noise, aliasing, phase inconsistencies, measurement errors, shadows, or no phase data at all, these parts of the pattern can be appropriately weighted. Weighting may be binary into 0 and 1 or continuously inverse proportional to the probability of noise or error. An algorithm based on Picard iteration is given for weighted least squares phase unwrapping. Another algorithm which provides faster convergence is developed on the basis of the method of preconditioned conjugate gradients. Several examples in [513] show how phase noise, data inconsistencies and other degradations are automatically accommodated by this least squares approach. A least-squares modal estimation of wrapped phases by an algorithm specially designed to be used as part of an iterative algorithm can be found in [537].

tors is mounted. With such a system undesired fringe motion is compensated for, whether due to laser frequency drift or interferometer path difference perturbations [553].

The easiest way to perform a compensation of unwanted object motions by modulating the phase of the reference wave is to direct the reference beam over a small mirror fixed to the object [554–556]. The reference wave thus is reflected out of the object illumination wave by wavefront division. The mirror undergoes the same motions as the point of the object to which it is fixed. A mathematical analysis is based on the fact that the mirror is specularly reflecting while the object surface is diffusely reflecting [554]. The analysis shows that with such a reference mirror not only the rigid body motions of the object are compensated but additionally we have one point in the fringe pattern where the interference order is known to be zero, namely the point of the mirror reflecting the illumination source point to the observation point.

The consequent continuation of these ideas leads to the fixation of the whole holographic arrangement onto the object whose surface deformations instead of the rigid body motions should be measured. Especially for large scale objects this is feasible. In such a way the deformation of a 4.5 m high *pressure vessel* with steel walls of 45 mm thickness has been measured holographically [546]. Some interferograms produced in this experiment have been shown in Fig. 6.2. The whole holographic arrangement including the laser was attached to the vessel by magnetic feet. Moreover, good results have been obtained, if the object is sprayed with a retroreflective paint and only the hologram plate is rigidly attached to it [557].

Although there exist effective methods to stabilize the microinterference fringes to be recorded in the hologram there still may remain an unknown rigid body motion between the two exposures of a double-exposure hologram. A compensation for this motion is possible during the reconstruction when the holographic interference pattern is generated. If the two states are recorded on separate hologram plates, these two plates after development may be tilted and shifted in the reference wave, a procedure called *sandwich hologram interferometry* [558–565]. The relative motion of the plates leads to a relative motion of the reconstructed wave fields, with the consequence that an object translation can be compensated. It is helpful if an area on the object, on the frame, or else in the arrangement, is visible through the holograms, where it is known that no displacement must have occurred. While in sandwich holography the two plates are close together in a common reference wave – thus the name of the method – the two plates can be fully separated with individual reference waves acting on them [566, 567]. Then one of the two plates is fixed to a hologram holder that has all degrees of freedom to move with a precision better than 1 μm. This method is suitable for compensating stress-induced translations in non-destructive testing leading to defect fringes on an infinite fringe background [568]. A detailed theoretical analysis of fringe modifications in the holographic interferometric measurement of large deformations is given in [569, 570].

In holographic measurements of phase objects the idea of dividing the reference wave out of the object wave can be realized not only by using a reference mirror at or near the object but also by extracting the zero frequency out of the whole field in the focal plane of a lens, while the object wave is contained in the first diffraction order [571]. The phase object for this approach is illuminated by a plane wave that has passed a Ronchi grating. To separate the reference and the object wave a spatial filter with two holes must be placed in the focal plane of a lens. A second lens images these two fields onto a plane where they interfere. Since object beam and reference beam take a common path, this interferometer is proof against external vibrations.

6.2 The Sensitivity Matrix

In the preceding section it has been shown how the displacement vectors are determined from the evaluated interference phase distributions. Connecting elements are the sensitivity vectors which in all evaluations of more than one dimension are combined in the *sensitivity matrix*. This section now will discuss how to obtain the components of the sensitivity matrix and how its condition can be quantified, which is the main property defining the achievable accuracy of the measurement results.

6.2.1 Determination of the Sensitivity Vectors

The *sensitivity vectors* $e(P)$ are calculated by (4.20) with the help of the unit vectors $s(P)$ in the illumination direction and $b(P)$ in the observation direction as defined in (4.18) and (4.19). To find out the sensitivity vectors of a specific experiment in an arbitrary but fixed Cartesian coordinate system one needs the coordinates of the *illumination point* S, which is the focal point of the optics used for expanding the illuminating laser beam, the coordinates of the *observation point* B, which is the center of the entrance pupil of the observing optical system, and the coordinates of the object points P of interest. For collimated illumination or observation the unit vectors are colinear with the illumination or observation directions and are identical for all object surface points.

Although the choice of the origin of the coordinate system is arbitrary, two options yield easier mathematics, since some components then will be zero. First, if the surface to be measured is plane, an origin at the surface with two axes lying in the surface plane should be chosen; second, for arbitrarily shaped object surfaces, a coordinate system with one axis colinear to the line connecting the illumination with the observation point and the origin halfway between these points is advantageous. Nevertheless the determination of the sensitivity vectors remains involved, especially for objects of complicated three-dimensional shape, for wide-angle or multiple illumination directions, or when mirrors are used for back or side views of the tested objects.

The straightforward approach is to make simple yardstick measurements between representative points of the holographic arrangement and then to calculate the necessary coordinates. Often the geometry or the contour of the object surface is given by design data, otherwise it has to be measured. There are optical methods for *contour measurement*, like triangulation, projected fringe, moiré, or holographic methods, which may be applied [164,572]. In the following only the holographic interferometric methods of these will be discussed.

In the preceding section methods were treated as to how to determine the unknown displacement vectors of surface points from known sensitivity vectors and measured interference phase values. The methods for the determination of the sensitivity vectors to be presented here employ known displacements or known rotations together with measured phases to determine the a priori unknown sensitivity vectors [573].

6.2 The Sensitivity Matrix

Consider three separate double exposure holograms of the object undergoing three different, but known, *rigid body displacements* d^1, d^2, and d^3, which must not be effected in coplanar directions. Assume the same recording and observation geometry for the three holograms with respect to the object. This ensures an invariant sensitivity vector from one recording to the other. Now evaluate the interference phases $\Delta\phi^1(P)$, $\Delta\phi^2(P)$, and $\Delta\phi^3(P)$ corresponding to the three displacements and the sensitivity vector $e(P)$ is obtained as the solution of

$$\begin{pmatrix} \Delta\phi_1(P) \\ \Delta\phi_2(P) \\ \Delta\phi_3(P) \end{pmatrix} = \begin{pmatrix} d^1_x & d^1_y & d^1_z \\ d^2_x & d^2_y & d^2_z \\ d^3_x & d^3_y & d^3_z \end{pmatrix} \begin{pmatrix} e_x(P) \\ e_y(P) \\ e_z(P) \end{pmatrix}. \tag{6.15}$$

The alternative approach is to use three rigid body rotations. Let the rotation vectors defining direction and rotation angles be θ^1, θ^2, and θ^3. We employ the so called *fringe-vector* $K_f(P)$, whose magnitude is inversely proportional to the normal distance between the spatial fringe shells intersecting the object surface and whose direction coincides with the direction of this normal pointing to the fringe shell of higher interference order. The fringe vector for rotations is given by

$$K_f(P) = -\theta \times e(P) \tag{6.16}$$

where $\times$ here denotes the vectorial product. The threefold evaluation of $K_f(P)$ for each P of interest with θ^1, θ^2, and θ^3 and the identical sensitivity vectors enables one to set up the system of equations

$$\begin{pmatrix} K^1_{fx}(P) & K^1_{fy}(P) & K^1_{fz}(P) \\ K^2_{fx}(P) & K^2_{fy}(P) & K^2_{fz}(P) \\ K^3_{fx}(P) & K^3_{fy}(P) & K^3_{fz}(P) \end{pmatrix}$$
$$= \begin{pmatrix} \theta^1_x & \theta^1_y & \theta^1_z \\ \theta^2_x & \theta^2_y & \theta^2_z \\ \theta^3_x & \theta^3_y & \theta^3_z \end{pmatrix} \begin{pmatrix} 0 & e_z(P) & -e_y(P) \\ -e_z(P) & 0 & e_x(P) \\ e_y(P) & -e_x(P) & 0 \end{pmatrix} \tag{6.17}$$

which is then solved to obtain $e(P)$ for each P. With these methods the problem of the determination of the geometry data is transferred to the task of performing precisely the necessary displacements or rotations and to evaluate the interference phase or fringe-vector unambiguously [574].

Another holographic approach is proposed in [575], where the object contours are measured by the holographic contouring using displaced illumination points, see Section 6.6.3. The main advantages are that the fringe pattern now is located on the surface and shows good visibility even for large illumination point displacements, and that the method works contactless with respect to the object.

6.2.2 Correction of Perspective Distortion

The determination of more than one component of the displacement vector necessitates the generation of several interference patterns which in turn are produced with different configuration geometries. If the observation direction varies between the recordings, the fringe patterns suffer from different *perspective distortions*. Especially when measuring at non-plane objects, errors induced by defocusing of the object surface and distortion due to perspective have to be taken into account [576]. For an oblique sight of the inspected surface, this surface appears deformed in the recorded image and often does not fill the full frame. A *spatial transform* has to map the pixels of the recorded interferogram to new pixels in an output interferogram, such that identical points of the object surface, which are imaged to different pixels in the various recorded images, are mapped to identical pixels in all output images.

In the following, an algorithm is given for the spatial transform which corrects perspective distortion in the case of plane surfaces. Without loss of generality, we assume a rectangular surface area to be evaluated, which by perspective distortion is deformed to an arbitrary convex quadrangle, Fig. 6.3. The spatial transform is performed by a *bilinear interpolation*, which

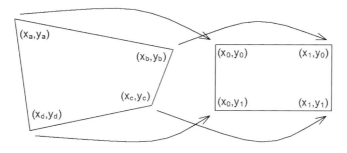

Figure 6.3: Bilinear mapping of a quadrangle onto a rectangle.

on the one hand is fast and computationally simple, and on the other hand produces a smooth mapping that preserves continuity and connectivity.

Let $I(x, y)$ be the recorded and stored perspectively distorted interferogram with pixel coordinates (x, y). Then the corrected interference pattern is $I'(x, y)$:

$$I'(x, y) = I(x', y') = I(ax + by + cxy + d, ex + fy + gxy + h). \qquad (6.18)$$

This bilinear transformation is defined by the values of the eight coefficients a through h. By specifying the mapping of the four vertices (x_a, y_a), (x_b, y_b), (x_c, y_c), (x_d, y_d) of the quadrangle to the four vertices (x_0, y_0), (x_1, y_0), (x_1, y_1), (x_0, y_1) of the output rectangle, we create a system of four equations

$$\begin{pmatrix} x_a \\ x_b \\ x_c \\ x_d \end{pmatrix} = \begin{pmatrix} x_0 & y_0 & x_0 y_0 & 1 \\ x_1 & y_0 & x_1 y_0 & 1 \\ x_1 & y_1 & x_1 y_1 & 1 \\ x_0 & y_1 & x_0 y_1 & 1 \end{pmatrix} \begin{pmatrix} a \\ b \\ c \\ d \end{pmatrix}. \qquad (6.19)$$

After inversion of the matrix we get the four coefficients a, b, c, and d as

$$\begin{pmatrix} a \\ b \\ c \\ d \end{pmatrix} = \frac{1}{(x_0 - x_1)(y_1 - y_0)} \begin{pmatrix} y_1 & -y_1 & y_0 & -y_0 \\ x_1 & -x_0 & x_0 & -x_1 \\ -1 & 1 & -1 & 1 \\ -x_1 y_1 & x_0 y_1 & -x_0 y_0 & x_1 y_0 \end{pmatrix} \begin{pmatrix} x_a \\ x_b \\ x_c \\ x_d \end{pmatrix}. \qquad (6.20)$$

6.2 The Sensitivity Matrix

The y_a through y_d are given by a system of equations similar to (6.19)

$$\begin{pmatrix} y_a \\ y_b \\ y_c \\ y_d \end{pmatrix} = \begin{pmatrix} x_0 & y_0 & x_0 y_0 & 1 \\ x_1 & y_0 & x_1 y_0 & 1 \\ x_1 & y_1 & x_1 y_1 & 1 \\ x_0 & y_1 & x_0 y_1 & 1 \end{pmatrix} \begin{pmatrix} e \\ f \\ g \\ h \end{pmatrix} \quad (6.21)$$

where the four coefficients e, f, g, and h are given by

$$\begin{pmatrix} e \\ f \\ g \\ h \end{pmatrix} = \frac{1}{(x_0 - x_1)(y_1 - y_0)} \begin{pmatrix} y_1 & -y_1 & y_0 & -y_0 \\ x_1 & -x_0 & x_0 & -x_1 \\ -1 & 1 & -1 & 1 \\ -x_1 y_1 & x_0 y_1 & -x_0 y_0 & x_1 y_0 \end{pmatrix} \begin{pmatrix} y_a \\ y_b \\ y_c \\ y_d \end{pmatrix}. \quad (6.22)$$

Thus the coefficients of the bilinear transform are defined. The transform can be implemented by (6.18), but a computationally more efficient algorithm is based on a line by line processing of the output image. Each new pixel coordinate is calculated by an increment from the foregoing one. The increments are constant along one line and are raised by a constant increment from line to line.

The spatial transform for each pixel calculates the original position of this pixel in the input image. In most cases it stems from a fractional position in the input pattern, meaning that its origin is between four adjacent pixels. So an interpolation is necessary to determine the gray level of the output pixel. The simplest interpolation is the nearest neighbor interpolation. A better solution, especially if significant gray level changes occur over one unit of pixel spacing, consists of applying the *bilinear interpolation*. Let us look for the intensity $I(n + x, m + y)$ at the fractional position $(n + x, m + y)$ with integers n and m and $x, y \in (0., 1.)$, Fig. 6.4.

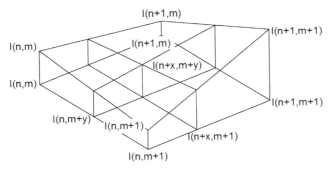

Figure 6.4: Bilinear mapping of the intensity.

The bilinear interpolation then gives

$$\begin{aligned} I(n+x, m+y) = I(n,m) &+ [I(n+1,m) - I(n,m)]x \\ &+ [I(n,m+1) - I(n,m)]y \\ &+ [I(n+1,m+1) + I(n,m) - I(n,m+1) - I(n+1,m)]xy \end{aligned} \quad (6.23)$$

An example of a correction of perspective distortion of a holographic interference pattern of a plane rectangular plate under deformation is given in Fig. 6.5.

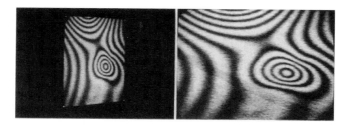

Figure 6.5: Correction of perspective distortion.

6.2.3 Condition of the Sensitivity Matrix

The determination of three-dimensional displacement fields requires the solution of linear systems of equations, as e. g. given in (6.3), (6.10), or (6.11). It is well known that if the equations are linearly dependent, leading to a *singular matrix* whose determinant is zero, the system is not solvable. However, even if the matrices on the right-hand side of (6.3), (6.10), or (6.11) are not singular, the solutions may show errors of up to some 100 percent in magnitude and direction if we have an *ill-conditioned matrix*. Therefore a holographic arrangement with sensitivity vectors oriented in a way yielding a well-conditioned matrix must be striven for.

A figure to measure the *condition of a system of linear equations*, or equivalently that of the matrix E of this system, is the *Hadamard condition number* $K_H(E)$ defined as

$$K_H(E) = \frac{|\det E|}{\prod_{i=1}^{n} \alpha_i} \tag{6.24}$$

with

$$\alpha_i = \sqrt{\sum_{k=1}^{n} e_{ik}^2} \tag{6.25}$$

where n is the rank of matrix E, whose elements are given by e_{ik}; det denotes the determinant. The system is ill-conditioned if $K_H(E) \ll 1.0$, and is optimum for $K_H(E) = 1.0$. A condition number of $K_H(E) > 0.1$ is considered to be desirable.

The elements of the matrix are the components of the sensitivity vectors, these are derived from measurements of the geometry of the holographic arrangement. A good condition means that the directions of the sensitivity vectors are linearly independent and as different as possible. As a rule of thumb the sensitivity vectors should have as different directions in the three dimensions as possible.

Of course, the geometry has to be measured precisely. The precision of the geometry data influences the achievable accuracy of the measurement. However, even more important than a precise measurement of the geometry is a proper separation of the sensitivity vectors. With an ill-conditioned system even the smallest errors in the geometry values would cause severe measurement errors. The use of an overdetermined system of equations, which is solved by Gaussian least squares, yields on principle a higher accuracy due to the averaging property [577–580], but it will cause nearly the same errors if the sensitivity vectors are not separated far enough.

6.3 Holographic Strain and Stress Analysis

In experimental mechanics the deformation of test objects in response to a mechanical or thermal load is studied in order to determine strains, stresses, or bending moments. These quantities are of interest because they affect the strength, safety, and lifetime of mechanical structures or components [170, 336]. A structure most likely will fail where the strain or stress is maximal.

Microstructures such as *microelectromechanical systems (MEMS)* have found numerous applications [239]. The reliability of these structures is determined by the assessment of the micromachining processes in realizing MEMS with the geometry and mechanical properties as required by the proper design. Since MEMS fabrication needs different materials and technological processes involving high temperature treatments, residual stresses can affect the final shape of the MEMS appearing in the form of undesired out-of-plane deformations. Thus it is instrumental for fabricating reliable MEMS to measure the effect of residual stress on the deformation of the single microstructures [113, 114, 239, 581, 582].

In this section the derivation of strains, stresses, and bending moments from holographically measured displacements of objects with opaque diffusely reflecting surfaces will be discussed. After the fundamental definitions, the most common structures, namely beams and plates, are treated in more detail, followed by a more general approach using the fringe-vector theory.

6.3.1 Definition of Elastomechanical Parameters

Let $P = (x_P, y_P, z_P)$ be the Cartesian coordinate description of a point of a solid object which is displaced by $\boldsymbol{d}(P) = (d_x(P), d_y(P), d_z(P))$. This notation is preferred here for reasons of consistency, while in the standard literature the components of the displacement vector often are written $(u, v, w) = (d_x, d_y, d_z)$. If there is no confusion, the argument P will be omitted in the following.

Strain generally is a tensor completely specified by nine components, of which six are independent [170, 583]. At any point in a solid body the three components of *normal strain* are

$$\varepsilon_x = \frac{\partial d_x}{\partial x}, \qquad \varepsilon_y = \frac{\partial d_y}{\partial y}, \qquad \varepsilon_z = \frac{\partial d_z}{\partial z} \tag{6.26}$$

and the three independent *shear strains* are

$$\gamma_{xy} = \frac{\partial d_x}{\partial y} + \frac{\partial d_y}{\partial x}, \quad \gamma_{yz} = \frac{\partial d_y}{\partial z} + \frac{\partial d_z}{\partial y}, \quad \gamma_{zx} = \frac{\partial d_z}{\partial x} + \frac{\partial d_x}{\partial z}. \tag{6.27}$$

While the normal strains describe the change of length per unit length in each coordinate direction, the shear strains measure the decrease in the angle between two line segments initially orthogonal and parallel to the coordinate axes. This is depicted in two dimensions for a small rectangle with sidelengths Δx and Δy in Fig. 6.6. Since the angles are assumed to be only small, the arctangent in the exact definition of the shear strain is approximated by its argument in the above definition.

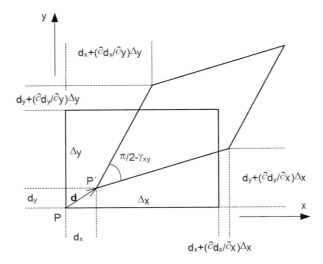

Figure 6.6: Elastic deformation of a small solid rectangle.

The strains are expressed as derivatives of the displacement components. Also the *rotations* of an object can be written by these derivatives: The components ω_x, ω_y, and ω_z describing the rotations about the x-, y-, and z-axes, respectively, are

$$\omega_x = \frac{1}{2}\left(\frac{\partial d_z}{\partial y} - \frac{\partial d_y}{\partial z}\right), \quad \omega_y = \frac{1}{2}\left(\frac{\partial d_x}{\partial z} - \frac{\partial d_z}{\partial x}\right), \quad \omega_z = \frac{1}{2}\left(\frac{\partial d_y}{\partial x} - \frac{\partial d_x}{\partial y}\right). \quad (6.28)$$

If external forces or moments affect a solid body, internal reactive forces maintain the equilibrium. As long as the masses are homogeneously distributed in the body, the reactive forces are distributed in planes. At each point of the solid an arbitrary cut dA can be defined. The *force* $\boldsymbol{F}$ related to the unit plane element dA is the *stress* $\boldsymbol{s}$

$$\boldsymbol{s} = \frac{d\boldsymbol{F}}{dA} \quad (6.29)$$

which is composed of the *normal stress* $\sigma = dF_n/dA$ and the *tangential* or *shear stress* $\tau = dF_t/dA$, Fig. 6.7. The description of the complete stress state at a point requires three planes or equivalently a cubic element to define the *stress tensor*, Fig. 6.8.

$$\boldsymbol{S} = \begin{pmatrix} \sigma_{xx} & \tau_{xy} & \tau_{xz} \\ \tau_{yx} & \sigma_{yy} & \tau_{yz} \\ \tau_{zx} & \tau_{zy} & \sigma_{zz} \end{pmatrix}. \quad (6.30)$$

Since $\tau_{xy} = \tau_{yx}$, $\tau_{xz} = \tau_{zx}$, and $\tau_{yz} = \tau_{zy}$, we need three normal stresses and three shear stresses to describe the stress state of a point of a solid body. Some relations between strains and stresses for special cases will be discussed in the next subsection.

6.3 Holographic Strain and Stress Analysis

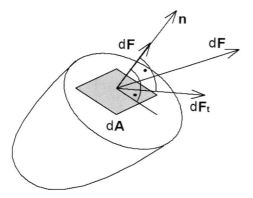

Figure 6.7: Normal and shear stress.

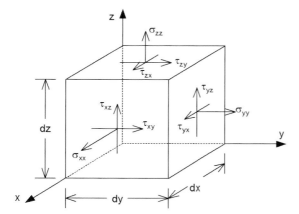

Figure 6.8: Elements of the stress tensor.

The aim of holographic strain and stress analysis is the determination of the strain or stress state of a tested component, which normally is done by measuring the displacement vector field and then calculating the strains and stresses according to the above introduced equations. But holographic interferometry does not provide sufficient information to process all the derivatives of (6.26) to (6.28). In particular the displacement derivatives in the direction normal to the surface of the opaque body cannot be evaluated. By applying the methods of Section 6.1 for a plane surface in the x-y-plane we may evaluate the displacement vector components $d_x(x, y)$, $d_y(x, y)$, $d_z(x, y)$ or precisely $d_x(x, y, z = 0)$, $d_y(x, y, z = 0)$, $d_z(x, y, z = 0)$, but we have no access to the general $d_x(x, y, z)$, $d_y(x, y, z)$, $d_z(x, y, z)$ in the interior $z \neq 0$ of the body.

Nevertheless what we have is sufficient to calculate the in-plane strains ε_x, ε_y and γ_{xy} in the plane surface as well as the in-plane rotation ω_z about an axis normal to the surface. We notice that the rotation components (6.28) are arithmetic averages of the rotations of two orthogonal faces of a cubic element, but this averaging is not required to evaluate the out-of-plane rotation of object surface points, $z = 0$ [170]. So the out of plane rotations at the surface are

$$\omega_x = \frac{\partial d_z}{\partial y} \qquad \omega_y = \frac{\partial d_z}{\partial x}. \tag{6.31}$$

A classic method for determining stresses in models of the interesting component fabricated from birefringent material or in other birefringent media, e. g. biological objects is *polarization imaging*. Polarization sensitive digital holography is presented in [122, 125]. A hologram of a specimen is created by the interference between object wave and two reference waves that have perpendicular polarization states. The reconstruction of such a hologram produces two wavefronts, one for each perpendicular polarization state. While this method takes advantage of only the reconstructed amplitude distributions, also the phase information is reconstructed in the approach published in [123].

6.3.2 Beams and Plates

The holographically measurable displacements $d(x, y, z = 0)$ are sufficient to describe the strains in the case of *plane stress*. We speak of a state of plane stress if a thin flat specimen is affected only by stresses parallel to its surface. Typical examples are the stretching of thin sheets or membranes and the tearing of flat tensile specimens commonly used to measure the mechanical properties of materials [170]. The in-plane strains for these objects are calculated from the measured $d_x(x, y)$- and $d_y(x, y)$-components of the displacement vector field.

The strains in another class of objects like beams, plates, or shells are determined from the measured displacement component $d_z(x, y)$ normal to the object surface.

A *beam* is a long slim solid component with a constant cross section subjected to transverse point loads, distributed or area loads, axial loads, and bending moments, Fig. 6.9. The length L of a beam generally is large compared with the width l and the thickness h: $L \gg l, L \gg h$. A beam is a common model for a number of technical components, like supporting beams, shafts, connecting rods, or turbine blades to name only a few.

A popular test object among holographers is the *cantilever beam* showing only deflections $d_z(x)$ in the z-direction. The reasons for the popularity are that these deflections are easy to measure holographically – they are in the direction of highest sensitivity – most often we have $d_z = 0$ at the base of the beam thus avoiding the absolute phase problem. Furthermore the theory of these beams is simple and well elaborated to allow a comparison of experimental and theoretical results.

If the beam consists of a material which behaves elastically for the applied load amplitudes, which means stress and strain are proportional, then for example in the x-direction we have

$$\sigma_{xx} = E\varepsilon_x \tag{6.32}$$

where E is the *modulus of elasticity*. Also the shear stresses and the shear strains are proportional

$$\tau_{zx} = G\gamma_{zx} \tag{6.33}$$

6.3 Holographic Strain and Stress Analysis

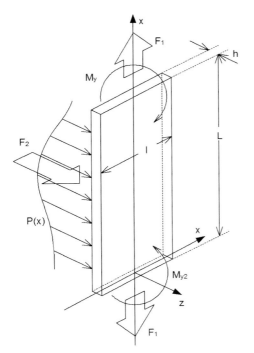

Figure 6.9: Elastic beam.

with G the *shear modulus of elasticity*. G and E are connected via

$$G = \frac{E}{2(1+\nu)} \qquad (6.34)$$

where ν is the *Poisson ratio*, which in a specimen subjected to tensile or compressive load in the x-direction describes the constant ratio of lateral strain to longitudinal strain

$$\varepsilon_z = \nu \varepsilon_x. \qquad (6.35)$$

E, G, and ν all are properties of the material of which the beam is fabricated.

If the beam of Fig. 6.9 is deflected responding to the transverse point load F_z, the distributed area-load $P_z(x)$, longitudinal forces F_x and bending moments M_{y1}, M_{y2} about axes parallel to the y-axis, the longitudinal strain at the observable surface is [170]

$$\varepsilon_x = \frac{\partial d_{x0}(x)}{\partial x} - \frac{1}{2}h\left(\frac{\partial^2 d_z(x)}{\partial x^2}\right). \qquad (6.36)$$

Here $d_{x0}(x)$ is the displacement in the x-direction of the central plane at $z = 0$ of the beam due to the longitudinal forces F_x and h denotes the beam thickness. The longitudinal stress σ_{xx} at the surface of the beam then is

$$\sigma_{xx} = E\left[\frac{\partial d_{x0}(x)}{\partial x} - \frac{1}{2}h\left(\frac{\partial^2 d_z(x)}{\partial x^2}\right)\right]. \qquad (6.37)$$

As long as the deflections remain small the bending moment at any x is

$$M_y = -\frac{Eh^3}{12}\left(\frac{\partial^2 d_z(x)}{\partial x^2}\right). \tag{6.38}$$

A *plate* is a component where the width l is in the same range as the length L, $l \approx L$, but the constant thickness h is small compared to L and l: $h \ll l$, $h \ll L$, Fig. 6.10. The plate is

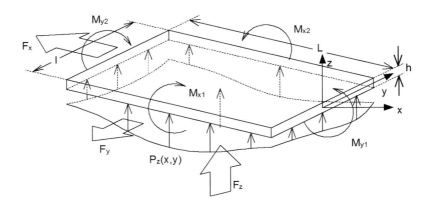

Figure 6.10: Elastic plate.

deformed e. g. by a distributed area-load $P_z(x,y)$, transverse forces F_z, axial forces F_x, F_y applied at its edges, as well as moments, M_x and M_y, acting at the edges. Then the strains at the observable surface are

$$\begin{aligned}
\varepsilon_x &= \frac{\partial d_{x0}(x,y)}{\partial x} - \frac{1}{2}h\left(\frac{\partial^2 d_z(x,y)}{\partial x^2}\right) \\
\varepsilon_y &= \frac{\partial d_{y0}(x,y)}{\partial y} - \frac{1}{2}h\left(\frac{\partial^2 d_z(x,y)}{\partial y^2}\right) \\
\gamma_{xy} &= \left(\frac{\partial d_{x0}(x,y)}{\partial x} + \frac{\partial d_{y0}(x,y)}{\partial y}\right) - h\left(\frac{\partial^2 d_z(x,y)}{\partial x \partial y}\right)
\end{aligned} \tag{6.39}$$

where $d_{x0}(x,y)$ and $d_{y0}(x,y)$ are the components of the in-plane translation of the central plane at $z=0$ of the plate due to the axial forces F_x or F_y. The corresponding stresses are again $\sigma_{xx} = E\varepsilon_x$, $\sigma_{yy} = E\varepsilon_y$, $\tau_{xy} = G\gamma_{xy}$ and the bending moments per unit length are

$$\begin{aligned}
M_x &= -D\left(\frac{\partial^2 d_z(x,y)}{\partial x^2} + \nu\frac{\partial^2 d_z(x,y)}{\partial y^2}\right) \\
M_y &= -D\left(\frac{\partial^2 d_z(x,y)}{\partial y^2} + \nu\frac{\partial^2 d_z(x,y)}{\partial x^2}\right) \\
M_{xy} &= D(1-\nu)\frac{\partial^2 d_z(x,y)}{\partial x \partial y}
\end{aligned} \tag{6.40}$$

6.4 Hybrid Methods

The accurate measurement of displacements by holographic interferometry and computer-aided quantitative evaluation allows a subsequent numerical differentiation or other processing steps [325]. Thus these measurements can be combined with structural analysis methods like the finite element methods, boundary-element methods, or the techniques known under the heading of fracture mechanics. The refined analysis methods resulting from such combinations may be categorized as *hybrid methods*. As an example in car production one is highly interested in the behavior of brakes, especially in their noise and vibration. Therefore the results of various metrologies such as CW and pulsed laser holographic interferometry, speckle methods, and laser Doppler vibrometry are combined to identify the root causes of brake concerns and verify engineering solutions [325, 591, 592].

6.4.1 Finite Element Methods and Holographic Interferometry

The *finite element method* is a structural analysis method, where a technical structure is divided into a number of discrete elements which have a simple geometry. All these elements are connected via the nodes by which they are defined. The mechanical behavior in each element can be calculated due to its simple geometry, then the program produces an overall solution that is compatible to all the nodes. Thus displacements, strains and stresses or the thermal behavior of a structure under mechanical and/or thermal load can be approximated with good accuracy.

Holographic deformation analysis [592–596] or holographic vibration analysis [597] and finite element calculations can effectively be combined to reach a number of goals. One of the most important is the holographic verification of the finite element model [598, 599]. Especially for complex shaped or composite structures it is not an easy task to find a proper discretization and to choose the right material parameters. A comparison of the measured displacements with the calculated displacements can confirm the finite element model, which can then be used for, say, strain and stress calculations.

This strategy has been successfully followed in the investigation of vibration modes [600, 601] or stress distributions [547]. For the analysis of *adhesive bondings* it has been determined to what proportions the metal layers and the adhesive layers contribute to the deformation of the specimen [602]. A complete representation of the strain and stress conditions within the specimen has been obtained. The combination of an electro-optic holographic microscope measuring displacements of vibrating microbeams and FEM to describe the behavior of microelectromechanical systems (MEMS) is employed by Brown and Pryputniewicz [603].

The temperature distribution and the deformation of a thermally loaded overlap adhesive bond with a local void in the adhesive layer was calculated by the finite element method and measured by holographic interferometry [604]. The combined evaluations have shown that the characteristic surface deformation above the defect in the internal adhesive layer is not caused by thermal expansion of the enclosed gas. The inhomogeneities in the surface deformation arise from the disturbed heat transfer in the defect area. So in this region we get locally higher temperature differences, which in consequence leads to locally different thermal deformations. These results explain the relevance of *thermal load* for *defect detection* in holographic non-destructive testing.

The combination of holographic interferometry and finite element methods not only enables a defect detection but also a *defect validation*. In systematic calculations a catalog of surface deformations in the region of internal cracks and voids in, say, steel with variations of defect type, length, orientation, volume or position, is compiled. For a given holographically measured displacement field above a defect, one starts with the best fitting displacement field from this catalog. In an iterative process in a finite element model, the parameters of the simulated defect are varied. The displacement field is calculated over and over again with varied defects until the agreement with the measured displacement field is sufficiently good [547].

Of course, the calculation of a displacement field for a given discrete structure and loading by the finite element method is not a one-to-one mapping. Thus the inverse process of determining the discrete structure from the loading and the measured displacement field is not possible. Nevertheless, with the iterative method we get a defect which is representative of the equivalent class of all defects producing the same deformation under the specific applied load.

6.4.2 Boundary Element Methods and Holographic Interferometry

In the finite element method, the whole body to be tested is discretized into finite volumes connected at the nodes. Continuity of parameters which are not explicit variables is only warranted at the nodes and not at the borders between the elements. If we have other functions which fulfill exactly the differential equation in the whole region, we have no discretization errors in the interior of the body. Since only the boundary conditions have to be satisfied, the requirement that the boundary is discretized is in itself sufficient. The relating methods are called *boundary element methods*.

The boundary element methods can be combined advantageously with experimental methods like holography. Especially, if in practical applications the boundary conditions are too complicated to be described theoretically, they have to be measured. Having measured the displacements by, e.g., holographic interferometry, the boundary of two- or three-dimensional regions is then divided into segments on which the displacements and strains are approximated by polynomials of the first degree. The stress components at prescribed internal points of the region are then calculated by means of the boundary element method.

In [605] this method is applied to transparent models manufactured from PMMA with roughened faces. The in-plane components of the displacement vectors are measured by double exposure double aperture speckle interferometry. The objects considered are a three-point loaded beam with an edge crack and a model of a large slab wall stiffened by a frame. Based on the measurements, exact values of stresses σ_x, σ_y, and σ_{xy} in the neighborhood of the crack tip are determined. In another application, the friction between the wall and its base is measured by applying the hybrid evaluation method of coherent optical measurement combined with the boundary element method.

6.4.3 Fracture Mechanics

In linear elastic *fracture mechanics* the influence of a *crack* or another defect on the damage of a technical structure is estimated. An important figure is the *stress intensity factor* K_I and its critical value K_{Ic}, the *fracture toughness*, which is a material property. The K_I-value can be determined holographically by first measuring the deformation field of the structure exhibiting a crack [606] followed by the determination of the boundary of the *plastic zone* arising above the crack during *tensile loading*. This is functionally related to the K_I-value [607–609]. Another method is based on the integration of the strain equations of Sneddon or Williams-Irwin. This method only requires a displacement measurement along a line perpendicular to the crack propagation direction [610]. In conjunction with these methods holographic interferometry has been used to determine the K_I-value for a CT 500 specimen with high accuracy and without any previous knowledge of the specific defect properties like its size and location.

A further criterion for *crack propagation* is the so-called *J-integral*, which is a figure independent of the path of integration. Crack propagation occurs if the J-integral exceeds a critical material parameter [609]. In [611], the determination of the J-integral is based on measurements of the displacement field by holographic interferometry. Power series estimations up to quadratic terms fulfilling the Lame-Navier equations are set up for the three displacement components. From the coefficients of the power series the integration can then be performed along a rectangular path to yield the J-integral.

6.5 Vibration Analysis

The measurement of vibrations is an important task in engineering, on the one hand to ascertain the operation of components, which should vibrate, like loudspeakers, ultrasonic transducers etc., and on the other hand to check the behavior of components, which have natural frequencies of response within the range of frequencies excited by the operation of an engine of which the component may be part. The aims can be the prevention of fatigue failure or the detection of noise-generating parts or areas, to name just two [591]. Of course non-contacting measurement methods which do not affect and bias the vibration are recommended. Since furthermore the amplitudes to be measured normally are in the range of the wavelengths of laser light, and since one is interested in the simultaneous measurement at a manifold of points, holographic interferometry is a suitable tool for analyzing vibrations [612, 613]. It is worth noting that the first holographic interferograms were made of diffusely reflecting surfaces under vibration [16].

6.5.1 Surface Vibrations

The fundamental surface vibration is the *sinusoidal vibration* also called *harmonic vibration*. Here the displacement of each point of the vibrating surface is

$$\boldsymbol{d}(P,t) = \boldsymbol{d}(P)\sin\omega t \tag{6.61}$$

where $\boldsymbol{d}(P)$ is the vector amplitude and ω the circular frequency of the vibration. The vibrations of continua are described by partial differential equations. Generally the boundary conditions yield transcendental equations for the eigenvalues. Approximate solutions use the Rayleigh quotients or the Ritz method. For some special cases like vibrating cantilever beams, circular or rectangular shells or plates, closed solutions are known.

The solutions differ in the distribution of $\boldsymbol{d}(P)$ over the points P, which defines the *mode-shapes* of the specific vibration. This can be explained with the example of a vibrating rectangular plate being jointed at the edges. The differential equation is

$$\frac{\partial^2 d_z}{\partial t^2} = -\frac{N}{\rho h}\left(\frac{\partial^4 d_z}{\partial x^4} + 2\frac{\partial^4 d_z}{\partial x^2 \partial y^2} + \frac{\partial^4 d_z}{\partial y^4}\right) \tag{6.62}$$

with the stiffness of the plate $N = Eh^3/[12(1-\nu^2)]$, the thickness h, and the specific mass ρ. If the sidelengths are a and b, solutions are given by

$$d_z(x,y,t) = d_{max}\sin(\omega t + \phi_0)\sin\left(\frac{j\pi x}{a}\right)\sin\left(\frac{k\pi y}{b}\right) \tag{6.63}$$

with the eigenvalues $\omega_{jk} = (j^2/a^2 + k^2/b^2)\pi^2\sqrt{N/(\rho h)}$ for $j,k = 1,2,\ldots$. The two extreme deformation states of a plate vibrating in the $(j=2, k=2)$-mode is shown in Fig. 6.12. The

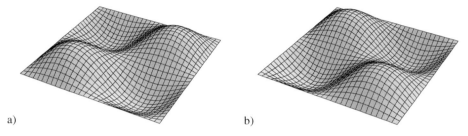

Figure 6.12: Vibrating rectangular plate.

maximum amplitude at each point (x,y) is $|d_{max}\sin(j\pi x/a)\sin(k\pi y/b)|$. The loci where the maximum amplitude is zero all the time are the *nodes*, on the other hand the loci of local extrema of the amplitude are called *antinodes*. In Fig. 6.12 we have two nodal lines parallel to the edges and intersecting in the center of the plate. It is one task of holographic vibration analysis to identify the *mode-shape* (j,k) and the maximum amplitude of the amplitude distribution of a vibration.

In Fig. 6.12 all points vibrate in-phase, there is no phase shift between the oscillations of each point, except that when one antinode reaches its positive maximum another takes its negative maximum, which may be interpreted as a phase shift by π. The phase relation between the vibrating points is another item to be analyzed holographically. Figure 6.13 shows some oscillatory states of a vibrating circular disc where the phase varies over 6π around the circumference. The phase is constant along each radius, while each radius of constant phase is continuously rotating. Figure 6.13a – d indicate how a, say, phase maximum moves around the circumference of the disc. There is only one nodal point in the center of the disc.

The sinusoidal vibration of (6.61) is an example for a separable object motion. *Separable object motions* are such motions which can be separated into a product of a displacement vector $\boldsymbol{d}(P)$ and a real temporal function $f(t)$. This $f(t)$ is $\sin(\omega t)$ in (6.61). Here the point P moves along a line defined by $\boldsymbol{d}(P)$. If the point moves along an arc, this motion can be described by the sum of two or three separable motions. Besides the sinusoidal vibration, $f(t)$ may represent damped harmonic vibrations [614] or other nonlinear vibrations [615–617].

6.5 Vibration Analysis

of a wave field proportional to

$$M_T = \frac{1}{T} \int_0^T f_{obj}(t) E_P f_{ref}^*(t) dt. \tag{6.73}$$

The expression M_T often is called the *characteristic function* [170, 172, 376, 628], but the concept of characteristic functions will not be stressed extensively in this book. The resulting intensity is $I = |M_T|^2$. The M_T for some ways of modulating the reference wave now should be investigated in more detail.

Let the object surface vibrate harmonically with frequency ω, then the object wave is proportional to $\exp[i\mathbf{d}(P) \cdot \mathbf{e}(P) \sin \omega t]$, while $f_{obj}(t) = 1$. In *frequency translated holography* the frequency of the reference wave is modulated by an integer multiple $n\omega$ of the object vibration frequency ω

$$f_{ref}(t) = \exp(in\omega t). \tag{6.74}$$

The resulting characteristic function is proportional to

$$M_T = \frac{1}{T} \int_0^T \exp[i\mathbf{d}(P) \cdot \mathbf{e}(P) \sin \omega t] \exp(-in\omega t) dt. \tag{6.75}$$

Using the identity (C.4) and reversing the order of integration and summation we get

$$\begin{aligned} M_T &= \frac{1}{T} \int_0^T \sum_{m=-\infty}^{\infty} J_m[\mathbf{d}(P) \cdot \mathbf{e}(P)] \exp(im\omega t) \exp(-in\omega t) dt \\ &= \sum_{m=-\infty}^{\infty} J_m[\mathbf{d}(P) \cdot \mathbf{e}(P)] \frac{1}{T} \int_0^T \exp[i(m-n)\omega t] dt. \end{aligned} \tag{6.76}$$

If the exposure time is long compared with the object vibration period, $T \gg 2\pi/\omega$, the integral vanishes for all m except for $m = n$, so that

$$M_T = J_n[\mathbf{d}(P) \cdot \mathbf{e}(P)] \tag{6.77}$$

or

$$I(P) = J_n^2[\mathbf{d}(P) \cdot \mathbf{e}(P)]. \tag{6.78}$$

This result is consistent with the unmodulated time average case having $n = 0$, see (6.69). Proportionality factors influencing only the overall brightness have been omitted for convenience.

Frequency translated holography is used to increase the sensitivity for vibrations with small as well as with large amplitudes: Small amplitudes here are such that the interference phase $\Delta\phi(P) = \mathbf{d}(P) \cdot \mathbf{e}(P)$ remains small, $\Delta\phi(P) \ll 1$. In time average holography with no modulation $J_0^2(0)$ is unity and has slope zero. So no significant intensity variation in the bright field will result. On the other hand $J_1^2(0)$ has a positive slope in the dark field, see Fig. C.1, yielding visible intensity variations even for small amplitudes. The smallest detectable amplitude was estimated as $2.7 \times 10^{-4}\lambda$ [629]. In the case of large vibration amplitudes we take advantage of the fact that the locations of the zeros of the Bessel functions are spread apart for increasing order n, Fig. C.1. This results in a decreasing number of fringes for the same amplitudes with increasing n, meaning too high fringe densities can be effectively avoided [630].

A general temporally periodic object wave may be considered as composed of its Fourier series terms, having frequencies $\omega_0 + m\omega$, $m = 0, \pm 1, \pm 2, \ldots$ with ω_0 the frequency of the used laser light. The reference wave in frequency translated holography has the frequency $\omega_0 + n\omega$. So only the frequency component with $m = n$ will produce a time average hologram, because only this one is coherent with the reference wave. Thus we have a method for temporal filtering by selecting single frequency components from a periodic object motion [170, 630, 631].

In *amplitude modulation holography* the amplitude of the reference wave is modulated with the same frequency as the object vibrates, but with a controllable phase difference ψ with respect to the vibration of a selected object point P, $f_{ref}(t) = \cos(\omega t - \psi)$. In this case the n of (6.74) is 1. The resulting intensity is proportional to

$$I(P) = J_1^2[\boldsymbol{d}(P) \cdot \boldsymbol{e}(P)] \cos^2 \psi. \tag{6.79}$$

A sequence of holograms, recorded with amplitude modulation at varied phase ψ, will display contours of constant relative phase [99].

A further method to obtain phase information about the object vibration is *phase modulation holography*. The phase of the reference beam is modulated at the frequency ω of the vibrating object with a modulation depth of Ω_R. The modulation function is

$$f_{ref}(t) = \exp(i\Omega_R \sin \omega t) \tag{6.80}$$

which gives the characteristic function [170]

$$\begin{aligned} M_T &= \frac{1}{T} \int_0^T \exp[i\boldsymbol{d}(P) \cdot \boldsymbol{e}(P) \sin(\omega t - \phi_0)] \exp(-i\Omega_R \sin \omega t) dt \\ &= J_0\{[(\boldsymbol{d}(P) \cdot \boldsymbol{e}(P))^2 + \Omega_R^2 - 2\boldsymbol{d}(P) \cdot \boldsymbol{e}(P)\Omega_R \cos \phi_0]^{1/2}\}. \end{aligned} \tag{6.81}$$

The implication of this characteristic function is that the relative phase ϕ_0 of the vibration at each point is encoded in the fringe pattern. Relative phases [632] as well as amplitudes [633] are measured based on this principle. For simplicity of discussion assume $\phi_0 = 0$. Then we have an intensity

$$I(P) = |M_T(P)|^2 = J_0^2(\boldsymbol{d}(P) \cdot \boldsymbol{e}(P) - \Omega_R). \tag{6.82}$$

Now the loci of bright zero fringes are controllable by the user, since they appear where $\boldsymbol{d}(P) \cdot \boldsymbol{e}(P) = \Omega_R$. Further insight is gained when the modulation (6.80) includes a phase term ϕ_R. The dark fringes in the interferogram then are characterized by

$$b_m^2 = \boldsymbol{d}(P) \cdot \boldsymbol{e}(P) + \Omega_R^2 - 2\boldsymbol{d}(P) \cdot \boldsymbol{e}(P)\Omega_R \cos(\phi_0 - \phi_R) \tag{6.83}$$

6.5 Vibration Analysis

where b_m denotes the m-th zero of J_0 [634]. If two holographic interferograms are produced in this way with phases ϕ_{R1} and ϕ_{R2}, they may be superimposed. At the intersections of fringes of equal order m we have

$$\phi_0 = \frac{\phi_{R1} + \phi_{R2}}{2} \pm m\pi. \tag{6.84}$$

Using this approach one gets phase contours by simple visual inspection. This work has been extended [635] to bright fringes so that also bright-bright and bright-dark combinations can be used. Hereby phases intermediate to those of (6.84) can be inspected. Objects excited with swept sinusoidal vibration or randomly have been investigated by time average holography with a mechanically excited reference beam. Modes and amplitudes of randomly excited objects were analyzed by a method called *spectroscopic holography* [636].

The holographic interferometry using phase modulation described by (6.82) is closely related to general motion compensation by reference waves modulated by the object motion [554, 556]. If the vibration amplitudes in the vicinity of a point P are large, a reference wave modulated by a mirror fixed at P and undergoing the same motions will yield a vibration measurement relative to this point.

6.5.5 Numerical Analysis of Time Average Holograms

Numerical analysis of cosine fringes by phase shifting or Fourier transform methods allows the determination of interference phases even between the fringe intensity maxima and minima with high accuracy. While a lot of research has been performed to automate the evaluation of cosine fringes, not a great deal has been done to automate Bessel-type fringe interpretation.

One approach is to convert J_0-fringes into sinusoidal fringes by stroboscopic techniques, and to apply heterodyning [621] or phase stepping [637]. The real-time method of vibration analysis was combined with heterodyne [622] and phase step evaluation [638]. In those methods vibration amplitudes must still be interpolated from integer fringe orders. The direct numerical extraction of vibration amplitudes from time average interference patterns as presented in [639] is given in the following.

The intensity of a time average hologram is (4.10)

$$I(x,y) = I_0(x,y) J_0^2 (\Delta\phi(x,y)) \tag{6.85}$$

where $I_0(x,y)$ is the irradiance of the object surface. The interference phase is $\Delta\phi(x,y) = \boldsymbol{d}(x,y) \cdot \boldsymbol{e}(x,y)$ with $\boldsymbol{d}(x,y)$ the vectorial displacement and $\boldsymbol{e}(x,y)$ the sensitivity vector in (x,y). As we have seen in Section 6.5.4, the J_0-fringes can be shifted by modulating the phase of either the object or the reference beam sinusoidally at the same frequency as the object vibration. This adds a phasor bias Ω_R to the argument of the Bessel function. If the object is vibrating in only one vibration mode and if the phase of the sinusoidal beam modulation is adjusted to coincide with that of the object vibration the phasor bias becomes an additive term and the irradiance of a time average hologram reconstruction together with the unavoidable distortions is (6.82)

$$I(x,y) = a(x,y) + b(x,y) J_0^2 (\Delta\phi(x,y) - \Omega_R) \tag{6.86}$$

with $a(x, y)$ containing the background irradiance and $b(x, y)$ the multiplicative object surface irradiance. Equation (6.86) has a form analogous to the phase sampling equation (5.15) for cosine fringes. Unfortunately the Bessel function of a sum cannot be expressed as a sum of terms as done in (5.19) for the cosine, so straightforward solutions are not possible. An iterative process for calculation of the phase $\Delta\phi$ of each (x, y) is outlined in [639], but this requires quite lengthy calculations.

A solution that makes use of the nearly periodic nature of the J_0-function employs the formula

$$\Delta\phi^* = \arctan \frac{(1 - \cos\Omega_R)(I_3 - I_1)}{\sin\Omega_R(I_1 - 2I_2 + I_3)}. \tag{6.87}$$

see Table 5.1, where we use three reconstructions with the bias terms $-\Omega_R$, 0, $+\Omega_R$ having known values. The interference phase $\Delta\phi^*$ computed by (6.87) differs from the correct argument of the J_0^2-function. Due to the approximating function $J_0^*(x)$, see (C.6), this error approaches $\pi/4$ for large arguments. The error can be computed for any phase angle $\Delta\phi$ and any value of Ω_R to create a lookup table to convert the incorrect answers obtained from the use of (6.87) into the correct phase values. In practice, since the influence of $a(x, y)$ and $b(x, y)$ is eliminated inherently in the evaluation by (6.87), we can proceed with $a = 0$ and $b = 1$ when compiling the lookup table. For each specific Ω_R the intensities I_1, I_2, and I_3 corresponding to $-\Omega_R$, 0, and $+\Omega_R$ are computed by (6.86) for the desired values of $\Delta\phi$. Then (6.87) is solved for these I_1, I_2, I_3, resulting in a false interference phase $\Delta\phi^*$. These $\Delta\phi^*$ are tabulated with the corresponding correct values of $\Delta\phi$.

The phase data obtained by evaluation using (6.87) with subsequent correction employing the lookup table still have to be demodulated. The one pattern with no bias vibration now is used to identify the loci of the zero-order fringes. These are recognized by their high brightness relative to the rest of the fringes. So when working with time average hologram data we even have access to the absolute fringe order.

Difficulties still occur when two or more vibration modes lie sufficiently close in frequency so that both are excited at the same time. It seems reasonable to suppose, that data recorded with the bias phase at a number of phase angles could be used to extract both the amplitude and the phase of the vibration [639]. In this direction work still has to be done.

6.5.6 Vibration Analysis by Digital Holography

The first experiments in which holographic interference was observed were in holographic vibration measurement [17], and still today holography is a frequently used tool for displaying and evaluating two-dimensional vibration modes [237]. The digital approach to holography also can be employed in vibration analysis. The simplest way is the use of a pulsed laser triggered to a time instant when the modes have maximum amplitude [65]. The reconstructed phase then is compared to the evaluated phase distribution related to a hologram of the object at rest with no vibration. This method does not differ from the digital holographic interferometric measurements of surface deformations already treated in Chapter 5.8.

The use of an image-intensifier system is reported in [640]. The advantage of the image-intensifier is that it can be gated, meaning the electronic shutter action produced by controlling its photocathode voltage. This allows one to record holograms with short exposure times.

6.6 Holographic Contouring

Transient vibrations can be measured with multipulse digital holography [641]. Hereby a ruby laser producing four pulses is used, the corresponding digital holograms are captured with three different CCD sensors.

But also the *time average method* for holographic vibration analysis can be performed in the digital mode. Now the CCD-array integrates over a time interval T which is long compared to the period of the vibration, $T \gg 2\pi/\omega$. The intensity of the wave field reconstructed from such a time average hologram obeys a squared Bessel function as is explained in Section 6.5.3.

An example for the digital realization of the time average method in holographic vibration analysis is shown in Fig. 6.15. Here a piezo-membrane as is used in micro-pumps was excited to vibration [262]. The size of the membrane was 12 mm × 12 mm. We observe the fundamental mode with increasing amplitude over the six intensity reconstructions. The membrane is clamped at the four edges where we have maximum brightness. The higher order bright fringes are much darker than the zero order fringe, as is predicted by the squared Bessel function J_0^2.

6.6 Holographic Contouring

The holographic interference pattern arises from the superposition of at least two states of a reflected or refracted wave field. In holographic deformation measurements the phase differences between the two states are given by the scalar product of the displacement vector field and the sensitivity vectors, (4.21). While the displacement vectors describe the variation of the surface point positions between the interferometrically measured states, the sensitivity vectors, which are given by the used laser wavelength, the directions of illumination and observation, and the geometry of the measured surface, remain constant.

This concept is inverted for *holographic contouring*. Contouring in general means the modulation of the image of a three-dimensional object by fringes corresponding to contours of constant elevation with respect to a reference plane [386, 642, 643]. For holographic contouring the object is not displaced, but the two states which by superposition form the fringe pattern are produced by variation of other factors in (4.13) and (4.20), respectively.

6.6.1 Contouring by Wavelength Differences

The length of the sensitivity vector depends on the wavelength λ, see (4.20). Of course a change in the wavelength varies the sensitivity vector and thus the optical path length. The wavelength can be changed in predetermined discrete steps with e. g. an argon-ion laser or continuously with a dye laser. A further method is the quick variation of the etalon in a pulsed ruby laser by altering the distance between the etalon's plates. One plate of the etalon may be mounted on a vibrating piezoelectric element [644]. For contouring by the *two-wavelength method* [23–26, 28, 645–647] the optical arrangement shown in Fig. 6.16 is employed. It uses plane illumination and reference waves, and a telecentric viewing system with an image plane hologram.

Let λ be the wavelength utilized in recording of the hologram and then apply the real-time method with wavelength λ'. The points P' of the reconstructed image are shifted relative to the really existing point P according to (2.151) – (2.153). The lateral displacements of the

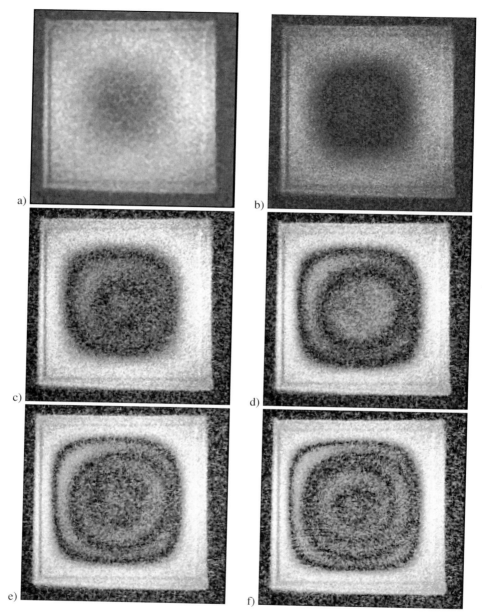

Figure 6.15: Reconstructed intensities in digital time-average holographic interferometry, vibration amplitudes increasing from (a) to (f) (Courtesy of S. Seebacher, BIAS).

6.6 Holographic Contouring

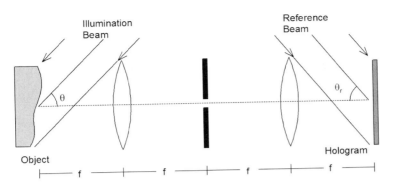

Figure 6.16: Arrangement for wavelength-difference contouring.

points can be eliminated by means of plane recording and reconstructing reference beams and shifting back the reconstructed image by tilting the reference beam by an appropriate amount such that the condition

$$\lambda \sin \Theta'_r = \lambda' \sin \Theta_r \tag{6.88}$$

is satisfied [309]. Θ_r and Θ'_r represent the initial and final angular positions of the reference beam. Another way of eliminating the lateral shift requires the object to be brought very close to the hologram plate, meaning to record an image plane hologram of the object, Fig. 6.16. Under these assumptions we have the situation of (2.157) – (2.159), so that there is a fictitious displacement vector

$$\boldsymbol{d} = P' - P = \begin{pmatrix} x'_P - x_P \\ y'_P - y_P \\ z'_P - z_P \end{pmatrix} = \begin{pmatrix} 0 \\ 0 \\ \frac{1}{\mu} z_P - z_P \end{pmatrix} \tag{6.89}$$

with $\mu = \lambda'/\lambda$. By the special choice of the configuration the observation unit vector is $\boldsymbol{b}(P) = (0, 0, 1)^T$ and the illumination unit vector is $\boldsymbol{s}(P) = (-\sin \Theta, 0, -\cos \Theta)^T$, Fig. 6.16. Thus the sensitivity vector is

$$\boldsymbol{e}(P) = \frac{2\pi}{\lambda} (\sin \Theta, 0, 1 + \cos \Theta)^T \tag{6.90}$$

and the resulting phase difference, (4.21), is

$$\begin{aligned} \Delta\phi(P) &= \boldsymbol{d}(P) \cdot \boldsymbol{e}(P) \\ &= \frac{2\pi}{\lambda} \left(\frac{1}{\mu} z_P - z_P \right) (1 + \cos \Theta) \\ &= 2\pi \frac{\lambda - \lambda'}{\lambda \lambda'} (1 + \cos \Theta) z_P. \end{aligned} \tag{6.91}$$

The two wavefronts interfere and produce fringes corresponding to *contours* of constant altitude. The fringe planes intersect the object in a direction parallel to the hologram plane. The contour sensitivity is given by the depth difference Δz

$$\Delta z = \frac{\lambda \lambda'}{(\lambda - \lambda')(1 + \cos \Theta)} \qquad (6.92)$$

which induces a change of 2π in the phase difference $\Delta \phi$.

As an example, Fig. 6.17 shows a statuette that was contoured by the wavelength difference method using a dye laser.

Figure 6.17: Wavelength difference contouring.

6.6.2 Contouring by Refractive Index Variation

A method closely related to the two-wavelength method is the *contouring by refractive index variation*, also called the *immersion method* [27, 309, 648]. Here the wavelength is not modified by changing the frequency of the laser, but by a change in the speed of light according to

$$\nu \lambda = nc \qquad (6.93)$$

where n is the refractive index of a transparent material the light is passing through. Equation (2.10) is the special case for vacuum, $n = 1$. The schematic of the immersion method is illustrated in Fig. 6.18. The object is placed in a glass tank filled with a transparent gas or liquid having refractive index n. While the first recording of the object is performed with the medium at refractive index n, the second exposure of the double exposure method is taken when the medium is replaced by another having refractive index n'. In the real-time method during reconstruction of the first holographically stored wave field the object rests in

6.6 Holographic Contouring

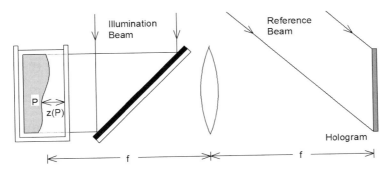

Figure 6.18: Arrangement for contouring by the immersion method.

the medium of refractive index n'. In both methods the produced interference fringes depend on the difference of the refractive indices and the distance $z(P)$ between the object surface and the tank wall.

The object is assumed to be illuminated by a plane wave in a direction perpendicular to the glass plane. Observation is carried out using a telecentric imaging system. Thus different deflections of the rays by the different refractive indices are avoided. The optical phase difference giving rise to fringe formation is

$$\Delta\phi(P) = \frac{4\pi}{\lambda}(n - n')z(P). \qquad (6.94)$$

The contour interval Δz producing a change in the phase difference $\Delta\Delta\phi = 2\pi$ is

$$\Delta z = \frac{\lambda}{2(n - n')}. \qquad (6.95)$$

If the liquid is a mixture of water and alcohol, and the refractive index is adjusted by the mixture ratio, the method is called the *grog method* [196].

6.6.3 Contouring by Varied Illumination Direction

In the preceding subsections the sensitivity vector between the two interfering states was changed by a variation of the wavelength. Of course the sensitivity vector can be changed by altering the directions of illumination or observation.

If the illumination point S is changed to S' between the two exposures of the double exposure method, the resulting optical path length change δ and the interference phase $\Delta\phi$ can be calculated analogously to (4.13) – (4.21). For $\boldsymbol{d}_S(S) = S' - S$, we get

$$\Delta\phi(P) = -\frac{2\pi}{\lambda}\boldsymbol{d}_S(S) \cdot \boldsymbol{s}(P). \qquad (6.96)$$

This means that the object is intersected by fringe surfaces which consist of a set of rotationally symmetric hyperboloids, their common foci being the two points of illumination. The fringe pattern is independent of the point of observation. The longer the distance between object and illumination sources the flatter are the intersecting surfaces that are approximately parallel to the illuminating beams. Collimated beams produce equidistant parallel flat surfaces. The distance Δh of two such surfaces is

$$\Delta h = -\frac{\lambda}{2\sin\frac{\Theta}{2}} \tag{6.97}$$

where Θ is the angle between the two illumination directions. The analogy to the results of Section 2.2.1, (2.130), is obvious. Exactly the same result would be produced if holography was not used at all, but rather the object illuminated from the two points simultaneously, this procedure being called *projected fringe contouring*.

If the angle of illumination is changed by translating the object between the two exposures in the proper direction, contouring surfaces of nearly any orientation can be produced [649, 650]. This approach may be combined with a variation of the illumination direction [651–653].

Theoretically the double exposure method with two observation points might be performed by moving the hologram slightly between the two exposures. But since the intersecting surfaces now are parallel to the line of sight, this attempt is useless for contouring purposes. However, if the two states are recorded on different holograms in *sandwich hologram interferometry*, contour lines can be obtained by mutually shifting the plates [654].

6.6.4 Contouring by Light-in-Flight recording

A conceptually different approach to contouring of three-dimensional objects is provided by the holographic *light-in-flight recording* and reconstruction [655–661]. The idea behind holographic light-in-flight recording is the equivalence of a short temporal coherence and a fictitious extremely fast shutter or short light-pulse to produce a motion picture of a propagating optical wavefront. In each small region of a hologram only those parts of an object will be recorded for which the path length from the laser to this small region via the object does not differ from the path length of the reference beam by more than the temporal coherence length of the laser light used for the recording [655, 662].

If the coherence length is short, only those parts of a large object are recorded and appear brightly reconstructed for which a near-zero path difference holds. The object surface seems to be intersected by imaginary interference surfaces in the form of ellipsoids. One focus of these ellipsoids is the source point of the illuminating spherical wave, point A in Fig. 6.19, the other focus is the point in the holographic plate used for observation, point H in Fig. 6.19. To each observation point in the hologram plate there corresponds an ellipsoid representing zero path length difference between object and reference beams, namely $\overline{AP}+\overline{PH} = \overline{AM_1} + \overline{M_1M_2} + \overline{M_2H}$ in Fig. 6.19.

When the point of observation H is moved along the holographic plate, the bright fringe of zero path length can move across the object. Only if H is varied in a way that does not change the zero path length for a specific object point P, this P remains stable. Thus to each object point P there exists a hyperboloid in space representing this zero path length. The two foci of these hyperboloids are the studied object point P and the virtual point source A' of the reference beam. By moving the observation point along the hologram in a way that it crosses

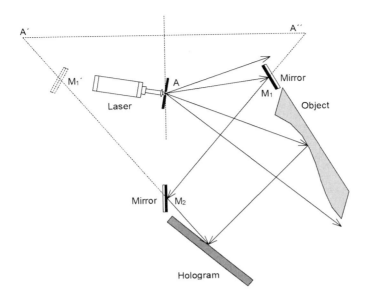

Figure 6.19: Arrangement for holographic light-in-flight recording.

a number of hyperboloids, one sees the fringe intersecting different parts of the object surface as a continuous high-speed motion picture displayed at arbitrary low speeds.

For *contouring* by holographic light-in-flight recording the geometry of the holographic setup has to be optimized in such a way that the ellipsoids in the volume of the three-dimensional object can be approximated by flat surfaces, Fig. 6.20. The object is seen intersected by one flat interference surface S, whose depth position is varied during reconstruction as the observation point is shifted, Fig. 6.20. Here when looking from H_i, one only sees fringes corresponding to the intersection of surface S_i with the object.

The method seems especially suited for a computer-aided evaluation. There is just a single fringe corresponding to one ellipsoid, no zero-order-fringe problem arises. The depth resolution can be controlled by the shift of the observation point during reconstruction. The residual deviations from flatness of the ellipsoids can be taken into account numerically. An evaluation of the contouring lines of light-in-flight holography utilizing an image processing system is described in [117]. Holographic light-in-flight methods also can be used advantageously for visualization and velocimetry in three-dimensional flows [663].

6.7 Contour Measurement by Digital Holography

In Section 6.6 it was shown how holographic interference patterns can be produced, whose local interference phase depends uniquely on the coordinates of the object's surface. These methods can be combined with digital holography, as will be shown. Moreover, the digital approach offers some important advantages over conventional holographic contouring. Because

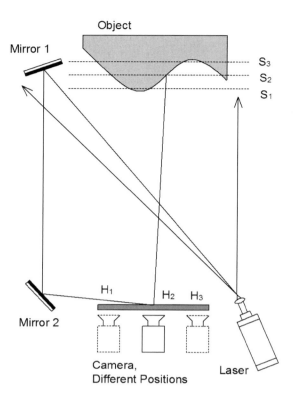

Figure 6.20: Contouring by holographic light-in-flight recording and reconstruction.

now the interference phases are compared numerically they do not have to be reconstructed with the same wavelength as in the optical interferometric comparison. This section presents the methods of Section 6.6 in their digital form.

A special aspect of *digital holographic contouring* is the potential of integrating deformation and contour measurement in a common holographic setup, performing both measurements simultaneously and having an integrated software with common formats [664].

The application of ultrashort laser pulses or laser light beams with only short coherence length enables holographic reconstructions of only those parts of the object which are intersected by specific curves in space.

The presented computer simulations use an object which consists of a rectangular plane plate with a central cap of a sphere [490]. The sphere has a radius of 25 mm, the largest height of the cap over the plane is 5 mm.

6.7.1 Contouring by Digital Holographic Interferometry

If the wavelength between the two recordings is varied, the interference fringes can be interpreted as intersections of the object's surface with a set of ellipsoids. Figure 6.21a shows

6.7 Contour Measurement by Digital Holography

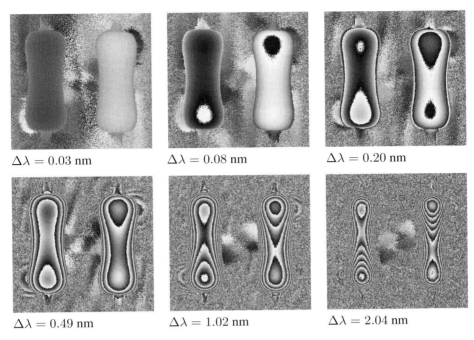

Figure 6.28: Interference phase distributions modulo 2π resulting from digital multi-wavelength holography.

distributions the contours are calculated with high precision, Fig. 6.29, if we demodulate and take into account the variation of the sensitivity vector.

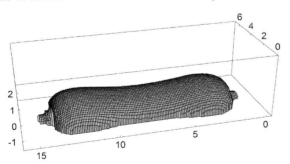

Figure 6.29: Measured contour of the resistor.

6.7.3 Holographic Contouring by Digital Light-in-Flight Measurement

Holographic light-in-flight measurement as a contouring method was already introduced in 6.6.4. It is ideally suited to be combined with digital holography. The main problem in adapting digital holography to this concept is the oblique incidence of the reference wave onto the hologram. The large angle to the object wave direction produces high frequency holographic structures, higher than can be resolved by a CCD array.

The first attempt to solve this problem was the installation of delay plates in the passage of the plane reference wave impinging normally onto the CCD array, Fig. 6.30. Behind the delay

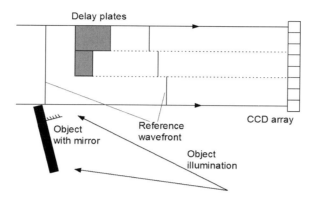

Figure 6.30: Light-in-flight arrangement with delay plates.

plates the wavefront was delayed according to a step function [668]. In this experiment three different light paths over the object to the CCD are recorded. Reconstructions are performed from small stripes in the digital hologram. The laser used in this experiment is an Ar-ion laser pumped CW dye laser with no frequency selecting elements. The resulting coherence length was about 2.3 mm.

A significant progress over the delay of the reference wave with only a few discrete steps was achieved by Carlsson and Nilsson [115, 116, 118, 669]. They use a blazed reflection grating in a Littrow configuration that introduces a continuous linear delay to the reference beam across its diameter while still having constant phase, Fig. 6.31. The resulting reference

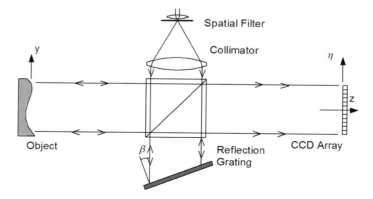

Figure 6.31: Littrow setup for light-in-flight contouring.

6.7 Contour Measurement by Digital Holography

field in the hologram plane, the (ξ, η)-plane of the CCD array, can be described by

$$r(\xi, \eta) = \gamma \left(\frac{2\pi}{\lambda} \rho_r - \omega t \right) E_r e^{i(\phi - \omega t)}. \tag{6.98}$$

Here $\gamma \left(\frac{2\pi}{\lambda} \rho_r - \omega t \right)$ describes the properties of the short laser pulse of duration τ_C or short coherence length L_C. We still have a plane wave, therefore $\phi(\xi, \eta) = \phi$ is constant, and the varying delay of the reference wave is contained in $\rho_r(\xi, \eta)$, the path length of the reference wave. It can be expressed by

$$\rho_r(\xi, \eta) = 2\eta \tan \beta + 2z_r \tag{6.99}$$

with z_r the distance between grating and CCD along the z-axis and β the grating angle. The factor 2 comes from the reference wave's double pass to and from the grating. With this reference wave the hologram

$$\begin{aligned} h(\xi, \eta) &= \iiint E_r e^{-i(\phi - \omega t)} \gamma(k\rho_r - \omega t) \frac{E_O}{\rho_O} \gamma(k\rho_O - \omega t) e^{i(k\rho_O - \omega t)} \, dx \, dy \, dt \\ &= \iint \frac{E_r e^{-i\phi}}{\rho_O} \Gamma(\Delta \rho) E_O e^{ik\rho_O} \, dx \, dy \end{aligned} \tag{6.100}$$

results. The time integration here is over the pulse length τ_C. The ρ_O is the distance between the object points and the hologram points

$$\rho_O(x, y, \xi, \eta) = \sqrt{d^2 + (x - \xi)^2 + (y - \eta)^2}, \tag{6.101}$$

and $\Gamma(\Delta \rho)$ is the coherence function, i. e. the autocorrelation of $\gamma(k\rho - \omega t)$. It represents the overlapping between the pulses, depending on the path difference $\Delta \rho = \rho_r - \rho_O$.

The reconstruction of the whole object wave field now can be done by using the normally impinging plane wave of unit intensity in the usual way, but then the depth information is lost. Instead one should employ the reconstruction beam

$$r_E(\eta, \eta_0) = \exp\left[-\frac{(\eta - \eta_0)^2}{2\Lambda^2} \right]. \tag{6.102}$$

This constitutes a Gaussian formed strip centered on the line $\eta = \eta_0$ where $\Lambda = L_C/(2 \tan \beta)$ represents the η-width of coherence length L_C in the CCD array. The center η_0 is varied between the reconstructions to permit evaluations of different depths.

Since $\Gamma(\Delta \rho)$ is an autocorrelation, its maximum is $\Gamma(\Delta \rho) = \Gamma(0)$. The maximum intensities in the reconstructed field therefore are at object points for which $\Delta \rho = \rho_r - \rho_O = 0$ is fulfilled. Thus

$$\rho_O = 2\eta_0 \tan \beta + 2z_r. \tag{6.103}$$

It was shown [115] that the curves for equal optical path length form paraboloids in space. There remains an ambiguity in the second, the ξ-coordinate. This can be resolved by use of two exposures with rotated reference gratings.

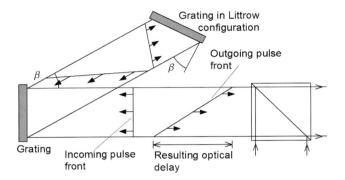

Figure 6.32: Increase of optical delay by gratings in series.

An improvement of the depth measurement range and resolution can be obtained by using larger grating angles. One way to increase the resulting optical delay is to put two gratings in series [118], Fig. 6.32. Up to now only a single digital hologram has been recorded. By recording two pulsed holograms the method can be combined with digital holographic interferometry to allow the simultaneous measurement of shape and deformation.

The concept of holographic light-in-flight contouring also can be realized by using ultrashort pulse lasers [262]. In an arrangement related to a Michelson interferometer, Fig. 6.33, only wave fronts reflected from those parts of the object form holographic structures on the CCD, where their light-paths are of the same length as the path over the piezo-mirror. The agreement must be in the range of the way the wave front travels during this pulse.

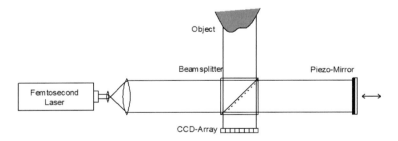

Figure 6.33: Arrangement for contouring using a femtosecond laser.

In the experiments described in the following a Ti:sapphire laser with pulses of 80 fs has been used, this corresponds to a layer of a thickness of about 24 µm the object is intersected by. In a measurement of the contour the optical path length difference is changed in constant steps by the piezo-mirror. Intersections of different depth are recorded and reconstructed by digital holography. There is an advantage of this method over the use of continuously emitting short-coherence lasers, because now the whole CCD records each hologram and not only narrow strips. This improves the quality of the reconstructed images.

6.8 Comparative Holographic Interferometry

The intensity $I(P)$ observed at any point in the image plane during observation of the reconstructed wave fields E_{1m} and E_{1t} together with the scattered fields E_{2m} and E_{2t} is

$$\begin{aligned}
I(P) &= \langle |E_{1m}(P) + E_{2m}(P) + E_{1t}(P) + E_{2t}(P)|^2 \rangle \\
&= \langle |E_{1m}|^2 + |E_{2m}|^2 + |E_{1t}|^2 + |E_{2t}|^2 + E_{01}^2 \, e^{-i\Delta\phi(P)} + E_{01}^2 \, e^{i\Delta\phi(P)} \\
&\quad + E_{01}^2 \, e^{-i\Delta\phi'(P)} + E_{01}^2 \, e^{i\Delta\phi'(P)} \rangle \\
&= 4I_1(P) + 2I_1(P)\cos\Delta\phi(P) + 2I_1(P)\cos\Delta\phi'(P) \\
&= 4I_1(P)\left(1 + \cos\frac{\Delta\phi(P)+\Delta\phi'(P)}{2}\cos\frac{\Delta\phi(P)-\Delta\phi'(P)}{2}\right). \quad (6.113)
\end{aligned}$$

The last line is the expression for an additive moiré. The high frequency fringes described by the cosine of the sum of the interference phases are modulated by the low frequency difference phase $\Delta\phi(P) - \Delta\phi'(P)$. For identical mechanical behavior, $\Delta\phi(P) = \Delta\phi'(P)$, the resulting intensity is

$$I(P) = 4I_1(P)(1 + \cos\Delta\phi(P)). \quad (6.114)$$

No moiré fringes are recognized in this case.

Generally the master and test surfaces are illuminated from different directions, s and s', but observed in common direction b. That means

$$\begin{aligned}
\Delta\phi(P) &= d(P) \cdot \frac{2\pi}{\lambda}[b(P) - s(P)] \\
\Delta\phi'(P) &= d'(P) \cdot \frac{2\pi}{\lambda}[b(P) - s'(P)]
\end{aligned} \quad (6.115)$$

and thus

$$\Delta\phi(P) - \Delta\phi'(P) = \frac{2\pi}{\lambda}\{[d(P) - d'(P)] \cdot [b(P) - s(P)] - d'(P) \cdot [s(P) - s'(P)]\}. \quad (6.116)$$

In practice it is advantageous to have common illumination directions s and s', as in Fig. 6.37, then the second term in the right-hand side of (6.116) vanishes. The resulting moiré fringe pattern provides information about the difference of the displacement vectors along the direction of the sensitivity vector $b(P) - s(P)$ [680].

Comparative holographic interferometry and comparative holographic moiré interferometry enable the measurement of displacement differences or contour differences of distinct objects. This meets many requirements of holographic non-destructive testing. An improvement of the precision of quantitative evaluation and of the quality of the fringe patterns can be achieved by employing phase shifting techniques [683, 684] or Fourier transform evaluation [685] in these techniques.

6.8.2 Digital Comparative Holography

We have seen the various advantages of digital holography, Chapter 3 and Section 5.8, so this approach seems promising when used with the comparative holographic interferometry concept. Now the difference phases (6.110) can be calculated in a computer. However the

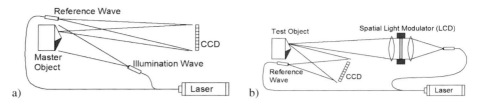

Figure 6.38: Digital comparative holography, a) Recording of coherent mask, b) Coherent illumination of test object.

comparison of the intermediate states is only possible optically. Hereby a spatial light modulator is used to reconstruct the conjugated wavefront of the master object. The arrangements of the optical components for digital comparative holography are shown in Fig. 6.38. The digital holograms of the undeformed and the deformed master object are recorded with the setup of Fig. 6.38a. For comparison purposes then the test object is replaced into the same or an identical arrangement according to Fig. 6.38b which enables the illumination of the test object with the reconstructed conjugated wavefront of the master object. Therefore in the experiment described here a liquid crystal display (LCD) is controlled by the stored digital hologram and optically generates the desired wavefronts [212, 667]. These wavefronts now act as an adaptive illumination of the test object. For a complete cancellation of the fringes, the sensitivity vectors of the master recording arrangement and that of the test arrangement must be the same except for the sign. In the optical realization of comparative holography, Section 6.8.1, we had to record undeformed and deformed states of the objects onto individual hologram plates or we had to separate by different reference waves. In *digital comparative holography* the separate recording and storing of all holograms is delivered automatically, so no additional reference waves are necessary. Thus the whole process is simplified significantly.

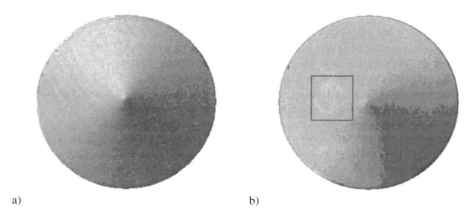

Figure 6.39: Master object (a) and test object with defect in indicated area (b) (Courtesy of T. Baumbach, BIAS).

On the other hand also a digital comparison of phase distributions as in comparative holographic moiré interferometry is possible. The incoherent superposition of intensity recon-

6.8 Comparative Holographic Interferometry

structions of the optical realization is now replaced by the subtraction of the two interference phase distributions calculated by computer.

Digital comparative holography is demonstrated in the following using two macroscopically identical aluminum cylinders with a cone at their upper end, Fig. 6.39. One of the two cylinders has a dent of some micrometers depth in its cone, Fig. 6.39b. In a preliminary test both objects are measured in the same holographic arrangement by holographic contouring using the wavelength difference method of Secs. 6.6.1 and 6.7.2. The wavelengths in the experiment are $\lambda_1 = 746.898$ nm and $\lambda_2 = 749.459$ nm, so the synthetic wavelength of (6.121) is $\lambda_{eq} = 0.218$ mm. The test object's interference phase distribution according to (6.91) is displayed in Fig. 6.40a. The difference of the related interference phase distributions of mas-

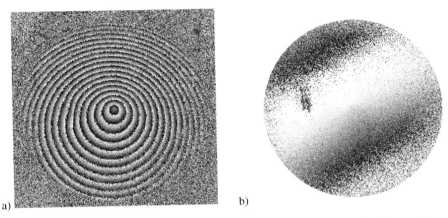

Figure 6.40: Interference phase modulo 2π (contour lines) (a) and result of comparison of interference phases of master and test object (b) (Courtesy of T. Baumbach, BIAS).

ter and test object is given in Fig. 6.40b. While in the difference image, Fig. 6.40b, the dent is clearly visible, in the phase distribution of Fig. 6.40a the dent is hardly recognizable. The remaining two fringes in Fig. 6.40b result from repositioning errors.

The experiments in digital comparative holography on the basis of coherent adaptive illumination presented in Figs. 6.41 and 6.42 use no deformation of the objects, but instead a contouring with the two-wavelength method. The recorded digital holograms have been adjusted to the pixel size of the LCD and were fed into the LCD. The experimental setup of the lensless Fourier transform type, schematically shown in Figs. 6.38a and b is given in Fig. 6.41. The LCD has 1024×768 pixels of size 18 μm$\times$18 μm. Now the master object is the cone with the dent, and the test object is the same cylinder rotated by $180°$ around the cylinder's axis. The rotated cone is illuminated by the reconstructed real image of the cone in the unrotated state. The resulting digital holograms are recorded and the phase distributions belonging to the different wavelengths used for recording are calculated. The differences of these phase distributions are displayed in Figs. 6.42a and b, first for a synthetic wavelength of $\lambda_{eq} = 0.330$ mm, second with $\lambda_{eq} = 0.254$ mm. We see that the smaller wavelength exhibits a higher noise level but a better recognition of the dents. Further experiments have demon-

Figure 6.41: Experimental setup for digital comparative holography (Courtesy of T. Baumbach, BIAS).

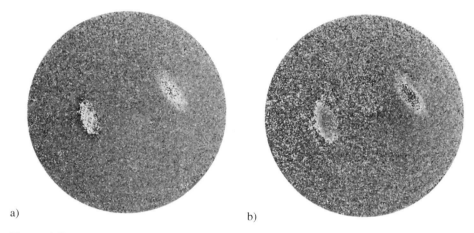

Figure 6.42: Resulting phase differences obtained with coherent illumination using synthetic wavelengths of (a) 0.330 mm, (b) 0.254 mm (Courtesy of T. Baumbach, BIAS).

strated that the noise is diminished and the detection quality is increased by using LCDs with a higher space bandwidth product [212].

The digital approach to comparative holography offers a lot of advantages and benefits: Misalignment between master and sample can be compensated numerically [212,686], there is no cross reconstruction problem with the second wavelengths in the two-wavelength method, and the double exposures are performed in the computer using the individually recorded digital holograms. If master and test object are at different places, the data about the holographic setup and the digital holograms can be easily transmitted using modern communication channels like the internet [687].

6.9 Measurement Range Extension

There are metrologic problems to be solved by holographic interferometry, where the methods introduced so far exhibit some drawbacks. The lack of the correct sign and the absolute phase, see Section 5.1, may be such a disadvantage. One would wish to measure with a wavelength much larger than the common laser wavelengths to overcome these deficiencies. In other applications one is interested only in the in-plane displacements, while the main sensitivity of the measurement is for out-of-plane displacements. The object may be loaded by centrifugal forces during rotation, for holographic interferometric measurements the interfering wavefronts must be steady. Since in holography only points of a rough surface interfere whose microstructures are identical in a small neighborhood, with conventional holographic interferometry different objects of the same macrostructure cannot be compared.

Solutions to these problems exist as we have seen in the last section. The price one has to pay is generally a more complicated technical and optical arrangement and procedure to be used for the experimental measurements. But these efforts enable a substantial extension of the measuring range of holographic interferometry.

6.9.1 Two-Wavelength Holographic Interferometry

The holographic interferometric measurement of large deformations often leads to high fringe densities which cannot be resolved any more. A longer wavelength, e. g. by a factor 10 to 100, would be desirable. Another effect of a long wavelength would be that it partially solves the absolute phase problem, at least as long as the maximal displacement remains less than this wavelength. But a long wavelength, e. g. the 10.6 μm of a CO_2 laser, would cause experimental difficulties since ordinary optical elements are no longer transparent, hologram plates have no sensitivity in this range and the observation requires additional experimental efforts [92]. On the other hand these problems can be circumvented by using two-wavelength holography [688].

In *two-wavelength holographic interferometry* two double-exposure interferograms of the same object undergoing the same deformation are taken with different wavelengths, say λ_1 and λ_2. In a first approach [688], which is recommended if high fringe densities are expected, the fringe pattern belonging to λ_1 is photographed, the developed transparency is replaced into the holographic arrangement and is illuminated by the interference pattern resulting from λ_2. A moiré pattern will be obtained, which we get by multiplication of the two overlayed and filtered intensities.

Let the displacement lead to a phase difference $\Delta\phi(x, y)$, then the complex amplitudes are

$$E_j(x, y) = e^{i\frac{2\pi}{\lambda_j}\phi(x,y)} + e^{i\frac{2\pi}{\lambda_j}[\phi(x,y) + \Delta\phi(x,y)]} \qquad j = 1, 2 \qquad (6.117)$$

and the intensities are

$$I_j(x, y) = E_j(x, y)E_j^*(x, y) = 2 + e^{i\frac{2\pi}{\lambda_j}\Delta\phi(x,y)} + e^{-i\frac{2\pi}{\lambda_j}\Delta\phi(x,y)} \qquad j = 1, 2 \quad (6.118)$$

where unit amplitudes were assumed. The moiré is

$$\begin{aligned}I_1(x,y)I_2(x,y) &= 4 + 2\mathrm{e}^{\mathrm{i}\frac{2\pi}{\lambda_1}\Delta\phi(x,y)} + 2\mathrm{e}^{-\mathrm{i}\frac{2\pi}{\lambda_1}\Delta\phi(x,y)} + 2\mathrm{e}^{\mathrm{i}\frac{2\pi}{\lambda_2}\Delta\phi(x,y)} \\ &+ 2\mathrm{e}^{-\mathrm{i}\frac{2\pi}{\lambda_2}\Delta\phi(x,y)} + \mathrm{e}^{\mathrm{i}2\pi(\frac{1}{\lambda_1}+\frac{1}{\lambda_2})\Delta\phi(x,y)} \\ &+ \mathrm{e}^{-\mathrm{i}2\pi(\frac{1}{\lambda_1}+\frac{1}{\lambda_2})\Delta\phi(x,y)} + \mathrm{e}^{\mathrm{i}2\pi(\frac{1}{\lambda_1}-\frac{1}{\lambda_2})\Delta\phi(x,y)} \\ &+ \mathrm{e}^{-\mathrm{i}2\pi(\frac{1}{\lambda_1}-\frac{1}{\lambda_2})\Delta\phi(x,y)}.\end{aligned} \quad (6.119)$$

Besides the d.c.- and the high-frequency terms the last two terms lead to an intensity proportional to

$$\cos\left[2\pi\left(\frac{1}{\lambda_1}-\frac{1}{\lambda_2}\right)\Delta\phi(x,y)\right] \quad (6.120)$$

which would have resulted if only the wavelength

$$\lambda_{eq} = \frac{\lambda_1\lambda_2}{|\lambda_1-\lambda_2|} \quad (6.121)$$

had been used. This wavelength is called *equivalent wavelength* [688,689] or *synthetic wavelength* [690–692]. As an example, for the two lines of an Ar laser of $\lambda_1 = 0.5145$ μm and $\lambda_2 = 0.4880$ μm the equivalent wavelength is $\lambda_{eq} = 9.4746$ μm.

To avoid difficulties in the exact adjusting of the same deformation twice or to avoid spurious interference by air turbulence, the two interferograms can be recorded simultaneously on the same hologram plate. Two separated reference waves, one for each wavelength, are recommended to avoid the disturbing cross-reconstructions [693], see also Section 4.2.2. Then the interferograms are reconstructed and evaluated separately for the two wavelengths. According to (4.20) and (4.21) we have

$$\Delta\phi_j(P) = \frac{2\pi}{\lambda_j}\boldsymbol{d}(P)\cdot[\boldsymbol{b}(P)-\boldsymbol{s}(P)] \qquad j=1,2. \quad (6.122)$$

The difference of the two evaluated interference phase distributions is

$$\begin{aligned}\Delta\phi_1(P) - \Delta\phi_2(P) &= 2\pi\left(\frac{1}{\lambda_1}-\frac{1}{\lambda_2}\right)\boldsymbol{d}(P)\cdot[\boldsymbol{b}(P)-\boldsymbol{s}(P)] \\ &= \frac{2\pi}{\lambda_{eq}}\boldsymbol{d}(P)\cdot[\boldsymbol{b}(P)-\boldsymbol{s}(P)].\end{aligned} \quad (6.123)$$

Thus we have a means to extend the range of unambiguity [694] or to reduce the sensitivity. Simultaneous recording avoids the influence of air turbulence, the remaining effect due to the different wavelengths can be neglected since air dispersion is small. Also chromatic aberration can be neglected, for only small wavelength differences are used. Theoretically the method can be expanded to multiple wavelengths yielding still larger equivalent wavelengths.

6.9.2 Holographic Moiré

The main sensitivity of holographic interferometry is in the direction of the bisector between the illumination and the observation direction, described by the sensitivity vector, Section 4.2,

6.9 Measurement Range Extension

and thus for out-of-plane displacements. For measurement of the in-plane components without multiple observation from different directions and subsequent numerical analysis of the resulting interferograms one can employ the *holographic moiré* method [124, 309, 695, 696]. Here the object is illuminated by collimated waves along two directions which are mutually coherent and symmetric to the surface normal, Fig. 6.43.

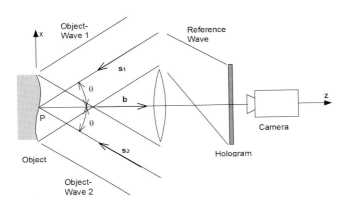

Figure 6.43: Arrangement for holographic moiré.

Let the directions of the two object illumination waves in object point P be described by the unit vectors $s_1(P) = (-\sin\theta, 0, \cos\theta)$ and $s_2(P) = (\sin\theta, 0, \cos\theta)$. The observation direction is $b(P) = (0, 0, 1)$. The complex amplitudes of the two illuminating object waves at P are A_1 and A_2 before the deformation and A'_1 and A'_2 after P has undergone the displacement $d(P) = (d_x(P), d_y(P), d_z(P))$. This displacement causes the phase differences (4.20) and (4.20)

$$\Delta\phi_1(P) = \frac{2\pi}{\lambda} d(P) \cdot [b(P) - s_1(P)]$$
$$\Delta\phi_2(P) = \frac{2\pi}{\lambda} d(P) \cdot [b(P) - s_2(P)] \qquad (6.124)$$

so that $A'_1(P) = A_1(P)e^{i\Delta\phi_1(P)}$ and $A'_2(P) = A_2(P)e^{i\Delta\phi_2(P)}$. The intensity observed at point P when using the real-time method is [309]

$$\langle I(P) \rangle = \langle |e^{i\pi} A_1 + e^{i\pi} A_2 + A'_1 + A'_2| \rangle. \qquad (6.125)$$

The reflected wave fields of the two illumination waves are uncorrelated, meaning $\langle A_i A_j^* \rangle = \langle A'_i A_j^{*'} \rangle = \langle A_i A_j^{*'} \rangle = \langle A'_i A_j^* \rangle = 0$ for $i \neq j$. Without restriction of generality we can assume equal real amplitudes $\langle |A_i|^2 \rangle = \langle |A'_i|^2 \rangle = \langle I_1 \rangle$ for $i = 1, 2$. Then (6.125) reduces to

$$\langle I(P) \rangle = 4\langle I_1(P) \rangle - 2\langle I_1(P) \rangle (\cos \Delta\phi_1(P) + \cos \Delta\phi_2(P)). \qquad (6.126)$$

Introducing $\Phi(P) = (\Delta\phi_1(P) - \Delta\phi_2(P))/2$ and $\Psi(P) = (\Delta\phi_1(P) + \Delta\phi_2(P))/2$ we obtain

$$\langle I \rangle = 4 \langle I_1 \rangle (1 - \cos\Phi \cos\Psi). \tag{6.127}$$

The argument (P) is omitted for clarity. From (6.124) it follows that

$$\Phi(P) = \frac{2\pi}{\lambda} d_x(P) \sin\theta$$
$$\Psi(P) = \frac{2\pi}{\lambda} d_z(P)(1 + \cos\theta). \tag{6.128}$$

The loci of the moiré fringes resulting from the multiplication of the two patterns in (6.127) represent contour lines of the projection of the displacement vector $\boldsymbol{d}(P)$ onto the object plane in the direction containing the two beams [309]. The spacing of the fringes corresponds to an incremental displacement of $\lambda/(2\sin\theta)$.

To get clearly visible moiré patterns, generally the two multiplied patterns must have high density and have to be oriented in nearly the same direction. Their planes of maximum contrast must match. In practice these requirements can be fulfilled by the introduction of additional phase differences much larger than those generated by the object deformation to be measured. To obtain parallel equidistant fringes of high density localized on the object surface one may rotate the holographic plate around an axis parallel to its plane [697, 698], translate the hologram plate in its plane [699], rotate the reference beam accompanied by a translation of the hologram [670], or rotate the reference beam only [699–701].

Although the auxiliary fringes are necessary to generate the holographic moiré at all, they disturb the visual appearance of the moiré fringes. Therefore an optical or digital low-pass filter eliminating the high-frequency carrier fringes and passing only the low-frequency moiré fringes is applied in the preprocessing step of the evaluation. After this the fringe pattern can be evaluated by one of the diverse quantitative evaluation methods. An interesting combination of numerical preprocessing and quantitative evaluation by Fourier transform fringe pattern analysis which determines instantaneously both phase functions of the pattern is presented in [685].

Another option is phase shifting holographic moiré [696] where the reference beam is directed over a mirror that is mounted on a piezoelectric transducer. From several phase shifted moiré holograms, recorded before and after deformation, the phase difference $\Phi(P)$ is calculated, from which $d_x(P)$ is determined using (6.128).

6.9.3 Holographic Interferometry at Rotating Objects

The *vibration modes* of spinning components excited by *centrifugal forces* generally are different from those of stationary objects excited conventionally. Wheels, propellers, or turbine blades are just a few engineering components where the knowledge of the actual dynamic behavior under real operating conditions helps to optimize the design and performance. Although the circumferential speed often reaches more than 100 m s^{-1}, holographic interferometry can be adapted to measure the deformation and vibration of such *rotating objects* [702–705].

Consequently the holographic recording of moving objects, especially rotating objects, normally requires the use of a pulsed laser. To produce time-resolved holographic interferograms, Carlsson et al. [706] describe a system consisting of a multiple-pulsed Q-switched ruby laser and a rotating disk having radial slits with a constant angular separation. The disk is used to scan the reference beam along the holographic plate, thereby achieving spatial multiplexing. This system is a tool for full-field dynamic measurements.

If we intend to measure the displacements and deformations excited by the rotation, e. g. by centrifugal forces, or if we want to use a CW laser, e. g. for time average vibration analysis, we need a recording configuration which is insensitive to the rotational motion but sensitive to radial and normal displacements. Such an arrangement must have its illumination point as well as the observation point on the axis of rotation, Fig. 6.44, which here is the z-axis.

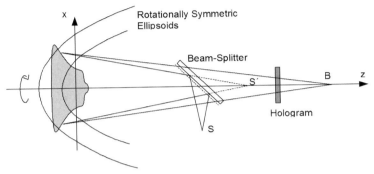

Figure 6.44: Holographic arrangement insensitive to in-plane rotations.

Rotationally symmetric ellipsoids with illumination point S' and observation point B as focal points define the loci of constant optical path length. Since object surface points undergoing only the in-plane rotation do not change the optical path lengths they do not contribute to the generation of the interference pattern. If furthermore the (virtual) illumination point S' and the observation point B coincide, the sensitivity vectors $e(P)$ for all surface points P are parallel to the rotation axis, see (4.20), and in this case the arrangement is insensitive to all in-plane displacements, even to radial displacements. For an arrangement optimized in the described way the maximum rotation speed is only restricted by motion blur and resolution. That means the lateral movement in the image plane is limited to about half a speckle size, (2.117) and (2.118), to obtain an acceptable visibility.

A common method to keep the unwanted contributions of the rotational motions to the interferogram small is the *object related triggering* of the double pulse laser [707]. By encoder disks or other non-contacting optical measurements the angular position of the object during the first of a double pulse is registered. The second pulse is triggered so that the object is in the same angular position, although during a different revolution, as with the first exposure. Pulse separation with this method must be rather high, i. e. some milliseconds, compared to conventional double pulse techniques. This object related triggering was successfully applied to measure forced vibrations of a rotating turbine blade model with maximum circumferential speed at the blade tip of 235 m s^{-1}. The measured deflection behavior was compared to numerical results [708].

A holographic interferometer spinning synchronously with the object is shown in Fig. 6.45. In this *rotating interferometer* two holograms and two reference waves are employed to achieve a better mechanical balance [707]. The object illumination comes from the

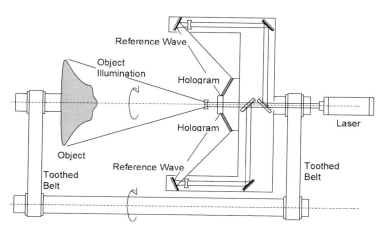

Figure 6.45: Rotating holographic interferometer.

laser beam passing through the hollow shaft of the interferometer. The axes of rotation of the object as well as of the holographic setup must be colinearly aligned. The synchronous rotation is achieved by mechanical coupling via toothed belts, Fig. 6.45, or electrooptical registration of the object rotation and electronic control of the interferometer rotation. A rotating interferometer as described allows one to record double exposure holograms in any angular position in contrast to the stroboscopic technique of the object related triggering method.

An approach to produce a stationary wave field reflected from the rotating object is the compensation of the rotation by an *image derotator* [352, 709, 710]. Its principle is shown in Fig. 6.46. If one observes an image reflected from a roof edge prism, this image appears to

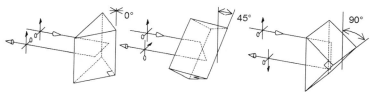

Figure 6.46: Image derotation by a roof edge prism.

rotate as soon as the prism is turned around its optical axis. The image rotates in reverse order to the prism but at twice the angular velocity of the prism rotation. So if the prism is rotating with half the number of revolutions and in the same direction as the rotating object then the reflected image appears stationary to an observer [711]. Of course the axes of rotation must be colinearly aligned. A holographic arrangement with an image derotator operating according to this principle is displayed in Fig. 6.47. The object is illuminated by a divergent wave field coming from a virtual point source located at the common axis of rotation. An encoder may be mounted on the shaft of the rotating component, or the object's rotational speed is recorded by a photocell detector or a tachogenerator to adapt the speeds of the object and the derotator prism. This prism is driven by a servo-controlled motor; if the object rotates with n revolutions

6.9 Measurement Range Extension

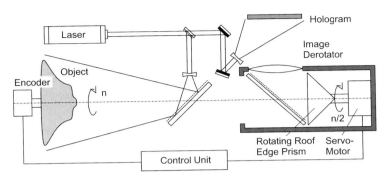

Figure 6.47: Holographic interferometric arrangement with image derotator.

per minute then the prism has to rotate with $n/2$ revolutions per minute. A fixed beamsplitter allows unhindered observation of the derotated image as well as the recording of holograms with double pulsed or continuous exposures. Using the image derotator, the rotating object can be viewed continuously in arbitrary angular positions: we have a non-stroboscopic system for freezing the wave field [712].

One application of holographic interferometry with rotation compensation by an image derotator was the investigation of the blade vibrations of the impeller of a radial compressor [713]. The radial impeller of 290 mm diameter consisted of 20 blades with every second blade cut back at the impeller inlet. Therefore 10 blades are observed along the rotational axis. The double exposure holographic interferogram of Fig. 6.48a was recorded at a rotational speed

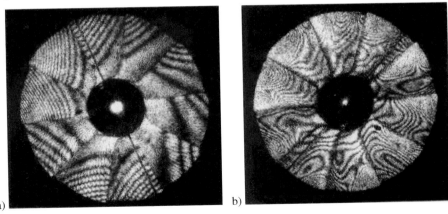

Figure 6.48: Holographic interferograms of a rotating radial impeller (a) at 2,935 rpm, (b) at 13,450 rpm (Courtesy of J. Geldmacher, BIAS).

of 2,935 rpm, and the one of Fig. 6.48b at 13,450 rpm. The resulting circumferential speed in this example reaches 204 m s^{-1}. The interferogram taken at 2,935 rpm was recorded with open compressor inlet. All blades vibrate in the first bending mode. At higher speeds, as in Fig. 6.48b, a closed compressor inlet had to be used. So the derotator was placed outside the inlet tube, the impeller was viewed via a deflection mirror installed inside the tube. The resulting interferogram displays a superposition of different vibration modes.

The aforementioned experiments still used holographic plates. Nowadays more flexibility is offered by digital holography. So pulsed digital holographic interferometry using a derotator for measuring dynamic deformations of rotating discs is described in [714]. Three-dimensional measurements of all displacement components of a rotating disc combining information from three different illumination directions are presented in [715]. Here pulsed digital holography without image derotator, but synchronized recording is employed.

6.9.4 Endoscopic Holographic Interferometry

Holographic interferometric measurement of form, deformation, vibration, etc. combined with endoscopy enhances the versatility of these methods by giving access to regions which are optically not accessible in the direct easy way. Thus hidden parts or cavities with only small access apertures like body cavities in medicine or the inside of machine components can be measured. Development of *endoscopic holographic interferometry* started with holographic film [716] but nowadays improved solutions using digital holographic interferometry are state of the art [717–720]. Different kinds of endoscopes enabling holographic interferometry are principally possible. So one can discriminate between rigid and flexible endoscopes. Modern rigid endoscopes are equipped with rod lens systems, so in a way a rigid endoscope can be viewed as a long objective. The flexible endoscopes on the other hand employ glass fibers for guiding the light. A further categorization can be introduced by dividing into the class of endoscopes with external interferometer head and the class where the interferometer is at the tip of the endoscope inside the cavity. While in the first named class the reflected wave field is guided over a bundle of fibers, e. g. 10,000 up to 50,000 fibers [718], in the second class illumination and reference light is transmitted over monomode fibers, and the interference between reflected object wave and the reference wave takes place on the 2D detector inside the cavity, so only the electronic data have to be transmitted to the outside world [717, 720].

The concept of an endoscope capable of full three-dimensional measurement of form and deformation by digital holographic interferometry is described in [717]. There reference wave and object illumination wave are guided into the sensor head by monomode glass fibers. Figure 6.49 shows only one object illumination arm although the real sensor has three arms. To

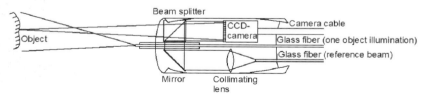

Figure 6.49: Sensor head of holographic endoscope.

obtain three significantly different sensitivity vectors, see Section 6.2, the illuminating fibers are within arms which are folded into the endoscope during induction into the cavity, but unfolded inside the cavity. The three unfolded arms are seen in the photograph of the holographic endoscope in Fig. 6.50. The beam splitter, Fig. 6.49, guides the collimated reference wave as

6.10 Refractive Index Fields in Transparent Media

which is approximated by

$$\delta = n_0 L \left(1 + \frac{1}{3} \left(\frac{m}{n_0} \right)^2 L^2 \right) \tag{6.143}$$

Even this simple example exhibits the bending of the rays when a spatial variation of the refractive index is present. This possibility has to be envisaged in the interpretation and evaluation of holographic interferograms which should measure refractive index variations of transparent media.

As mentioned before, each interpretation or evaluation of an interference pattern requires an imaging system to form a real image, at the least the lens of the observer's eye. The imaging system is represented by a single thin lens in Fig. 6.55. The refractive index field acting

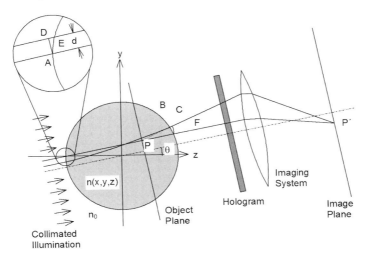

Figure 6.55: Imaging of a refractive index field.

as the object to be measured is assumed to be contained in a circular region as proposed in Appendix B. A ray parallel to the optical axis enters the refracting region in point A, is bent and leaves the region at B, it passes through the surrounding homogeneous medium of refractive index n_0, meets the thin lens and is imaged onto point P' in the image plane. In double exposure holographic interferometry of course this role is played by the holographic reconstruction of the ray belonging to the exposure during the presence of the refracting object. At P' the ray interferes with another holographically reconstructed ray, this second one stemming from the exposure when no refracting object was present. To find the corresponding point P in the object plane, we can trace back with the common thin lens techniques. Another way is to extend the first ray from the refracting medium to the lens further along a straight line to the object plane [722], Fig. 6.55. This straight ray enters the refracting region at E and leaves it at F. These two rays of the collimated illumination are separated by the distance d.

To find the optical path length changes δ, or equivalently the optical phase difference $\Delta\phi$ between the rays interfering at P', we notice that no relative changes occur left of points A and D or to the right of points C and F. D is the intersection of the straight ray with the perpendicular to the first ray through A. C is at the same distance from P as F, Fig. 6.55. The optical pathlength difference $\Delta\delta$ therefore is

$$\Delta\delta = \int_A^B n(x,y,z)ds + n_0(\overline{BC} - \overline{DE} - \overline{EF}). \tag{6.144}$$

While the first term of (6.144) is the integral of the refractive index along the curved path $\widetilde{AB}$, the second term accounts for the different path length in the surrounding with refractive index n_0. For each viewing direction, defined by the angle θ to the arbitrary reference direction z, $\Delta\delta(\rho,\theta)$ or $\Delta\phi(\rho,\theta) = \Delta\delta(\rho,\theta) \cdot 2\pi/\lambda$ can be measured holographically. The ray equation (6.132) and the integral (6.144) together define the so called *path length transform*, which is a nonlinear integral transform of $n(x,y,z)$ and n_0. If the refractive index variation is sufficiently small the curved ray $\widetilde{AB}$ coincides with the straight line $\overline{EF}$, so that (6.144) reduces to

$$\Delta\delta = \int_E^F (n(x,y,z) - n_0)dl \tag{6.145}$$

with dl denoting the differential distance along line $\overline{EF}$. The *line integral transform* defined by (6.132) and (6.145) is mathematically equivalent to the *Radon transform* (B.2), see Appendix B.

6.10.2 Physical Quantities Affecting the Refractive Index Field

In most applications of holographic interferometry for measurements at transparent media it is not the refractive index distribution, Table 6.2, which is of main concern, but another physical quantity. The value of this physical quantity is determined by the effect it has on the refractive index field. So in the following some relations between such physical quantities and the refractive index are given [170].

In *aerodynamics* and *flow visualization* the flow of compressible gases is studied, e. g. in wind tunnels or shock tubes. The quantity of interest is the density ρ in a gas, the mass per unit volume. Its relation to refractive index n is given by the *Gladstone-Dale equation*

$$n - 1 = K\rho \tag{6.146}$$

with the *Gladstone-Dale constant* K, which is a property of the gas. The Gladstone-Dale constant is nearly independent of temperature and pressure under moderate physical conditions and it is a weak function of wavelength [170]. Some values are given in Table 6.3 [170]. The Gladstone-Dale constant of a mixture of gases can be calculated as the mass-weighted average of the values for the component gases

$$K = \sum_i a_i K_i \tag{6.147}$$

6.10 Refractive Index Fields in Transparent Media

Table 6.2: Index of refraction of various optical materials (at yellow) [220].

Material	Index of Refraction	Material	Index of Refraction
Air	1.0003	Sodium chloride	1.54
Water	1.33	Light flint glass	1.57
Methanol	1.33	Carbon disulfide	1.62
Ethanol	1.36	Medium flint glass	1.63
Magnesium flouride	1.38	Dense flint glass	1.66
Fused quartz	1.46	Extra-dense flint glass	1.73
Pyrex glass	1.47	Sapphire	1.77
Benzene	1.50	Heaviest flint glass	1.89
Xylene	1.50	Zinc sulfide (thin film)	2.3
Crown glass	1.52	Titanium dioxide (thin film)	2.4–2.9
Canada balsam (cement)	1.53		

Table 6.3: Gladstone-Dale constants of gases.

Gas	K (m³/kg) at $\lambda = 0.5145$ μm	at $\lambda = 0.6328$ μm
Ar	0.175×10^{-3}	0.175×10^{-3}
O_2	0.191×10^{-3}	0.189×10^{-3}
He	0.196×10^{-3}	0.195×10^{-3}
CO_2	0.229×10^{-3}	0.227×10^{-3}
N_2	0.240×10^{-3}	0.238×10^{-3}

with the mass fraction a_i and Gladstone-Dale constant K_i of the i-th component.

The density ρ of a gas in most cases of interest can be calculated from the pressure P, the molecular weight M and the absolute temperature T via the *ideal gas equation*

$$\rho = \frac{MP}{RT} \qquad (6.148)$$

with the universal gas constant $R = 8.3143$ J/(mol K). The combination with (6.146) yields

$$n - 1 = \frac{KMP}{RT}. \qquad (6.149)$$

This can be used for holographic *temperature measurements*. If the temperature changes remain small, a linear relation between the change of the refractive index and the change of the temperature can be adopted. As an example for air at 288° K and 0.1013 MPa the Gladstone-Dale constant at $\lambda = 0.6328$ μm is 0.226×10^{-3} m³/kg and the molecular weight is 28.97.

This results in

$$\frac{dn}{dT} = -0.9617 \times 10^{-6} \text{ K}^{-1}. \tag{6.150}$$

With even more precision the dependence of the refractive index of air from temperature at $\lambda = 0.6328$ µm is [170]

$$n = 1 + \frac{0.292015 \times 10^{-3}}{1 + 0.368184 \times 10^{-2}T} \tag{6.151}$$

and at $\lambda = 0.5145$ µm

$$n = 1 + \frac{0.294036 \times 10^{-3}}{1 + 0.369203 \times 10^{-2}T} \tag{6.152}$$

with T in degrees Celsius.

In *liquids* the refractive index is related to density ρ by the *Lorentz-Lorenz equation*

$$\frac{n^2 - 1}{\rho(n^2 + 2)} = \bar{r}(\lambda) \tag{6.153}$$

where $\bar{r}(\lambda)$ is the *specific refractivity*, which depends on the substance and the wavelength of light. There is no direct analog to the ideal gas equation (6.148) in the case of liquids, instead empirical relations between refractive index and temperature must be used. Some are given in Table 6.4 [170, 723].

Table 6.4: Dependence of refractive index on temperature in liquids.

Liquid	$-dn/dT(\text{K}^{-1})$ at $\lambda = 0.5461$ µm	at $\lambda = 0.6328$ µm
Water	1.00×10^{-4}	0.985×10^{-4}
Methyl alcohol	4.05×10^{-4}	4.0×10^{-4}
Ethyl alcohol	4.05×10^{-4}	4.0×10^{-4}
Isopropyl alcohol	4.15×10^{-4}	4.15×10^{-4}
Benzene	6.42×10^{-4}	6.40×10^{-4}
Toluene	5.55×10^{-4}	5.55×10^{-4}
Nitrobenzene	4.68×10^{-4}	4.68×10^{-4}
c-Hexane	5.46×10^{-4}	5.43×10^{-4}
Acetone	5.31×10^{-4}	5.31×10^{-4}
Chloroform	5.98×10^{-4}	5.98×10^{-4}
Carbon tetrachloride	5.99×10^{-4}	5.98×10^{-4}
Carbon disulfide	7.96×10^{-4}	7.96×10^{-4}

Quite accurate equations for water are

$$n = 1.3331733 - (1.936\,T + 0.1699\,T^2) \times 10^{-5} \tag{6.154}$$

each annular element [726] or employ representations of $f(r)$ by sampling series [727, 728]. Methods based on the inversion of formula (6.174) use interpolated phase data and numerical differentiation.

A fast and efficient method utilizes a Fourier decomposition of the interference phase and calculates the Abel inversion of each spatial frequency component [729]. The Fourier coefficients are obtained from an FFT-routine, Appendix A.10.

6.10.5 Multidirectional Recording of Asymmetric Refractive Index Fields

The determination of *asymmetric refractive index fields* requires the analysis of a large number of holographic interferograms by methods of *computer aided tomography*, see Appendix B. Each of the reconstructed holographic interference patterns has to correspond to a different viewing direction [198]. Therefore holographic interferometry is ideally suited for rendering tomographic data, since just a single hologram allows the realization of a number of viewing directions. Thus although many interferograms are required, the number of necessary holograms remains limited. Some arrangements to record multiple holograms for subsequent tomographic reconstruction are proposed in Fig. 6.58 [730]. Figure 6.58a shows the diffuse illumination via a diffuser of e. g. ground glass. The range of viewing directions is only limited by the size of the diffuser and the aperture of the holograms. But speckles may become a problem due to complicated localization of the fringes as soon as the interferograms have to be observed with a small aperture, see Section 4.3. This possible disadvantage is circumvented by using a phase grating, Fig. 6.58b, which diffracts several plane waves out of the impinging plane wave. In Fig. 6.58c the individual plane waves are produced separately. A fixed plane object wave and a rotating object field are used in the arrangement of Fig. 6.58d. In all cases one has to consider whether a transient or a steady or at least repeatable refractive index distribution is present. The arrangements of Figs. 6.58a, b, and c allow a simultaneous recording of all holograms; the holograms in Fig. 6.58d are recorded sequentially. The system of Fig. 6.58c also enables a sequential registration.

Quite another way of registration of the refractive index variation may be taken by using the holographic *light-in-flight recording* and reconstruction of optical wavefronts, see Section 6.6.4. With this method wavefronts can be visualized as they pass a transparent object [655]. The spatial distortion of a plane or spherical wavefront induced by the refractive index field can be measured this way, Fig. 6.59.

In the treatment of reconstruction of asymmetric refractive index fields we have to consider two cases. The first is when ray bending due to refraction is minor and can be neglected. Then the integral defining the pathlength difference

$$\delta(x,y) = \int_s f(x,y,z)\, ds \qquad (6.175)$$

with $f(x,y,z) = n(x,y,z) - n_0$ can be evaluated along straight lines s. We speak of the *refractionless limit*. The second case is when we have to take into account ray bending by strongly refracting fields.

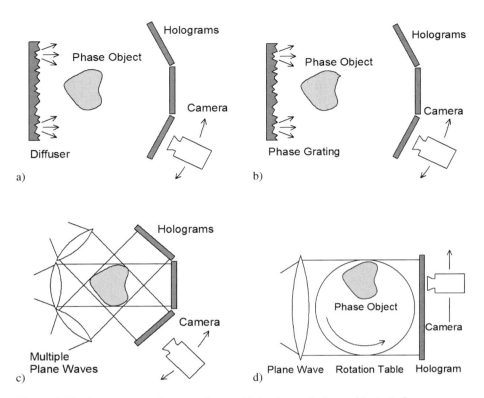

Figure 6.58: Arrangements for recording multiple views of phase objects (reference waves omitted for clarity).

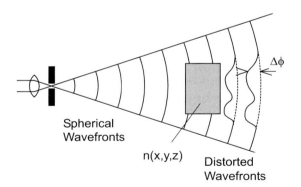

Figure 6.59: Registration of wavefront distortion by light-in-flight recording.

6.10.6 Tomographic Reconstruction in the Refractionless Limit

Reconstructions for both aforementioned cases are based on the tomographic methods introduced in Appendix B [731]. While most of the methods work well with multidirectional data collected along views subtending the whole angle of $180°$, in holographic interferometry the

6.10 Refractive Index Fields in Transparent Media

angular range is often less than 180°. The angle of view is restricted by the aperture of the hologram, the angular aperture of the illuminated diffuser, and the extent of the transparent object, Fig. 6.58.

Several reconstruction procedures have been introduced and compared with regard to their capability to work with restricted angles of view in [30, 732]. The first approach in [30] is the *Fourier synthesis*, which fills the spatial frequency plane with values along lines through the origin, as they are given by the *Fourier slice theorem*, (B.7). But this method suffers from the necessity of interpolation when in practice data are collected only for a finite number of views and positions, as pointed out in Appendix B.2.

The problem of the values not uniformly distributed in the frequency plane are avoided by the *direct inversion* of (6.175) as introduced by [733, 734] and adopted by [30]. This approach is equivalent to the one introduced in Appendix B.3. Its practical implementation, when discrete data are at hand, corresponds to the methods of Appendix B.4.

These two methods require optical pathlength data collected over the whole 180° angle of view. The methods described next can be applied even when data are available only for viewing angles less than 180°. A number of such methods represent the $f(x,y)$ of (6.175) in each plane z =const by a series expansion

$$f_e(x,y) = \sum_{m=0}^{M-1} \sum_{n=0}^{N-1} a_{mn} H_{mn}(x,y). \tag{6.176}$$

Together with (6.175) we get after interchanging the order of integration and summation

$$\delta(x,y) = \sum_{m=0}^{M-1} \sum_{n=0}^{N-1} a_{mn} \int_s H_{mn}(x,y) ds. \tag{6.177}$$

This shows that it is advantageous to choose generating functions $H_{mn}(x,y)$ which are easily integrated along arbitrary straight lines.

A convenient series expansion is directed by the Whittaker-Shannon *sampling theorem*, which states that a properly sampled band-limited function can be exactly represented by a linear combination of sinc-functions, see Appendix A.7

$$f_e(x,y) = \sum_{l=-\infty}^{\infty} \sum_{k=-\infty}^{\infty} f\left(\frac{l}{2B_x}, \frac{k}{2B_y}\right) \mathrm{sinc}\left[2B_x\left(x - \frac{l}{2B_x}\right)\right] \mathrm{sinc}\left[2B_y\left(y - \frac{k}{2B_y}\right)\right] \tag{6.178}$$

where $\mathrm{sinc}(x) = [\sin(\pi x)]/(\pi x)$ and B_x, B_y are the bandwidths in the x- and y-direction. The sinc-function approach leads to a set of algebraic equations [30]

$$\delta(\rho_i, \theta_j) = \sum_{m=0}^{M-1} \sum_{n=0}^{N-1} W_{mn}(\rho_i, \theta_j) f_e(\Delta x\, m, \Delta y\, n) \tag{6.179}$$

where the $W_{mn}(\rho_i, \theta_j)$ are defined by

$$W_{mn}(\rho_i,\theta_j) = \begin{cases} \sqrt{1+\tan^2\theta_j}\,\Delta x\,\text{sinc}[(\rho_i\sec\theta_j + \Delta xm\tan\theta_j - \Delta yn)/\Delta y] \\ \qquad\qquad\qquad \text{for } 0 \leq |\tan\theta_j| \leq \Delta y/\Delta x \\ \sqrt{1+\tan^2\theta_j}\,\frac{\Delta y}{\tan\theta_j}\text{sinc}[(\rho_i\sec\theta_j + \Delta xm\tan\theta_j - \Delta yn)/(\Delta y\tan\theta_j)] \\ \qquad\qquad\qquad \text{for } \Delta_y/\Delta x < |\tan\theta_j| < \infty \\ \Delta y\,\text{sinc}[(\rho_i + \Delta xm)/\Delta x] \qquad \text{for } |\tan\theta_j| = \infty \end{cases} \qquad (6.180)$$

The θ_j represent the different angular views, the ρ_i the pathlength values in each projection. Δx and Δy denote the spacing of the pathlength values in each direction. The system of equations (6.179) can be solved if at least $M \times N$ pathlength data are measured. Tests have shown that reliable results are obtained when redundant data are used, meaning much more than $M \times N$ values of $\delta(\rho_i, \theta_j)$ are present. This is particularly true if only restricted angles of view are possible. The overdetermined system of linear equations then is solved by the method of Gaussian least squares.

The next method investigated in [30], the so called *grid method*, uses a set of rectangular elements, with constant refractive index in each element. A similar approach is described in Section B.5. As with the sinc-method, we get a set of algebraic equations whose solution represents an approximation to the actual $f(x,y)$.

A Fourier transform approach to reconstruction from data over a restricted angular view is given by the *frequency plane restoration*. If multidirectional data are given only for directions between $-\gamma$ and $+\gamma$, Fig. 6.60a, the resulting frequency samples also are only within an angular range of 2γ, the shaded region in Fig. 6.60b. The determination of the Fourier

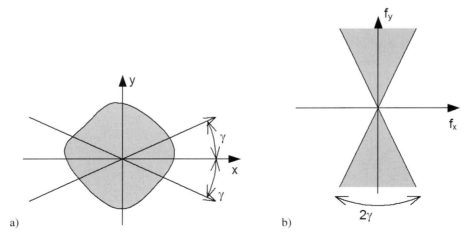

Figure 6.60: Restricted angular views, (a) spatial domain, (b) frequency domain.

6.10 Refractive Index Fields in Transparent Media

transform values outside of this known region is based on the fact that the Fourier transform of a continuous, bounded, and spatially limited function is analytic. If the Fourier transform can be uniquely determined over any finite domain, the entire transform can be determined using analytic continuation [30]. Therefore a truncated summation over sampling functions in the frequency domain is defined with the frequency values of the whole frequency plane as coefficients. This leads to a set of algebraic equations, which can be solved for the unknown coefficients.

The last reconstruction method addressed in [30] is an iterative method for solution of an underdetermined system of linear equations. Of the manifold of possible solutions the one with minimum variance is chosen. Nevertheless the method is also feasible for overdetermined systems of equations.

The six reconstruction techniques mentioned above have been compared with simulated data as they may arise in fluid temperature measurements [30]. These comparisons have given the following results. If a 180° angle of view is available, all techniques produce good reconstructions. Especially the sinc-method, the direct inversion and the frequency restoration achieve a good accuracy, while direct inversion requires the least computer time. If the angle of view is less than 180°, e. g. 45°, the frequency plane restoration provides the most accurate reconstruction. It has been observed that if the angle of view is decreased, the degree of redundancy necessary for a reliable reconstruction increases. On the other hand if in an experiment the amount of data is fixed, this relation limits the number of degrees of freedom, and therefore the achievable resolution.

Independent from the specific technique used for the reconstruction one has to reflect upon the data sampling rates. Theoretically only band-limited functions can be reconstructed exactly, although in practice we have spatially limited refractive index fields which therefore are not band-limited. But if the Fourier components of this field are sufficiently small outside some finite region in the frequency plane, an *effective bandwidth* B can be defined. The sampling theorem now guarantees a reliable reconstruction, if the spacing between consecutive ρ_i is less than $1/(2B)$ in the corresponding θ_j direction. In defining the angular separation $\Delta\theta$ between the views one has to consider that the evaluated points in the frequency plane are lying along radial lines, so they have maximum separation at the effective band limits. If we consider two adjacent radial lines oriented near $\pi/2$ and data should be sampled at the Nyquist rate corresponding to the effective band limit B_y, we meet the condition

$$B_y \tan(\Delta\theta) \leq 1/L_x \tag{6.181}$$

where L_x is the object's extent in the x-direction. The angular spacing between views near $\pi/2$ must obey

$$\Delta\theta \leq \arctan\left(\frac{1}{B_y L_x}\right) \tag{6.182}$$

and similarly for the other direction. In the vicinity of $\pi/2$ the tan-function is steepest, so this is a conservative estimate.

Samples are more close near the origin of the frequency plane than near the band limits. So if the views are chosen to satisfy (6.182) the data will be oversampled. Altogether approximately $8B_x B_y L_x L_y$ samples will be used. Compared to the space-bandwidth product of the representation of the refractive index distribution, which is $4B_x B_y L_x L_y$, we determine a data redundancy of about 2 [30]. The space-bandwidth product is a measure of the total number of degrees of freedom, and is invariant under Fourier transformation.

6.10.7 Tomographic Reconstruction of Strongly Refracting Fields

In tomography codes have been designed to perform an inversion in the case of the refractionless limit. If these codes are applied to pathlength transforms of *strongly refracting refractive index fields* appreciable errors may result. Although an analytical solution for radially symmetric strongly refracting fields exists, Section 6.10.4, there is no analytical solution for the general asymmetric case. Therefore in the following an iterative algorithm for reconstruction in this instance is presented [722].

Let $\widetilde{\delta}(\rho, \theta)$ be the pathlength difference along the ray bent by refraction as given in (6.144). Correspondingly let $\overline{\delta}(\rho, \theta)$ be the pathlength along the straight ray of (6.145). θ is the angle of the projection and ρ the coordinate along the projection, Fig. 6.55. Let the operator $\overline{P}$ express the straight line integral transform and $\widetilde{P}$ the bent line integral transform which maps $n(r, \phi) - n_0$ onto $\overline{\delta}(\rho, \theta)$ and $\widetilde{\delta}(\rho, \theta)$, resp. The iterative method is based on successive estimation of the deviation $D(\rho, \theta)$ between the straight ray and bended ray pathlength transforms

$$D(\rho, \theta) = \widetilde{\delta}(\rho, \theta) - \overline{\delta}(\rho, \theta). \tag{6.183}$$

It is assumed that $\overline{\delta}(\rho, \theta)$ and $\widetilde{\delta}(\rho, \theta)$ are defined in the same domain [722]. The iterative algorithm begins with (1) setting an initial estimate of the deviation $D_0(\rho, \theta)$. Then (2) the estimate of the refractionless path length transform $\overline{\delta}(\rho, \theta)$ is calculated from the measured pathlength differences $\widetilde{\delta}(\rho, \theta)$ and the assumed deviation

$$\overline{\delta}_i(\rho, \theta) = \widetilde{\delta}(\rho, \theta) - D_i(\rho, \theta). \tag{6.184}$$

The corresponding refractive index field $n_i(r, \phi) - n_0$ is (3) reconstructed by a numerical inverse line integral transformation

$$n_i(r, \phi) - n_0 = \overline{P}^{-1}[\overline{\delta}_i(\rho, \theta)]. \tag{6.185}$$

Using computational ray tracing (4), according to (6.132) the pathlength transform of the estimated field is calculated

$$\widetilde{\delta}_i(\rho, \theta) = \widetilde{P}[n_i(r, \phi) - n_0]. \tag{6.186}$$

A new estimate of the deviation is calculated (5) by

$$D_i(\rho, \theta) = \widetilde{\delta}_i(\rho, \theta) - \overline{\delta}_i(\rho, \theta) \tag{6.187}$$

and the algorithm proceeds at step (2). This iterative procedure continues until the change in $D_i(\rho, \theta)$ or the difference between two successive reconstructed fields is smaller than a predetermined value.

6.10 Refractive Index Fields in Transparent Media

The line integral transform to be inverted in step (3) is expressed as a Fourier series within a circular domain of radius R, $\rho/R \leq 1$

$$\overline{\delta_i}(\rho, \theta) = \sum_{m=0}^{M-1} \sum_{n=0}^{N-1} A_{mn} g_{mn}(\rho) e^{im\theta}. \tag{6.188}$$

A finite number of coefficients A_{mn} are found by inverting an overdetermined system of algebraic equations of the form (6.188) with discrete ρ_j and θ_j by a least squares method. Once the coefficients A_{mn} have been calculated, the reconstructed $f(r, \phi) = n(r, \phi) - n_0$ can be presented as a Fourier series within the circular domain $r/R \leq 1$

$$f(r, \phi) = \sum_{m=0}^{M-1} \sum_{n=0}^{N-1} A_{mn} f_{mn}(r) e^{im\theta}. \tag{6.189}$$

Appropriate functions $g_{mn}(\rho)$ and $f_{mn}(r)$, which are line-integral transform pairs are discussed in [722].

The iterative algorithm was applied to numerically simulated data as well as tested in experiments measuring strongly refracting boundary layers. Refractive index fields with multiple maxima and quite steep gradients have been successfully reconstructed, even when refraction was strong enough to bend some rays by as much as $27°$. Nevertheless operator interaction was needed to detect computational ray crossing [170]. Path length data contaminated by this effect were eliminated [735].

The iterative algorithm for correction of errors caused by ray bending also is employed in [736], where optical tomography for flow visualization of the density field around a revolving helicopter rotor blade is investigated. The tomographic reconstruction there is based on the convolution backprojection employing a Shepp-Logan-kernel for filtering, see Appendix B.3.

Further approaches to reconstruction of asymmetric refractive index fields which bend rays are the application of perturbation analysis [737], which is feasible for mildly refracting objects or the approach via inverse scattering: If diffraction is mild, the Rytov or Born approximations to the wave equation can be used [738, 739].

More problems encountered in tomographic reconstruction of refractive index fields are caused by limited interferometric data. Physical constraints such as test section enclosures may restrict the angular views to less than the desired $180°$, and/or a portion of the probing rays may be blocked [736]. Limited data reconstructions are sensitive to noise and produce geometric distortions with various artifacts. An approach to reconstruct under such circumstances is published in [740]: The so called complementary field method is an iterative one and incorporates a priori information effectively. Another problem occurs when an opaque object is present within the field under study, e. g. the test model around which flow is being studied in aerodynamic testing. A solution by illuminating the embedded objects and holographic recording of the wave field scattered by the object's surface is presented in [741]. By this technique furthermore the fringe localization surface is compressed as well as displaced and the number of fringes is doubled. The compression of the localization surface enables the observer to use a larger numerical aperture than would be possible otherwise, see Section 4.3.5.

To summarize, it can be stated that the approaches to reconstruction of refractive index distributions with ray bending, published up to now, seem to be designed for specialized cases with no guarantee to work under general conditions. Much more needs to be done in the future in this important, challenging, and interesting field [724, 735].

6.10.8 Analysis of Transparent Media with Digital Holography

In the preceding sections the applications of digital holography were in measuring at opaque rough surfaces or in detecting particles. However as with ordinary optical holography and holographic interferometry digital holography also can be used for measuring refractive index distributions in transparent media. A typical application is the measurement of gas flow in front of a nozzle [102] or combustion experiments [151]. In Fig. 6.61 the modified Mach-Zehnder interferometer performing this task by digital holographic interferometry is shown. While the gas flow is injected into the probe arm, the reference beam is focused by a lens to form a point source. When viewed from the CCD-array through the beam splitter the reference point source seems to be at a location which is optically coplanar to the location of the gas flow, depicted as the virtual source of the reference wave in Fig. 6.61. Two digital holograms are recorded with and without the gas flow injection.

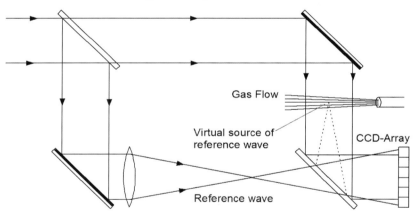

Figure 6.61: Modified Mach-Zehnder interferometer for digital recording of lensless Fourier transform holograms.

This lensless Fourier transform geometry adopted by Takeda et al. [102] has the main advantage that we do not need knowledge about the exact distance d between object and CCD-array in the hologram plane. While in the Fresnel case, d determines the plane of focusing, any error in the measurement of d will cause defocusing and may distort the phase distribution in the observation plane. Alternatively in the Fourier case the error in d influences only the spatial scale factor and not the relative phase of the reconstructed optical fields. A further but minor advantage is the possible saving in computation time and storage requirement: If we Fourier transform the real function $h(\xi, \eta)$ rather than the complex function $h(\xi, \eta) r(\xi, \eta) \exp\{\frac{\pi i}{\lambda d}(\xi^2 + \eta^2)\}$ we need essentially half the number of operations than re-

6.11 Defect Detection by Holographic Non-Destructive Testing

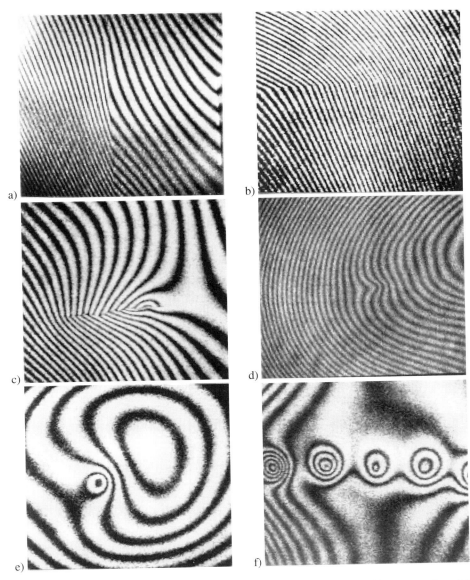

Figure 6.66: Experimentally generated defect induced fringe patterns: (a) compression, (b) bend, (c) displacement, (d) groove, (e) bull's eye, (f) eye chain (Courtesy of W. Osten, BIAS).

6.11.2 Data Reduction for Automatic Qualitative Evaluation

A central step in each solution of a pattern recognition problem is the reduction of the registered input data to a tractable number of parameters, which resemble the significant features of the input data. In our context we have to reduce from some 10^5 to 10^6 pixels of the recorded holographic interference pattern to a few parameters which still can discriminate interferograms with characteristic partial patterns indicating a material defect from those without. A first stage of data reduction may be the skeletonizing introduced in Section 5.3.

A data reduction strategy that often is successful in digital image processing is to take the 2D Fourier transform. The real amplitude spectrum already is translation invariant. If we describe the amplitude spectrum in polar coordinates and integrate all spectral values along the angles from $0°$ to $360°$ ($0°$ to $180°$ also suffice) at each radius, we get a rotation- as well as translation-invariant 1D spectrum. If needed, a further Fourier transform may produce even a scale-invariant set of features.

Since we are interested only in the existence of a defect induced partial pattern, but not primarily in its location, orientation, or size, translation-, rotation-, and scale-invariant features are exactly what we need. But the outlined Fourier transform based strategy here gives no significant results. The reason is the averaging effect of the global 2D Fourier transform. Small partial patterns may cover only a few pixels of all the pixels of the interferogram. During calculation of each spectral value a weighted sum of the intensities of all pixels is formed, the contribution of a small partial pattern is suppressed by the averaging effect. Especially if some broadband noise is present in the interferogram, a defect induced local pattern will not contribute significantly to the spectrum.

The few published approaches for automatic qualitative evaluation are based on dividing the whole pattern into a number of sections and comparing the fringe densities in these sections. In [752] the interference patterns of pressure vessels are partitioned into squares and the number of fringes in each square is counted. Based on holographic interferograms of proven intact specimens, each square gets a minimum and maximum acceptable fringe count. The fringes are counted automatically in each square, regardless of their orientation. If all fringe counts fall between the predetermined thresholds, the pressure vessel is accepted, otherwise it is rejected as a defective one.

The sections into which the patterns are divided, can even be degenerated rectangles, like rows or columns [244, 753]. In [753] fringe peaks are counted along horizontal lines. If the number of peaks exceeds a threshold in one line, fringes are counted along short vertical columns centering at this horizontal line until again a threshold is reached. To ensure that a closed ring pattern is detected, the fringes are counted along inclined vectors as well. A fault is considered to be detected if the fringe count in each direction is above the threshold value. The process can be repeated to detect multiple defects. The procedure is applied to the detection of debrazes in brazed cooling panels.

In [244] Fourier amplitude spectra are calculated one-dimensionally along lines and columns or two-dimensionally in small rectangles. From these spectra an average amplitude spectrum together with an acceptance band broader than its bandwidth is determined. If the spectrum along one line, column, or rectangle differs significantly from the average spectrum, then a locally higher or lower fringe density must have occurred, indicating a defect.

A feature selection scheme that associates parameters to sections of the pattern, which was successfully applied in an artificial neural network approach to qualitative evaluation, is presented in the following. It is based on the fact that the intensity distribution of the defect induced partial patterns, which are searched, is varying more rapidly than the interference pattern in its defect free neighborhood. But the local variation must not be compared to an

6.11 Defect Detection by Holographic Non-Destructive Testing

averaged global variation, since the fringe density changes continuously even in the non-defect case due to the loading and the varying sensitivity of the holographic arrangement. This change may lead to a higher fringe density in a non-defect area than the density of a defect in a low fringe density area [293]. So the comparison of intensity variation in a small area is restricted to its immediate neighborhood.

For the determination of the intensity variation $I(x,y) \in \{0, 1, \ldots, 255\}$ at each pixel (x, y) the slopes a and b of the two-dimensional plane

$$I(x,y) = ax + by + c \tag{6.191}$$

tangential to the intensity distribution $I(x,y)$ are calculated by Gaussian least squares based on its eight neighbors. The slopes are

$$\begin{aligned}
a &= \tfrac{1}{6}\big[I(x+1,y-1) - I(x-1,y-1) + I(x+1,y) - I(x-1,y) \\
&\quad + I(x+1,y+1) - I(x-1,y+1)\big] \\
b &= \tfrac{1}{6}\big[I(x-1,y+1) - I(x-1,y-1) + I(x,y+1) - I(x,y-1) \\
&\quad + I(x+1,y+1) - I(x+1,y-1)\big]
\end{aligned} \tag{6.192}$$

a and b may have different or equal signs, Fig. 6.67, so the maximum slope at (a,b) is

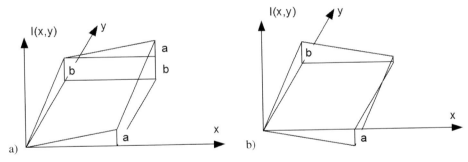

Figure 6.67: Local slope for equal signs of a and b (a), and different signs (b).

approximated by

$$s(x,y) = \max\{|a+b|, |a-b|\}. \tag{6.193}$$

To remain in the range of 8 bits in image processing, we truncate this expression

$$s(x,y) = \min\{\max\{|a+b|, |a-b|\}, 255\}. \tag{6.194}$$

The whole pattern now is partitioned into non-overlapping areas, e. g. the pattern of 512×512 pixels is divided into 8×8 areas of 64×64 pixels each. Each area E gets the parameter $k(E)$, which is the maximal slope of all its pixels

$$k(E) = \max\{s(x,y) : (x,y) \in E\}. \tag{6.195}$$

Figure 6.68a shows an interference pattern with a defect induced local variation in an 8-bit gray-scale display. The parameters $k(E)$ of a partition into 8×8 areas E are given as gray-values in Fig. 6.68b and as numbers in Fig. 6.68c. One can notice the effect of higher fringe density and thus higher slope at the defect-free lower margin of the pattern compared to the defect area in the center of the pattern.

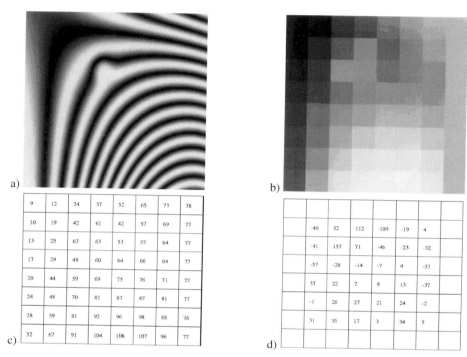

Figure 6.68: Feature selection: (a) interference pattern, (b) maximal slopes in 8×8 areas as gray-values, (c) maximal slopes in 8×8 areas as numbers, (d) Laplace values.

A defect induced local variation of the fringe density is present in area E, if the parameter $k(E)$ differs significantly from the $k(E')$ of the neighboring areas E'. To detect this, we apply a Laplace-filter to the image of area parameters. If the neighboring areas of E are denoted by A to I according to

A	B	C
D	E	F
G	H	I

6.11 Defect Detection by Holographic Non-Destructive Testing

two possible realizations of the Laplace-filter are

$$f_1(E) = 4k(E) - k(B) - k(D) - k(F) - k(H) \tag{6.196}$$

or

$$f_2(E) = 8k(E) - k(A) - k(B) - k(C) - k(D) - k(F) - k(G) - k(H) - k(I). \tag{6.197}$$

Figure 6.68d shows the Laplace values for all areas having 8 neighboring areas. A defect now is indicated by a high Laplace value. The highest Laplace values of the pattern, regardless where they occur, are translation-invariant as well as rotation-invariant features.

Defects which cause high density fringes even of low contrast are detected by this procedure as long as they are confined to one area or extend over only a few areas.

In a first attempt an automatic defect detection by a threshold comparison of the highest Laplace value over all areas of the tested pattern was carried out. To define the optimal threshold, a sample set of 1000 holographic interferograms was simulated on a computer, see Section 4.1.6, 500 with and 500 without defects. The threshold was chosen at the valley between the two modes of the bimodal histogram of the highest Laplace value of each pattern of the sample set. This approach was feasible as long as the partial patterns did not vary too much. All defect induced partial patterns had to fit into a single area.

Another approach to automatic defect detection is presented in [474]. There the variety of possible interference patterns is limited by experimental modifications, which produce only linear fringes. This holographic fringe linearization is obtained by swinging the object beam between the two exposures. Proper selection of the fringe frequency by adjusting the object beam swing and of the loading force creates a reconstructed image laced with linear fringes that have highly visible fringe shifts at the defect locations. These fringe shifts furthermore have characteristic Fourier signatures different from those of the linear fringe acting as a carrier.

6.11.3 Neural Network Approach to Qualitative Evaluation

The concept of *artificial neural networks* promises a reliable defect detection based on a number of typical examples because a neural network (1) has the intrinsic ability to learn from the input data and to generalize; (2) is nonparametric and makes weaker assumptions about the input data distributions than traditional statistical (Bayesian) methods; and (3) is capable of forming highly nonlinear decision bounds in the feature space.

Each neural network consists of a number of neurons which are the fundamental information processing units, Fig. 6.69. Every *neuron* has a number n of input paths numbered by $i = 1, \ldots, n$. The inputs x_i, $i = 1, \ldots, n$ are multiplied by *synaptic weights* w_{ji}, where j counts the neurons in the network. The weighted inputs are summed, the result is the internal activity level I_j

$$I_j = \sum_i w_{ji} x_i. \tag{6.198}$$

This activity is modified by a transfer function f which can be a binary (0 and 1) or bipolar (-1 and $+1$) hard limiter, a threshold function with a linear range, the *sigmoid function* $f(I) =$

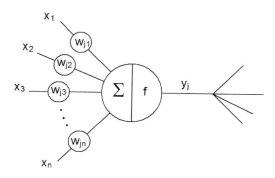

Figure 6.69: Processing element of an artificial neural network.

$[1 + \exp(-I)]^{-1}$, or any nondecreasing bounded function. The result

$$y_j = f\left(\sum_i w_{ji} x_i\right) \qquad (6.199)$$

is given to the output path which may be branched to be connected to the input paths of other neurons or it may present the results of the processing of the whole network to the outside.

Usually the neurons of an artificial neural network are organized into groups called layers. A typical network consists of an input layer of source elements, followed by one or more so-called hidden layers, and an output layer, Fig. 6.70. While the basic operation of a neuron

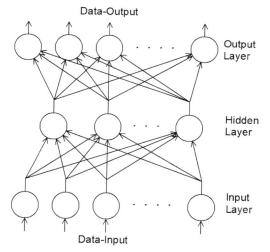

Figure 6.70: Artificial neural network with one hidden layer.

is always the same, the different concepts of neural networks differ in their architecture; that is, the number of inputs and outputs, the number of layers, the number of neurons in each layer, the number of weights in each neuron, the way the neurons and corresponding weights

6.11 Defect Detection by Holographic Non-Destructive Testing

are linked together within a layer or between the layers, or which neurons receive correction signals. Closely related to the architecture is the way information is fed to the network, its training. Although a single neuron is far from achieving relevant processing of information, it is the network of many interconnected neurons and the information stored in the synaptic weights that exhibits the far reaching problem solving capability. In the brains of living creatures information is processed in parallel by many neurons; this is actually performed sequentially in the computer software realizations of artificial neural networks.

To achieve an automatic detection of the characteristic local patterns of HNDT, a multilayer network trained by backpropagation learning at sample patterns has been implemented and tested [293, 531, 754]. Since it is not possible to generate the whole training set experimentally – one needs several thousands of holographic interferograms – the training samples are calculated by computer simulation, see Section 4.1.6. For a given application the material and shape of the object, the type of loading, and the typical defects to be detected have first to be examined practically and theoretically. Based on these examinations a restricted number of experiments are performed to check the validity of the simulation program and help to optimize it.

The features of each interference pattern to be presented to the input neurons are the four highest Laplace values introduced in Section 6.11.2, calculated for each partition of the 512×512 pixel pattern into 8×8, 16×16, 32×32, and 64×64 areas. Thus a total of 16 features were to be fed to a neural net with 16 input neurons. The reasons for the advantages of the multiple area partitions are explained with the help of Fig. 6.71. In Fig. 6.71a the defect

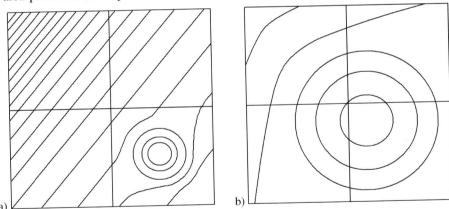

Figure 6.71: Area partitioning and defect size: (a) small defect and large areas, (b) large defect and small areas.

defines the maximum fringe density of its area, but neighboring areas, especially the upper left, also contain high fringe densities, leading to a small Laplace value. Contrary to this, the defect of Fig. 6.71b defines the maximum slopes of several neighboring areas, so the Laplace values representing the difference of slopes between adjacent areas remain small.

The neural network structure that showed optimal performance in this special application is shown in Fig. 6.72. Training of this network was performed by the backpropagation method for supervised learning. This technique, basically a gradient-descent method, propagates the

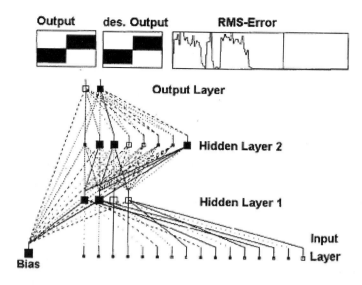

Figure 6.72: Neural network structure.

input through the layers, determining the output according to the actual weights. The output is compared with the output expected for this specific input, which quantifies an error. This error then is propagated back through the network from the output layer to the input layer while modifying the weights with the objective of minimizing the global error. This process is repeated with all input samples of the training set, which are taken each one several times in a stochastic order.

The evaluation of actually measured interference patterns, after the training phase was successfully finished, is shown in Fig. 6.73.

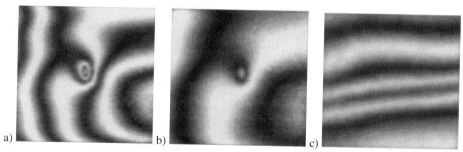

Figure 6.73: Evaluation by neural network: (a–c) interference patterns.

Three interference patterns were generated by digital holography. In a preprocessing step the speckle noise is reduced by low-pass filtering. The three patterns are given in Figs. 6.73a–c, the maximal slopes in gray scale display for 8×8 pixel areas in Fig. 6.73d-f, for 16×16 pixel areas in Figs. 6.73g–i, for 32×32 pixel areas in Figs. 6.73j–l, and for 64×64 pixel

6.11 Defect Detection by Holographic Non-Destructive Testing

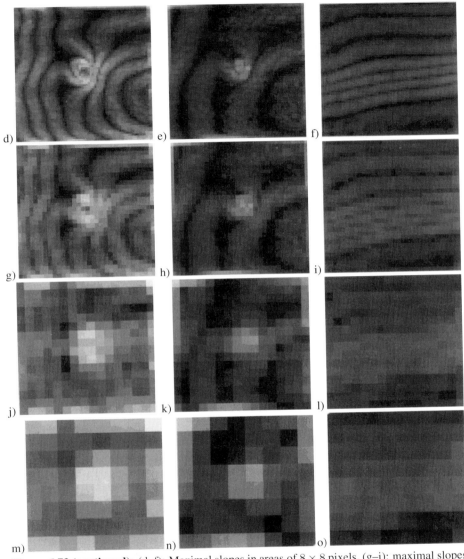

Figure 6.73 (continued): (d–f): Maximal slopes in areas of 8×8 pixels, (g–i): maximal slopes in areas of 16×16 pixels, (j–l): maximal slopes in areas of 32×32 pixels, (m–o): maximal slopes in areas of 64×64 pixels

areas in Figs. 6.73m–o. The corresponding Laplace values in gray scale coding are shown in Figs. 6.73p–A.

The neural network detected a defect in the patterns of Figs. 6.73a and b, and no defect in the pattern of Fig. 6.73c. The same result would have been achieved by a skilled human investigator.

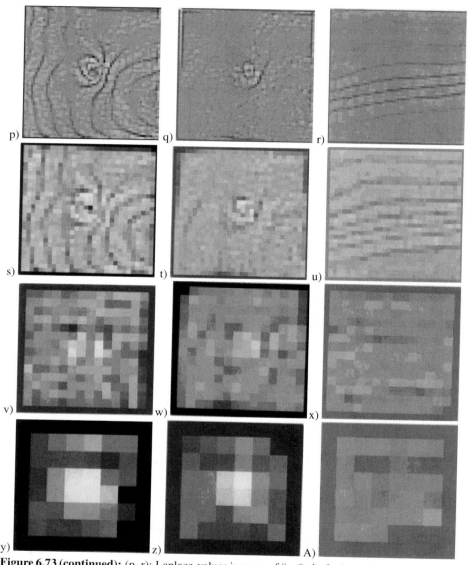

Figure 6.73 (continued): (p–r): Laplace-values in areas of 8×8 pixels, (s–u): Laplace-values in areas of 16×16 pixels, (v–x): Laplace-values in areas of 32×32 pixels, (y–A): Laplace-values in areas of 64×64 pixels

The artificial neural network so far has been applied to find characteristic partial patterns in the holographic interferograms. Another application of an artificial neural network is described in [755, 756] where it processes three-dimensional images obtained by digital holography, aiming at object recognition even for different orientations of the object.

7 Speckle Metrology

In the preceeding chapters the speckles which always appear when coherent light is diffusely scattered or reflected were treated as a disturbance to be suppressed or eliminated. On the other hand in coherent optical metrology the speckles can be viewed as the fundamental carriers of information, and thus can be used for specific measurement techniques. A number of methods of speckle metrology are closely related to methods of holographic interferometry. Therefore holographic interferometry and speckle metrology are often presented together in a closed form. A typical example for the close relation between holographic and speckle interferometry is the ESPI/DSPI method, which by several authors is regarded as image plane holography and therefore a holographic interferometric method. For these reasons in the following a brief introduction to the main techniques of speckle metrology is given.

Two principal approaches have to be distinguished in speckle metrology applied to e. g. deformation analysis of opaque diffusely reflecting surfaces: In speckle photography two reflected speckle fields are incoherently superposed to give information about an in-plane displacement; in speckle interferometry two interference fields are compared, each one generated by coherent superposition of the reflected wave field and a reference wave. The two fields to be compared correspond to the object states before and after the deformation.

The nature of the speckles and their statistics, which are of general interest also for holographic interferometry are described in detail in Section 2.5.

7.1 Speckle Photography

In *speckle photography* an opaque diffusely reflecting surface is illuminated by coherent light. The resulting speckle pattern is imaged by the lens of a photo-camera onto photographic film. The exposure results in a pointwise blackening of the film. When the surface point motion has a component in a direction normal to the optical axis, the speckle pattern follows this displacement component. A developed double exposure negative with the two exposures before and after the deformation will consist of a manifold of speckle pairs. The distance between the points of each pair is proportional to the displacement component normal to the optical axis of the corresponding object point, the direction of this lateral displacement component is the same as the direction of the shift of the related speckles.

For reconstruction the double exposure negative, often called *specklegram*, is illuminated in a pointwise manner by an unexpanded laser beam, Fig. 7.1. The point pairs in the small region where the beam passes the specklegram act like the two apertures in *Young's double aperture interferometer*, see Section 2.3.2. We get parallel equispaced fringes with a spacing of $\lambda L/d$ at a screen, which is placed at a distance L from the specklegram. d is the distance

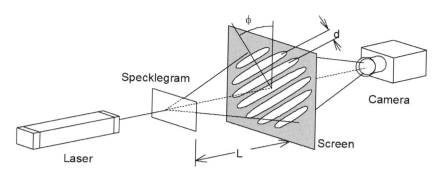

Figure 7.1: Reconstruction of Young's fringes in speckle photography.

in the pointpair in the specklegram, λ the wavelength used. The orientation of the fringes is orthogonal to the direction of the measured displacement component. By scanning the laser spot over the specklegram a two-dimensional field of displacement components can be measured. It should be stressed that from a single specklegram one can get only two pieces of information at each point: The modulus of the displacement projection onto a plane orthogonal to the optical axis, and the direction of this displacement component. The first is given by the distance of the Young's fringes, the second by the orientation of the fringes. In other words, we obtain the x- and y-components of the displacement vector in a coordinate system with the z-axis as the optical axis. There are a number of approaches to determine exactly the spacing and orientation of the fringes [757–759]. Good results have been obtained by locating the primary side lobes in the numerical 2D Fourier spectrum of the fringe pattern [760, 761].

Contrary to the pointwise evaluation and scanning there is a full field measurement method [164], where the specklegram is placed in an optical Fourier processor. A screen with a small hole is placed in the diffraction plane and passes only a small part of the spectrum, ideally a single spatial frequency. The resulting pattern in the image plane after such a filtering shows contours of equal displacement components in the direction given by the position of the filter. The filter position in the spatial frequency domain can now be varied by shifting the hole. It is obvious that in this way the measuring sensitivity can be varied, even after the speckle patterns have been originally recorded.

The main sensitivity of speckle photography is in the direction normal to the optical axis, that means for in-plane displacements of the surface points. The displacements generally have to be larger than the speckle size, which can be controlled by the aperture of the photo-camera.

7.2 Electronic and Digital Speckle Interferometry

The most important technique of speckle interferometry is the *digital speckle pattern interferometry (DSPI)*, originally called *electronic speckle pattern interferometry (ESPI)*, sometimes named *electronic holography* [762], *TV holography* [763–765], or *electro-optic holography* [516, 766, 767], or even *digital holography without wavefront reconstruction* [225] as well as *digital image plane holography* [768–771]. It was invented independently by several

7.2 Electronic and Digital Speckle Interferometry

groups [46–48]. The original aim was to overcome the time consuming wet-chemical processing of the silver halide holograms and to use electronic camera tubes instead. To adapt the micro-interference between object and reference wave to the resolution of the cameras, colinear reference and object waves have to be employed and an imaging system has to be used, Fig. 7.2. The object surface is focused onto the camera target, which in conjunction

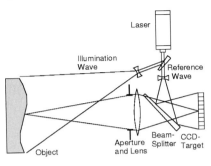

Figure 7.2: Arrangement for digital speckle interferometry.

with the colinear reference wave results in large speckles which now can be resolved by the camera but degrade the resulting interference pattern. This disadvantage is accepted due to the nearly real-time recording and reconstruction, and becomes less severe as CCD-targets get more and smaller pixels. Due to the focusing of the object's surface onto the camera target we record an image plane hologram, contrary to digital holography where only the Fresnel or Fraunhofer diffraction field of the object wavefront is registered, as defined in Chapter 3.

The object wave field in the image plane (x, y), the plane of the camera target, Fig. 7.2, can be described by

$$E^{(\text{ob})}(x, y) = E_0^{(\text{ob})}(x, y)\, e^{i\phi^{(\text{ob})}(x, y)} \tag{7.1}$$

where $E_0^{(\text{ob})}(x, y)$ is the real amplitude and $\phi^{(\text{ob})}(x, y)$ is the random phase due to the surface roughness, Fig. 7.3a. The colinear reference wave field

$$E^{(\text{ref})}(x, y) = E_0^{(\text{ref})}(x, y)\, e^{i\phi^{(\text{ref})}(x, y)} \tag{7.2}$$

is superposed. This reference wave may be a plane wave, a spherical wave, or an arbitrary reflected one, Fig. 7.3b. Only intensities are recorded by the TV target, Fig. 7.3d

$$\begin{aligned} I_A(x, y) &= |E^{(\text{ob})}(x, y) + E^{(\text{ref})}(x, y)|^2 \\ &= I^{(\text{ob})}(x, y) + I^{(\text{ref})}(x, y) + 2\sqrt{I^{(\text{ob})}(x, y)\, I^{(\text{ref})}(x, y)} \cos\psi(x, y). \end{aligned} \tag{7.3}$$

This is the recorded, digitized, and stored speckle pattern with the stochastic phase difference $\psi(x, y) = \phi^{(\text{ob})}(x, y) - \phi^{(\text{ref})}(x, y)$, Fig. 7.3c. A deformation changes the phase $\phi^{(\text{ob})}(x, y)$ of each point by $\Delta\phi(x, y)$, Fig. 7.3e, so that the wave field after deformation is

$$E^{(\text{ob})\prime}(x, y) = E_0^{(\text{ob})}(x, y)\, e^{i[\phi^{(\text{ob})}(x, y) + \Delta\phi(x, y)]}. \tag{7.4}$$

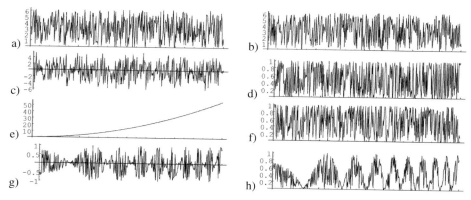

Figure 7.3: Digital speckle interferometry: (a) $\phi^{(\text{ob})}(x)$, (b) $\phi^{(\text{ref})}(x)$, (c) $\phi^{(\text{ob})}(x) - \phi^{(\text{ref})}(x)$, (d) $I_A(x)$, (e) $\Delta\phi(x)$, (f) $I_B(x)$, (g) $I_A(x) - I_B(x)$, (h) $|I_A(x) - I_B(x)|$.

Superposition with the colinear reference wave $E^{(\text{ref})}(x,y)$ leads to $I_B(x,y)$, Fig. 7.3f

$$I_B(x,y) = I^{(\text{ob})\prime}(x,y) + I^{(\text{ref})}(x,y) + 2\sqrt{I^{(\text{ob})\prime}(x,y)\,I^{(\text{ref})}(x,y)}\cos[\psi(x,y) + \Delta\phi(x,y)]. \quad (7.5)$$

In the digital image processing system this second speckle pattern $I_B(x,y)$ is subtracted in a pointwise manner in real time from the stored $I_A(x,y)$, where it is assumed that the deformation changes the phase but not the amplitude, meaning $I^{(\text{ob})\prime}(x,y) = I^{(\text{ob})}(x,y)$. The resulting difference is, Fig. 7.3g

$$\begin{aligned}(I_A - I_B)(x,y) &= 2\sqrt{I^{(\text{ob})}(x,y)\,I^{(\text{ref})}(x,y)}[\cos\psi - \cos\psi\cos\Delta\phi + \sin\psi\sin\Delta\phi](x,y) \\ &= 4\sqrt{I^{(\text{ob})}(x,y)\,I^{(\text{ref})}(x,y)}\sin\left[\psi(x,y) + \frac{\Delta\phi(x,y)}{2}\right]\sin\frac{\Delta\phi(x,y)}{2}.\end{aligned} \quad (7.6)$$

To display this result in real time on a monitor, positive intensities are obtained by taking the modulus $|I_A - I_B|$ or the square $(I_A - I_B)^2$, Fig. 7.3h. The square root in (7.6) describes the background illumination. The first sine-term gives the stochastic speckle noise which varies randomly from pixel to pixel. This noise is modulated by the sine of the half phase difference induced by the deformation. This low frequency modulation of the high frequency speckle noise is recognized as an interference pattern. The relation between the displacement vector $\boldsymbol{d}(x,y)$ and the phase difference $\Delta\phi(x,y)$ is as in holographic interferometry, see (4.20) and (4.21)

$$\Delta\phi(x,y) = \frac{2\pi}{\lambda}\boldsymbol{d}(x,y) \cdot [\boldsymbol{b}(x,y) - \boldsymbol{s}(x,y)]. \quad (7.7)$$

An alternative way for determining the phase difference from two speckle interferograms is the Fourier transform method first suggested in [457]. Here both intensities $I_A(x,y)$ and

7.2 Electronic and Digital Speckle Interferometry

$I_B(x,y)$ are Fourier transformed. The amplitude spectra in the first and minus first diffraction orders – the first two side-lobes – resemble the aperture, see Fig. 7.2. If the lateral placement of the aperture and the position of the virtual reference source point are properly chosen, the side-lobes do not overlap [306] and a bandpass filter as described in Section 5.6.1 can be applied. A virtual reference source point outside the aperture generates carrier fringes as described in Section 5.6.5. After one side-lobe is isolated by bandpass filtering an inverse Fourier transform is applied and gives complex $c_A(x,y)$ and $c_B(x,y)$. From these the interference phase $\Delta\phi(x,y)$ is calculated as defined in (5.71) by

$$\begin{aligned}\Delta\phi(n,m) &= \arctan\frac{\operatorname{Re}\{c_A(x,y)\}\operatorname{Im}\{c_B(x,y)\} - \operatorname{Re}\{c_B(x,y)\}\operatorname{Im}\{c_A(x,y)\}}{\operatorname{Re}\{c_A(x,y)\}\operatorname{Re}\{c_B(x,y)\} + \operatorname{Im}\{c_A(x,y)\}\operatorname{Im}\{c_B(x,y)\}} \\ &= \arctan\frac{\operatorname{Im}\{c_B(x,y)c_A^*(x,y)\}}{\operatorname{Re}\{c_B(x,y)c_A^*(x,y)\}}.\end{aligned} \quad (7.8)$$

This approach elucidates the close relationship to digital holography, therefore some authors use the term digital image plane holography [768–770].

Equation (7.7) shows that maximum sensitivity of DSPI/ESPI is for out-of-plane displacements. For measuring transversal displacements a modified setup using two illuminations from opposite directions having equal angles to the surface normal is recommended [582, 697, 772]. DSPI/ESPI patterns essentially contain the same information as the corresponding holographic interferograms. Thus their production requires the same precautions concerning vibration isolation and stability during the recording process. The results can be observed in real time, due to the electronic recording there is no problem with an exact repositioning of a hologram plate. The interference phase map quality can be improved by filtering in the spatial or spatial frequency domain [773].

While most applications of digital speckle interferometry are in deformation measurements of opaque surfaces these methods can also be successfully applied in fluid mechanics. Now the flowing transparent medium is seeded with tracer particles which constitute the reflecting object [267]. The particle field may be illuminated from any convenient direction; however, a useful approach is illumination by a thin light sheet as in particle image velocimetry (PIV).

The measurement of transient phenomena like impact studies, vibration analysis, or flow diagnostics is possible using double-pulsed illumination and recording. In this way mechanical amplitude and phase of surface acoustic waves have been measured [774] with interference phase determination by the Fourier transform method. While in this application the surface acoustic waves constitute a spatial carrier frequency by themselves, carrier fringes can be introduced by a tilt between object and reference wave. This allows the use of Fourier transform evaluation, e. g. in the double pulse measurement of brake squeal [775]. The separation of vibration modes using four pulses of a Q-switched ruby laser is described in [776, 777]. Here the method is called digital holography but since the "laser light scattered by the object is collected with an imaging lens that forms the image of the object on each of the three CCD camera faceplates" [776] there clearly an electronic speckle pattern interferometer with out-of-plane sensitivity is employed.

7.3 Electro-optic Holography

Electro-optic holography, also known as *electronic holography* or *TV holography*, is a combination of *phase stepping* and digital speckle interferometry [49, 50, 202]. For static measurements n phase stepped speckle patterns are recorded in the unstressed and n phase stepped speckle patterns in the stressed state. The recorded intensities are, see (7.3)

$$\begin{aligned}I_n(x,y) &= I^{(\text{ob})}(x,y) + I^{(\text{ref})}(x,y) + 2\sqrt{I^{(\text{ob})}(x,y)\,I^{(\text{ref})}(x,y)}\cos[\psi(x,y)+\phi_{Rn}] \\ I'_n(x,y) &= I^{(\text{ob})\prime}(x,y) + I^{(\text{ref})}(x,y) \\ &\quad + 2\sqrt{I^{(\text{ob})\prime}(x,y)\,I^{(\text{ref})}(x,y)}\cos[\psi(x,y)+\Delta\phi(x,y)+\phi_{Rn}].\end{aligned} \quad (7.9)$$

The notation is as in Section 7.2, ϕ_{Rn} are the phase shifts. While generally arbitrary phase shifts ϕ_{Rn} can be employed, see Section 5.5, the most used are $\phi_{R1}=0°$, $\phi_{R2}=90°$, $\phi_{R3}=180°$, and $\phi_{R4}=270°$. This results in

$$\begin{aligned}I_1(x,y) &= I^{(\text{ob})}(x,y) + I^{(\text{ref})}(x,y) + 2\sqrt{I^{(\text{ob})}\,I^{(\text{ref})}}\cos\psi(x,y) \\ I_2(x,y) &= I^{(\text{ob})}(x,y) + I^{(\text{ref})}(x,y) + 2\sqrt{I^{(\text{ob})}\,I^{(\text{ref})}}\sin\psi(x,y) \\ I_3(x,y) &= I^{(\text{ob})}(x,y) + I^{(\text{ref})}(x,y) - 2\sqrt{I^{(\text{ob})}\,I^{(\text{ref})}}\cos\psi(x,y) \\ I_4(x,y) &= I^{(\text{ob})}(x,y) + I^{(\text{ref})}(x,y) - 2\sqrt{I^{(\text{ob})}\,I^{(\text{ref})}}\sin\psi(x,y)\end{aligned} \quad (7.10)$$

and

$$\begin{aligned}I'_1(x,y) &= I^{(\text{ob})}(x,y) + I^{(\text{ref})}(x,y) + 2\sqrt{I^{(\text{ob})}\,I^{(\text{ref})}}\cos[\psi(x,y)+\Delta\phi(x,y)] \\ I'_2(x,y) &= I^{(\text{ob})}(x,y) + I^{(\text{ref})}(x,y) + 2\sqrt{I^{(\text{ob})}\,I^{(\text{ref})}}\sin[\psi(x,y)+\Delta\phi(x,y)] \\ I'_3(x,y) &= I^{(\text{ob})}(x,y) + I^{(\text{ref})}(x,y) - 2\sqrt{I^{(\text{ob})}\,I^{(\text{ref})}}\cos[\psi(x,y)+\Delta\phi(x,y)] \\ I'_4(x,y) &= I^{(\text{ob})}(x,y) + I^{(\text{ref})}(x,y) - 2\sqrt{I^{(\text{ob})}\,I^{(\text{ref})}}\sin[\psi(x,y)+\Delta\phi(x,y)]\end{aligned} \quad (7.11)$$

These systems of equations can be solved, see (5.24), yielding $\psi(x,y)$ and $\psi(x,y)+\Delta\phi(x,y)$, whose difference is the interference phase distribution $\Delta\phi(x,y)$. The advantage over conventional DSPI-patterns becomes obvious, when

$$\sqrt{[(I_1-I_3)+(I'_1-I'_3)]^2 + [(I_2-I_4)+(I'_2-I'_4)]^2} = 8\sqrt{I^{(\text{ob})}\,I^{(\text{ref})}}\cos(\Delta\phi/2)$$

is displayed. This interference pattern exhibits much less speckle noise than DSPI-patterns.

Besides static deformation measurement electro-optic holography also has been effectively applied to sinusoidally vibrating objects [376]. High quality time average interference patterns can be synthesized from phase stepped recordings. The argument of the Bessel function is determined with high accuracy if the phase of the object or reference wave is modulated at the same frequency and phase as the object vibration [376, 762].

In the described method the four phase shifted patterns were taken one after the other. To get several phase shifted patterns not sequentially in time but spatially separated, the in-line reference wave of digital speckle interferometry must be tilted, which is accomplished by

A Signal Processing Fundamentals

A.1 Overview

Fourier analysis is the most important and appropriate mathematical tool employed in the description of coherent optics and of digital holography. The reasons for this fact are manifold: The propagation of optical fields, e. g. from an object to the recording medium, here the hologram, or from an illuminated hologram to an image plane, which may be the target of an electronic recording device, the retina of an observer's eye, or a virtual plane whose field is calculated and displayed on a monitor, is described in the framework of linear systems theory, where Fourier analysis plays the central role. The optical wave fields used in this book are solutions of the Maxwell equations and as such are superpositions of harmonic functions. Fourier analysis is the adequate tool to deal with those harmonic functions.

Therefore the objective of this appendix is to review the most notable concepts of signal processing, as far as they find applications in optics and especially in digital holography. Of course this remains far from being exhaustive, so the interested reader is referred to a number of excellent textbooks on one- and two-dimensional signal processing, where the topic is introduced systematically, questions regarding the existence of the Fourier transform of classes of signals are discussed, many more theorems are presented, and where proofs, applications, and examples can be found.

The following treatment first introduces the concepts in one dimension, followed by an extension to the two dimensions typical for image processing. At a conceptual level, there is a great deal of similarity between one-dimensional signal processing and two-dimensional image processing. Although in this book we deal with images and two-dimensional patterns the concepts can be described and understood more clearly and easily in one dimension. After this preparation the extension to two dimensions is straightforward

and poses no additional difficulties. Nevertheless also differences already exist between one- and two-dimensional signal processing. For example, many one-dimensional systems are described by ordinary differential equations, while many two-dimensional systems involve partial differential equations, which generally are much more difficult to solve. Another fact is the absence of the fundamental theorem of algebra for two-dimensional polynomials. So two-dimensional polynomials cannot be factored generally into lower-order polynomials, with consequences on filter-design, issues related to system stability, etc. [396]. A basic problem in digital holography refers to sampling of optical two-dimensional signals. That topic is addressed in this appendix as well as the discussion of the chirp function, a function frequently used in digital holography. The following treatment furthermore distinguishes between analog signals, i. e. continuous time and space-signals, and discrete signals. While theory elegantly deals with continuous infinite functions, in metrologic practice we always have a finite number of sampled discrete data for processing. The differences between signals defined over infinite and finite domains are also identified.

A.2 Definition of the Fourier Transform

Let $f(x)$ be a one-dimensional complex-valued function. The real variable x may stand for a temporal coordinate, e. g. in acoustics or in electronics, or for a spatial coordinate, e. g. in image processing or optics. Then the *Fourier transform* of $f(x)$ is defined as

$$\mathcal{F}\{f(x)\} = F(\xi) = \int_{-\infty}^{\infty} f(x) e^{-i2\pi\xi x}\, dx. \tag{A.1}$$

The Fourier transform is a linear integral transformation that maps the complex-valued function $f(x)$ to another complex-valued function $F(\xi)$ of the variable ξ. This ξ stands for the *temporal frequency* or the *one-dimensional spatial frequency*.

The *inverse Fourier transform* of a complex-valued function $F(\xi)$ defined in the frequency domain is given by

$$\mathcal{F}^{-1}\{F(\xi)\} = f(x) = \frac{1}{2\pi} \int_{-\infty}^{\infty} F(\xi) e^{i2\pi\xi x} d\xi. \tag{A.2}$$

The *Fourier integral theorem* now states that

$$f(x) = \frac{1}{2\pi} \int_{-\infty}^{\infty} \left[\int_{-\infty}^{\infty} f(x) e^{-i2\pi\xi x}\, dx \right] e^{i2\pi\xi x} d\xi \tag{A.3}$$

or shortly $f(x) = \mathcal{F}^{-1}\{\mathcal{F}\{f(x)\}\}$, which means that the transformation is reciprocal

$$\mathcal{F}\{f(x)\} = F(\xi) \quad \Longrightarrow \quad \mathcal{F}^{-1}\{F(\xi)\} = f(x). \tag{A.4}$$

The two functions $f(x)$ and $F(\xi)$ satisfying (A.4) together are called a *Fourier transform pair*. For any $f(x)$, if the Fourier transform exists, $F(\xi)$ is unique and vice versa. From the reciprocity we see that the same amount of information is contained in $f(x)$ described in the spatial (or temporal) domain as is in the complex spectrum $F(\xi)$ represented in the spatial (or temporal) frequency domain.

Since the Fourier transform is an integral transformation, the question of the existence of the infinite limit integrals (A.1) and (A.2) arises. The infinite limits are not the problem, because the Cauchy principal value $\int_{-\infty}^{\infty} f(x)\, dx = \lim_{A\to\infty} \int_{-A}^{A} f(x)\, dx$ is taken, but on the other hand no direct and simple criterion is known, which is both sufficient and necessary in ensuring that a function has a valid Fourier transform. However in our applications the functions we process generally are digitized images. These are necessarily truncated to a finite spatial extent and furthermore are bounded in their values. Thus they belong to the

A.2 Definition of the Fourier Transform

class of *transient functions* which fall to zero rapidly enough for large positive and negative arguments so that the integrals in (A.1) and (A.2) exist. If the integral of the absolute value of the function $f(x)$ exists

$$\int_{-\infty}^{\infty} |f(x)| dx < \infty \tag{A.5}$$

and it is either continuous or has only a finite number of discontinuities and these discontinuities are not infinite and furthermore it has at most a finite number of extrema in any finite interval, then its Fourier transform $F(\xi)$ exists for all values of ξ [160, 783].

Periodic or *constant functions* do not belong to the class of these transient functions, nevertheless they may be transformable. For their treatment as well as for other purposes the *Dirac delta* or *impulse* $\delta(x)$ is useful. It can be defined as the limit of a sequence of *rectangular functions* $\delta_n(x)$ which are given by

$$\delta(x) = \lim_{n \to \infty} \delta_n(x) = \lim_{n \to \infty} n \operatorname{rect}(nx) \tag{A.6}$$

$$\operatorname{rect}(x) = \begin{cases} 1 & \text{for } |x| < \frac{1}{2} \\ 0 & \text{elsewhere.} \end{cases} \tag{A.7}$$

The first components of the sequence $\{\delta_n(x)\}$ are illustrated in Fig. A.1. The limit of this

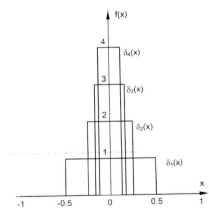

Figure A.1: Rectangular functions converging to the Dirac delta.

sequence of functions is of infinite height but zero width in such a manner that the area is still unity

$$\int_{-\infty}^{\infty} \delta(x) \, dx = 1. \tag{A.8}$$

The importance of the impulse function is based on the following *sampling property*:

$$\int_{-\infty}^{\infty} f(x) \delta(x - x') \, dx = f(x') \tag{A.9}$$

where $\delta(x - x')$ denotes an impulse shifted to location $x = x'$. It must be mentioned that from a mathematical standpoint the impulse is not a function in the classic sense. A rigorous treatment uses the calculus of distributions or generalized functions.

Based on the stated properties of the impulse, the Fourier transforms of some constant and periodic functions can be calculated. These transforms are given together with further frequently used Fourier transform pairs in Table A.1. The Fourier transforms of the Gaussian and of the infinite chirp function here are special cases of the correspondence [784]

$$\mathcal{F}\{e^{-sx^2}\} = \sqrt{\frac{\pi}{s}}\, e^{-\frac{\xi^2}{4s}} \qquad \text{for any complex } s \text{ with } \operatorname{Re}\{s\} \geq 0. \qquad (A.10)$$

In the first case we set $s = \pi$, and in the second $s = -i\pi$.

A.3 Interpretation of the Fourier Transform

Equation (A.2) indicates that the Fourier transform can be interpreted as a decomposition of the function $f(x)$ into a linear combination of elementary functions, here the complex exponentials of frequencies ξ. Due to the integral this combination consists of a continuum of terms. There are numerous applications where it is much easier to perform a specific processing of each of the simple complex exponentials and to superimpose the individual results to yield the overall result instead of processing the original function $f(x)$.

The frequency decomposition of the Fourier transform can be elucidated by the following argument. Suppose that $f(x)$ is expressed by a number of components of different frequencies

$$f(x) = F_0 + F_1 e^{i2\pi\xi_1 x} + F_2 e^{i2\pi\xi_2 x} + F_3 e^{i2\pi\xi_3 x} + \ldots \qquad (A.11)$$

Each of these terms rotates with a certain frequency ξ_i. In performing the Fourier transform (A.1) to $f(x)$, each term is multiplied with the factor $e^{-i2\pi\xi x}$. If in a product there is $\xi \neq \xi_i$, the product still rotates and its integral becomes zero. Only where $\xi = \xi_i$, which means that $e^{-i2\pi\xi x}$ rotates with the same frequency but in the other direction, the multiplication stops rotation and the integration gives a value proportional to F_i. Thus the Fourier transform picks the components rotating with specific frequencies. The vanishing integrals are due to the infinite limits and the fact that the real and imaginary parts of the complex exponentials are cosine- and sine-functions $e^{i2\pi\xi x} = \cos 2\pi\xi x + i\sin 2\pi\xi x$, which is the well known *formula of Euler*, Fig. A.2.

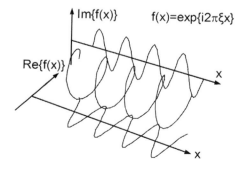

Figure A.2: Projections of complex exponential.

A.5 Linear Systems

can easily be determined, the output of a linear system to this input is just the weighted sum of the elementary outputs. This fact is called the *principle of superposition*.

The sampling property of the Dirac delta (A.9) expresses $f(x)$ as a linear combination of weighted and displaced delta functions. So here the delta functions act as the elementary functions of the decomposition. The response $g(x) = L\{f(x)\}$ of a linear system to the decomposed input $f(x)$ is

$$g(x) = L\{f(x)\} = L\left\{ \int_{-\infty}^{\infty} f(x')\delta(x-x')\,dx' \right\}$$

$$= \int_{-\infty}^{\infty} f(x') L\{\delta(x-x')\,dx'\}. \qquad (A.22)$$

Here we have taken the $f(x')$ as the weighting factors and extended the linearity property to the continuous case, say from the sum to the integral.

If the response of the system to a Dirac delta is expressed as

$$h(x,x') = L\{\delta(x-x')\} \qquad (A.23)$$

we can relate the input and output of the linear system by the so called *superposition integral*

$$g(x) = \int_{-\infty}^{\infty} f(x') h(x,x')\,dx'. \qquad (A.24)$$

Since the Dirac delta alone is used to describe an impulse, the function h is called the *impulse response*. In optics the equivalent to an impulse is a point, therefore the h describing the system that images this point is called a *point spread function*. The linear system now is completely characterized by the responses to the unit impulses for all locations of these impulses in the input plane.

A further important class of systems is that of the invariant systems. A *shift-invariant*, also called *space-invariant* or *isoplanatic system*, L fulfills

$$L\{f(x-x')\} = g(x-x') \qquad (A.25)$$

where $g(x) = L\{f(x)\}$ and x' can take any value. For a linear system described by the impulse response $h(x,x')$ the shift invariance gives

$$h(x,x') = h(x-x'). \qquad (A.26)$$

For different excitation points x' and response points x the impulse response only depends on the distance $x - x'$ between, but not on the specific location of these points on the x-axis. In this way the linear shift-invariant system is completely characterized by the function $h(x)$ depending on a single variable.

The superposition integral (A.24) now takes the simple form

$$g(x) = \int_{-\infty}^{\infty} f(x') h(x-x')\,dx'. \qquad (A.27)$$

The operation on two functions expressed on the right-hand side of (A.27) is called a *convolution* and is written shortly by

$$f(x) \star h(x) = \int_{-\infty}^{\infty} f(x')h(x-x')\,dx'. \tag{A.28}$$

The integrand in (A.27) is the product of $f(x')$ and $h(x')$ with the latter shifted by x giving $h(x'-x)$ and then rotated by 180° yielding $h(x-x')$. The calculation of the convolution of a triangular pulse $f(x) = \{x \text{ for } x \in [0.,1.],\ 2.-x \text{ for } x \in [1.,2.],\ 0.\text{ elsewhere}\}$ and a rectangular pulse $h(x) = \{1 \text{ for } |x| < 0.25,\ 0.\text{ elsewhere}\}$ is illustrated in Fig. A.3. Figures A.3a and b display the functions $f(x)$ and $h(x)$, and the shaded areas in Figs. A.3c

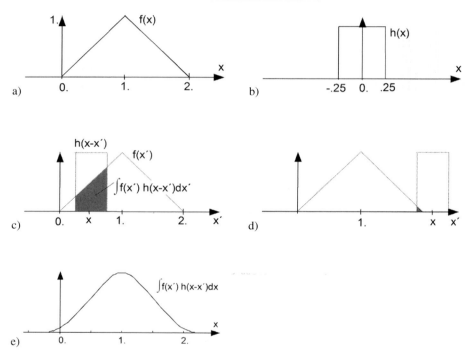

Figure A.3: Calculation of convolution.

and d indicate the values of the convolution for the specific x. Figure A.3d shows that even for $x > 2$, where $f(x) = 0$. and $h(x) = 0$. there is still a non-vanishing convolution. In fact $f(x) \star h(x)$ in this case has non-vanishing values in the range $-0.25 < x < 2.25$. The result shown in Fig. A.3e indicates that the rectangular pulse acts as a low-pass filter smearing out the sharp edges of the triangle. The 180° rotation is not explicitly shown because in this example $h(x)$ is an even function.

A.5 Linear Systems

The convolution operation has a number of useful properties such as *commutativity*

$$f(x) \star g(x) = g(x) \star f(x), \tag{A.29}$$

associativity

$$[f(x) \star g(x)] \star h(x) = f(x) \star [g(x) \star h(x)], \tag{A.30}$$

and *distributivity*

$$f(x) \star [g(x) + h(x)] = [f(x) \star g(x)] + [f(x) \star h(x)]. \tag{A.31}$$

Extremely useful is the *convolution theorem* which states that the Fourier transform of a convolution is the product of the individual Fourier transforms

$$\mathcal{F}\{f(x) \star g(x)\} = \mathcal{F}\{f(x)\} \cdot \mathcal{F}\{g(x)\}. \tag{A.32}$$

This theorem allows one to replace the often tedious operation of convolution by the much more simpler and more economic operations of Fourier transform and multiplication $f(x) \star g(x) = \mathcal{F}^{-1}\{\mathcal{F}\{f\} \cdot \mathcal{F}\{g\}\}$.

The linear shift-invariant system now can be characterized by its impulse response $h(x)$ with which the input function is convolved, or by the function $H(\xi) = \mathcal{F}\{h(x)\}$, the Fourier transform of the input is multiplied with to yield the Fourier transform of the result. The function $H(\xi)$ is called the system's *transfer function*. In other words, a harmonic input of complex amplitude $F(\xi)$ produces a harmonic output of the same spatial frequency with complex amplitude $G(\xi) = F(\xi) \cdot H(\xi)$. Either of the impulse response or the transfer function characterizes the linear shift-invariant system completely and enables us to determine the output corresponding to an arbitrary input.

A concept for analyzing mutual relationships or similarities between two deterministic as well as probabilistic functions is *correlation analysis*. Let $f(x)$ and $g(x)$ be two real- or complex-valued functions, then their *cross-correlation function* is defined as

$$\begin{aligned} \gamma_{fg}(x) &= \int_{-\infty}^{\infty} f^*(x' - x) g(x') \, dx' \\ &= \gamma_{gf}^*(-x). \end{aligned} \tag{A.33}$$

It is the average product of f and g for each shift x of the one relative to the other. Note that for real functions the cross correlation differs from the convolution in that before taking the product of the two mutually shifted functions no $180°$ rotation is performed.

The analog to the convolution theorem now reads

$$\mathcal{F}\{\gamma_{fg}(x)\} = \mathcal{F}\{f(x)\}^* \cdot \mathcal{F}\{g(x)\} \tag{A.34}$$

which allows the calculation of the correlation by $\gamma_{fg}(x) = \mathcal{F}^{-1}\{\mathcal{F}\{f\}^* \cdot \mathcal{F}\{g\}\}$. The * here denotes the complex conjugate.

If both function agree, we get the *autocorrelation function*

$$R(x) = \int_{-\infty}^{\infty} f^*(x' - x) f(x') \, dx'$$

$$= \int_{-\infty}^{\infty} f(x' + x) f^*(x') \, dx' \quad (A.35)$$

which indicates how much a function matches its own shifted replicas. The autocorrelation plays an important role in the investigation of periodic and of random functions. It is always even and has its maximum at $x = 0$. The *autocorrelation theorem* states that

$$\int_{-\infty}^{\infty} R(x) \, dx = \left[\int_{-\infty}^{\infty} f(x') \, dx' \right]^2. \quad (A.36)$$

Every function has a unique autocorrelation function but the converse is not true. From (A.34) directly follows the *Wiener-Khinchine theorem* which is expressed as

$$\mathcal{F}\{R(x)\} = \mathcal{F}\{f(x) \star f(-x)\} = |F(u)|^2. \quad (A.37)$$

For deterministic functions $f(x)$, $|F(u)|^2$ is the *power spectrum*. The Wiener-Khinchine theorem states that the power spectrum is the Fourier transform of the autocorrelation function.

A.6 Fourier Analysis of Sampled Functions

In technical applications, e. g. in measurements, we have no access to the continuous functions, but only to sampled values of a signal at discrete temporal or spatial instants. Furthermore we cannot measure from the infinite past to the infinite future, in the same way in the spatial domain we cannot record signals of infinite extent. Therefore we have to deal with a finite number of discrete signals. In the following the Fourier theory is extended to discrete sampled signals as well as to finite signals.

Let a continuous function $f(x)$ be defined for $X_1 \leq x \leq X_2$. This function can be expressed as

$$f(x) = \sum_{k=-\infty}^{\infty} a_k e^{i 2 \pi k f_0 x} \quad (A.38)$$

where $f_0 = 1/X$ with $X = X_2 - X_1$. Here $f(x)$ is represented by an infinite linear combination of sines and cosines, the so called *sinusoids*, which oscillate with $k f_0$ cycles per unit of x

$$e^{i 2 \pi k f_0 x} = \cos(2 \pi k f_0 x) + i \sin(2 \pi k f_0 x). \quad (A.39)$$

The frequencies of all sinusoids are integer multiples of the *fundamental frequency* f_0, the first four components of which are displayed in Fig. A.4. Each coefficient a_k is called the

A.6 Fourier Analysis of Sampled Functions

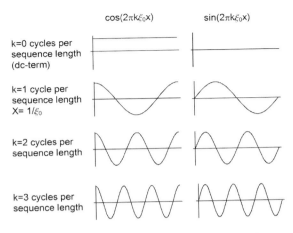

Figure A.4: First four components of a Fourier series.

complex amplitude of the k-th component, and these coefficients are obtained by

$$a_k = \frac{1}{X} \int_{X_1}^{X_2} f(x) e^{-i2\pi k \frac{x}{X}} \, dx. \tag{A.40}$$

The representation (A.38) with the a_k defined by (A.40) is called the *Fourier series*.

The difference between the continuous Fourier transform and the Fourier series is that we have all frequencies between $-\infty$ and ∞ in the sinusoids of the continuous Fourier transform but only integer multiples of the fundamental frequency f_0 for the Fourier series. The series representation is contained in the continuous Fourier transform representation especially if $f(x)$ is assumed to be zero outside $[X_1, X_2]$, which restricts the range of integration in (A.1). The additional information in $F(\xi)$ for $\xi \neq k f_0$ in the continuous reconstruction (A.2) is necessary to constrain the values of the reconstructed $f(x) = \mathcal{F}^{-1}\{F(\xi)\}$ outside the interval $[X_1, X_2]$. If we reconstruct $f(x)$ only from the a_k by (A.38) we will of course obtain the correct values of $f(x)$ within $[X_1, X_2]$, but we will get *periodic replications* of the original $f(x)$ outside the interval $[X_1, X_2]$. Contrarily, using $F(\xi)$ and (A.2) for reconstruction, we will obtain $f(x)$ exactly within $[X_1, X_2]$ and the correct zero everywhere outside.

As in the continuous case a discrete function may also be given a frequency domain representation. Let $\{f_n\}$ be the discrete samples of a continuous function $f(x)$ at all points $n\Delta x$, spaced by the sampling interval Δx

$$f_n = f(n\Delta x) \qquad n = \ldots, -1, 0, 1, \ldots \tag{A.41}$$

Then the *discrete Fourier transform* is defined by

$$F(\xi) = \sum_{n=-\infty}^{\infty} f_n e^{-i2\pi \xi n \Delta x}. \tag{A.42}$$

From the discrete Fourier transform the function $f(x)$ can be recovered by

$$f(n\Delta x) = \frac{\Delta x}{2\pi} \int_{-\pi/\Delta x}^{\pi/\Delta x} F(\xi) e^{i2\pi\xi n\Delta x} d\xi \qquad (A.43)$$

which gives the discrete function $f_n = f(n\Delta x)$ by integration over weighted sinusoids.

Although this discrete but still infinite Fourier transform is useful for many theoretical discussions, for practical purposes the following *finite Fourier transform* is actually calculated. Let the discrete function

$$f(0), f(\Delta x), f(2\Delta x), \ldots, f((N-1)\Delta x) \qquad (A.44)$$

be N elements long and be written as $\{f_0, f_1, f_2, \ldots, f_{N-1}\}$. Then the finite Fourier transform is defined as

$$F_k = \frac{1}{N} \sum_{n=0}^{N-1} f_n e^{-i2\pi\frac{kn}{N}} \qquad k = 0, 1, 2, \ldots, N-1. \qquad (A.45)$$

The F_k are samples of the continuous function $F(\xi)$ of (A.42) for $\xi = \frac{k}{N\Delta x}$ with $k = 0, 1, \ldots, N-1$, as can be proven by rewriting (A.45) with these ξ. This implies that if (A.45) is used to compute the frequency domain representation of a discrete function, a sampling interval of Δx in the x-domain corresponds to a sampling interval of $\Delta\xi = 1/(N\Delta x)$ in the frequency domain.

If $f(x)$ is a real function, or in the finite case f_n is a real sequence, then the d.c.-term F_0 is real. It is calculated by

$$F_0 = \frac{1}{N} \sum_{n=0}^{N-1} f_n. \qquad (A.46)$$

The value F_1 stands for the sinusoid of 1 cycle per sequence length. The value F_k represents k cycles per sequence length, as long as $k \leq N/2$. For higher values we must recognize the periodicity property $F_{-k} = F_{N-k}$, so that for $k > N/2$ the F_k represents $-(N-k)$ cycles per sequence length. But since an output where the negative axis information follows the positive axis information is somewhat unnatural to look at, the output should be rearranged. Normally looking finite Fourier transform outputs with the d.c.-component at the center, as for the well known optical diffraction patterns in the Fraunhofer region, can be produced by merely reordering the result or by multiplying the sequence to be transformed f_n with $(-1)^n$

$$f'_n = f_n(-1)^n \qquad (A.47)$$

prior to calculating the finite Fourier transform.

The *inverse finite Fourier transform* is

$$f_n = \sum_{k=0}^{N-1} F_k e^{i2\pi\frac{kn}{N}}. \qquad (A.48)$$

Both (A.45) and (A.48) define sequences that are periodically replicated due to the discrete nature. Since $e^{i(2\pi/N)Nm} = 1$ for all integers m we see that

$$F_{Nm+i} = F_i \quad \text{for all} \quad m \in \mathbb{Z}$$
$$\text{and} \quad f_{Nm+i} = f_i \quad \text{for all} \quad m \in \mathbb{Z}. \quad (A.49)$$

The four types of Fourier transforms are summarized in Table A.3 [785].

Table A.3: Types of Fourier transforms.

		Continuous Space/Time	Discrete Space/Time
Continuous Frequency	Name:	Continuous Fourier Transform	Discrete Fourier Transform
	Forward:	$F(\xi) = \int_{-\infty}^{\infty} f(x) e^{-i2\pi\xi x} dx$	$F(\xi) = \sum_{n=-\infty}^{\infty} f_n e^{-i2\pi\xi n \Delta x}$
	Inverse:	$f(x) = \frac{1}{2\pi} \int_{-\infty}^{\infty} F(\xi) e^{i2\pi\xi x} d\xi$	$f_n = \frac{\Delta x}{2\pi} \int_{-\pi/\Delta x}^{\pi/\Delta x} F(\xi) e^{i2\pi\xi n \Delta x} d\xi$
	Periodicity:	none	$F(\xi) = F(\xi + m(2\pi/\Delta x))$
Discrete Frequency	Name:	Fourier Series	Finite Fourier Transform
	Forward:	$F_k = \frac{1}{X} \int_{X_1}^{X_2} f(x) e^{-i2\pi k \frac{x}{X}} dx$	$F_k = \frac{1}{N} \sum_{n=0}^{N-1} f_n e^{-i2\pi \frac{kn}{N}}$
	Inverse:	$f(x) = \sum_{k=-\infty}^{\infty} F_k e^{i2\pi k \frac{x}{X}}$	$f_n = \sum_{k=0}^{N-1} F_k e^{i2\pi \frac{kn}{N}}$
	Periodicity:	$f(x) = f(x + mX)$	$f_i = f_{i+Nm}$ and $F_i = F_{i+Nm}$

A.7 The Sampling Theorem and Data Truncation Effects

In the definitions of the discrete infinite as well as finite Fourier transforms a sequence of numbers $\{f_n\}$ was used to approximate a continuous function $f(x)$. The question arises as to how finely must the data be sampled in order to accurately represent the original signal. The answer is given by the *sampling theorem*: If a signal $f(x)$ has a Fourier transform $F(\xi)$ such that

$$F(\xi) = 0 \quad \text{for} \quad \xi \geq \frac{\xi_N}{2} \quad (A.50)$$

then sampling of f at any rate greater than ξ_N is sufficient in order to ensure an exact reconstruction of $f(x)$ from the samples. That means if Δx is the interval between consecutive samples, it must be $2\pi/\Delta x > \xi_N$. Equation (A.50) implies that the sampling frequency ξ must be higher than twice the largest frequency contained in $F(\xi)$, which is estimated to ξ_N. The frequency ξ_N is known as the *Nyquist rate* or *Nyquist frequency*. Functions $f(x)$ for which an upper bound of frequencies as in (A.50) exists and which have finite energy, i. e. the energy E defined in (A.18) fulfills $E < \infty$, are called *band-limited functions*.

To consider the consequences of sampling at below the Nyquist rate, the sampled version $f_s(x)$ of a function $f(x)$ is expressed by a multiplication of the original continuous signal $f(x)$ with the sampling function $h(x)$ given by

$$h(x) = \sum_{n=-\infty}^{\infty} \delta(x - n\Delta x). \tag{A.51}$$

Since $h(x)$ is periodic, its Fourier transform is computed via the Fourier series to

$$H(\xi) = \frac{2\pi}{\Delta x} \sum_{n=-\infty}^{\infty} \delta\left(\xi - \frac{2\pi n}{\Delta x}\right). \tag{A.52}$$

Converting the multiplication into a convolution in the frequency domain we get the transform

$$F_s(\xi) = \frac{2\pi}{\Delta x} \sum_{n=-\infty}^{\infty} F\left(\xi - \frac{2\pi n}{\Delta x}\right) \tag{A.53}$$

which is a sum of shifted functions $F(\xi)$. The formula of the Fourier transform of a function f sampled at $n\Delta x$ is

$$F_s(\xi) = \frac{2\pi}{\Delta x} \sum_{n=-\infty}^{\infty} f(n\Delta x) e^{-i2\pi\xi n\Delta x}. \tag{A.54}$$

Figure A.5 shows the results of over- and undersampling. Figure A.5a gives the amplitude spectrum $|F(\xi)|$, Fig. A.5b displays the result after sampling at a rate faster than the Nyquist rate, i. e. $\Delta x < 2\pi/\xi_N$, meaning that the sampling theorem is fulfilled. Figure A.5c is the spectrum after sampling at exactly the Nyquist rate, and Fig. A.5d shows the overlap resulting from sampling at less than the Nyquist rate. An inverse Fourier transform of the spectrum in Fig. A.5d would produce an erroneous signal, the error is known generally as *aliasing* or as *moiré pattern* in two-dimensional image processing.

In many applications we have only a finite number of samples over a finite time or space while the signal extends beyond the limits of this interval. Nevertheless we assume that all the significant transitions of the signal occur in this base interval. Let the signal $f_n = f(x_n)$ be defined for all n. The true discrete Fourier transform of this signal is

$$F(\xi) = \sum_{n=-\infty}^{\infty} f_n e^{-i2\pi\xi n\Delta x}. \tag{A.55}$$

Suppose we take only an N-element transform meaning that of all the $x_n = n\Delta x$ we will retain only those going from $-(N/2 - 1)\Delta x$ to $(N/2)\Delta x$. It is assumed that N is even. The discrete Fourier transform of the *truncated data* is

$$\begin{aligned} F'(\xi) &= \sum_{n=-(N/2-1)}^{N/2} f_n e^{-i2\pi\xi n\Delta x} \\ &= \sum_{n=-\infty}^{\infty} f_n h_N(n) e^{-i2\pi\xi n\Delta x} \end{aligned} \tag{A.56}$$

A.9 Two-Dimensional Image Processing

For processing images like digital holograms or holographic interference patterns some of the aforementioned signal processing concepts are extended to two dimensions in the following. A *picture* or *image* is nothing more than a real-valued function $f(x,y)$ of two *spatial coordinates*. The values of this function can be interpreted as *gray-values*, so $f(x,y)$ gives the brightness distribution of, say, a black and white photograph.

Let L be an operation that maps an image f into another image $L[f]$. L is called *linear* if for all constants a, b and all images f, g

$$L[af + bg] = aL[f] + bL[g]. \tag{A.66}$$

For the analysis of two-dimensional linear operations we need the concept of a *point source*, the two-dimensional equivalent to the delta impulse. Over the *two-dimensional rectangular function*

$$\operatorname{rect}(x,y) = \begin{cases} 1 & : \text{ for } |x| \leq \tfrac{1}{2} \text{ and } |y| \leq \tfrac{1}{2} \\ 0 & : \text{ elsewhere} \end{cases} \tag{A.67}$$

and

$$\delta_n(x,y) = n^2 \operatorname{rect}(nx, ny) \qquad n = 1, 2, \ldots \tag{A.68}$$

the point source δ is defined by

$$\delta(x,y) = \lim_{n \to \infty} \delta_n(x,y) \tag{A.69}$$

which has the properties known from one dimension

$$\int_{-\infty}^{\infty} \int_{-\infty}^{\infty} \delta(x,y)\, dx\, dy = 1 \tag{A.70}$$

and

$$\int_{-\infty}^{\infty} \int_{-\infty}^{\infty} f(x,y) \delta(x-a, y-b)\, dx\, dy = f(a,b). \tag{A.71}$$

Although the same notation except of the two variables is used for the impulse and the point source, no confusion should arise.

A linear operation L is called *shift invariant* if

$$L[f(x-a, y-b)] = L[f](x-a, y-b); \tag{A.72}$$

in other words, if the input f is shifted by (a, b) then the output $L[f]$ is also merely shifted by (a, b). Using the representation (A.71) and the linearity of L we get

$$\begin{aligned} L[f(x,y)] &= L\left[\int_{-\infty}^{\infty}\int_{-\infty}^{\infty} f(x',y')\delta(x'-x, y'-y)\,dx\,dy\right] \\ &= \int_{-\infty}^{\infty}\int_{-\infty}^{\infty} f(x',y')L[\delta(x'-x, y'-y)]\,dx\,dy \\ &= \int_{-\infty}^{\infty}\int_{-\infty}^{\infty} f(x',y')h_L(x'-x, y'-y)\,dx\,dy \end{aligned} \qquad (A.73)$$

where the last equality uses the shift invariance of L, and h_L denotes the response of δ under L, the *point spread function*. The last expression in (A.73) defines the *two-dimensional convolution* $L[f] = f \star h_L = h_L \star f$. The *two-dimensional Fourier transform* $F(\xi, \eta)$ of the image $f(x, y)$ is defined by

$$F(\xi, \eta) = \int_{-\infty}^{\infty}\int_{-\infty}^{\infty} f(x,y)\mathrm{e}^{-\mathrm{i}2\pi(\xi x + \eta y)}\,dx\,dy. \qquad (A.74)$$

The two-dimensional Fourier transform can be considered as a one-dimensional transform with respect to, say, x, performed for all y, followed by the one-dimensional transform with respect to y, performed for all x

$$F(\xi, \eta) = \int_{-\infty}^{\infty}\left[\int_{-\infty}^{\infty} f(x,y)\mathrm{e}^{-\mathrm{i}2\pi\xi x}\,dx\right]\mathrm{e}^{-\mathrm{i}2\pi\eta y}\,dy. \qquad (A.75)$$

This *separability* is used in the *two-dimensional Fourier transform algorithm* where first all rows are replaced by their one-dimensional transforms and then all columns are transformed one-dimensionally or vice versa and the one-dimensional transforms are calculated by the effective FFT-procedure, see Section A.10. The discrete finite two-dimensional Fourier transform is

$$F(k, l) = \frac{1}{N^2}\sum_{m=0}^{N-1}\left[\sum_{n=0}^{N-1} f(n,m)W_N^{kn}\right]W_N^{lm} \qquad (A.76)$$

with $k = 0, 1, \ldots, N-1$, $l = 0, 1, \ldots, N-1$ numbering the sample points in the spatial frequency domain and $W_N = \exp\{-\mathrm{i}2\pi/N\}$.

The properties of the one-dimensional Fourier transform translate to two dimensions in a natural way, some additional properties due to the two dimensions arise. Some of these properties are summarized in Table A.4. Here $f(x, y)$, $f_1(x, y)$, $f_2(x, y)$ are functions in the spatial domain and $F(\xi, \eta)$, $F_1(\xi, \eta)$, $F_2(\xi, \eta)$ are the corresponding transformed functions in the spatial frequency domain. a and b are scalar numbers. (r, Θ) are polar coordinates in the

A.9 Two-Dimensional Image Processing

Table A.4: Properties of the two-dimensional Fourier transform.

Name	Function in the Spatial Domain	Transformed Function in the Spatial Frequency Domain
Linearity	$af_1(x,y) + bf_2(x,y)$	$aF_1(\xi,\eta) + bF_2(\xi,\eta)$
Scaling	$f(ax, by)$	$\frac{1}{|ab|} F\left(\frac{\xi}{a}, \frac{\eta}{b}\right)$
Shifting	$f(x - x_0, y - y_0)$	$e^{-i2\pi(\xi x_0 + \eta y_0)} F(\xi,\eta)$
	$e^{i2\pi(\xi_0 x + \eta_0 y)} f(x,y)$	$F(\xi - \xi_0, \eta - \eta_0)$
Differentiation	$\left(\frac{\partial}{\partial x}\right)^m \left(\frac{\partial}{\partial y}\right)^n f(x,y)$	$(i2\pi\xi)^m (i2\pi\eta)^n F(\xi,\eta)$
Laplacian	$\nabla^2 f(x,y) = \left(\frac{\partial^2}{\partial x^2} + \frac{\partial^2}{\partial y^2}\right) f(x,y)$	$-4\pi^2(\xi^2 + \eta^2) F(\xi,\eta)$
Rotation	$f(r, \Theta + \alpha)$	$F(p, \phi + \alpha)$
Rotational Symmetry	$f(r, \Theta) = f(r)$	$F(p, \phi) = F(p)$ $= 2\pi \int_0^\infty r f(r) \mathrm{J}_0(2\pi r p) dr$
Convolution	$f_1(x,y) \star f_2(x,y)$	$F_1(\xi,\eta) F_2(\xi,\eta)$
	$f_1(x,y) f_2(x,y)$	$F_1(\xi,\eta) \star F_2(\xi,\eta)$
Separability	$f(x,y) = f_1(x) f_2(y)$	$F(\xi,\eta) = F_1(\xi) F_2(\eta)$
Parseval's Theorem	$\int_{-\infty}^{\infty} \int_{-\infty}^{\infty} f_1(x,y) f_2^*(x,y)\, dx\, dy = \int_{-\infty}^{\infty} \int_{-\infty}^{\infty} F_1(\xi,\eta) F_2^*(\xi,\eta) d\xi d\eta$	
Conservation of Energy	$\int_{-\infty}^{\infty} \int_{-\infty}^{\infty} |f(x,y)|^2\, dx\, dy = \int_{-\infty}^{\infty} \int_{-\infty}^{\infty} |F(\xi,\eta)|^2 d\xi d\eta$	

spatial domain and (p, ϕ) polar coordinates in the spatial frequency domain. α is an angular coordinate, x_0, y_0, u_0, v_0 are fixed spatial coordinates and spatial frequencies, respectively. The transform $F(p)$ of the rotationally symmetric function $f(r)$ in line 7 of the table is called a *Hankel transform*. It contains the zero-order *Bessel function* of the first kind J_0.

If a two-dimensional Fourier transform of $f(x, y)$ is computed as described above (A.76), the zero peak at the spatial frequency $(0, 0)$ will not occur at the center of the array, as one is used to from the Fraunhofer diffraction patterns, but in the upper leftmost corner. A manipulation to force the frequency domain origin to approximately the center of the array (a precise center coordinate component does not exist if N is an even number) is the multiplication of the data with $(-1)^{m+n}$ before the transform

$$f'(m, n) = f(m, n)(-1)^{m+n}. \tag{A.77}$$

The *two-dimensional sampling theorem* states the following [789]: A function $f(x, y)$ whose Fourier transform $F(\xi, \eta)$ vanishes over all but a bounded region in the spatial frequency domain can be reproduced everywhere from its values taken over a lattice of points

$(m(\Delta x_1, \Delta y_1) + n(\Delta x_2, \Delta y_2))$, $m, n = 0, \pm 1, \pm 2, \ldots$ in the spatial domain provided the vectors $(\Delta x_1, \Delta y_1)$ and $(\Delta x_2, \Delta y_2)$ are small enough to ensure nonoverlapping of the spectrum $F(\xi, \eta)$ with its images on a periodic lattice of points $(l(\Delta \xi_1, \Delta \eta_1) + k(\Delta \xi_2, \Delta \eta_2))$, $k, l = 0, \pm 1, \pm 2, \ldots$ in the spatial frequency domain, where the vectors $(\Delta \xi_i, \Delta \eta_i)$, $i = 1, 2$, depend on the $(\Delta x_i, \Delta y_i)$, $i = 1, 2$, by

$$\Delta x_i \Delta \xi_j + \Delta y_i \Delta \eta_j = \begin{cases} 0 & : \quad i \neq j \\ 1 & : \quad i = j. \end{cases} \tag{A.78}$$

The definition of the sampling lattices over the two vectors reflects the fact that it is not necessary to sample in a rectangular grid. This is illustrated in Fig. A.8. Figure A.8a displays the sampling points in the spatial domain, Fig. A.8b depicts a boundary of the transform $F(\xi, \eta)$ of a band-limited two-dimensional function, and Fig. A.8c shows the nonoverlapping copies of $F(\xi, \eta)$.

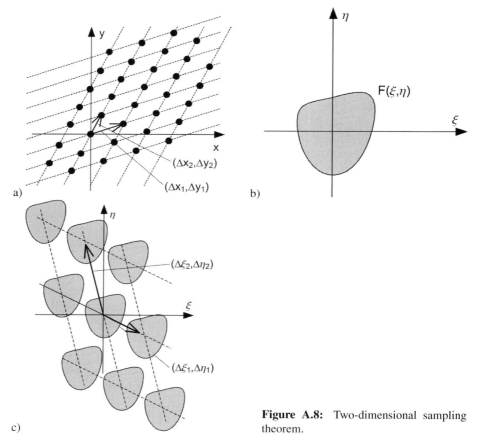

Figure A.8: Two-dimensional sampling theorem.

The effects of sampling without and with violation of the sampling theorem are shown in Figs. A.9 and A.10. The sinusoidal pattern of Fig. A.9a is sampled by discrete points at a rate

```
        J=I+L
        TEMPR=WR*FR(J)-WI*FI(J)
        TEMPI=WR*FI(J)+WI*FR(J)
        FR(J)=FR(I)-TEMPR
        FI(J)=FI(I)-TEMPI
        FR(I)=FR(I)+TEMPR
        FI(I)=FI(I)+TEMPI
4       CONTINUE
        L=ISTEP
        GOTO 3
        END
```

This should help for initial tests. More efficient programs should systematically avoid multiplications by zero and one, and should merge the multiplications by $\sqrt{2}$, or can be based on the Hartley transform. Detailed discussions of the FFT algorithms can be found in [790, 791].

A.11　Fast Fourier Transform for $N \neq 2^n$

Occasionally one needs to compute large DFTs of size $N \neq 2^n$ which can be done by combining several small DFTs of sizes $N_1, N_2, \ldots, N_l$, which are relative prime factors of N. Here we only consider the simple case of a DFT of size N, where N is the product of two mutually prime factors N_1 and N_2. The objective is to convert this one-dimensional DFT into a two-dimensional DFT of size $N_1 \times N_2$. In order to do this, one must convert the indices n and k in (A.80), defined modulo N, into two sets of indices n_1, k_1 and n_2, k_2, defined, respectively, modulo N_1 and modulo N_2. This can be done by the Chinese remainder theorem or by use of simple permutations [792]. Thus we arrive at the index transformations

$$
\begin{aligned}
n &\equiv N_1 n_2 + N_2 n_1 \text{ modulo } N \\
k &\equiv N_1 k_2 + N_2 k_1 \text{ modulo } N \\
&n_1, k_1 = 0, \ldots, N_1 - 1 \\
&n_2, k_2 = 0, \ldots, N_2 - 1.
\end{aligned}
\tag{A.81}
$$

This is valid only for $(N_1, N_2) = 1$. Now we substitute n and k in (A.80) and using the fact $N_1 N_2 \equiv 0$ modulo N we obtain

$$
F_{N_1 k_2 + N_2 k_1} = \frac{1}{N} \sum_{n_1=0}^{N_1-1} \sum_{n_2=0}^{N_2-1} f_{N_1 n_2 + N_2 n_1} W_1^{N_2 n_1 k_1} W_2^{N_1 n_2 k_2} \tag{A.82}
$$

with

$$
W_1 = e^{-i\frac{2\pi}{N_1}} \qquad W_2 = e^{-i\frac{2\pi}{N_2}}. \tag{A.83}
$$

In order to obtain a two-dimensional DFT in the conventional lexicographic order, it is convenient to replace k_1 and k_2 by their permuted values $t_2 k_1$ and $t_1 k_2$ such that $N_2 t_2 \equiv 1$ modulo N_1 and $N_1 t_1 \equiv 1$ modulo N_2. This is equivalent to replacing the mapping of k given by (A.82) with its Chinese remainder equivalent

$$
k \equiv N_1 t_1 k_2 + N_2 t_2 k_1 \text{ modulo } N. \tag{A.84}
$$

Now we have

$$F_{N_1t_1k_2+N_2t_2k_1} = \frac{1}{N}\sum_{n_1=0}^{N_1-1}\sum_{n_2=0}^{N_2-1} f_{N_1n_2+N_2n_1} W_1^{n_1k_1} W_2^{n_2k_2} \qquad (A.85)$$

which is the usual representation of a DFT of size $N_1 \times N_2$. Thus by using for n the permutation defined by (A.85), and for k the Chinese remainder correspondence defined by (A.84) (or vice versa), we are able to map a one-dimensional DFT of length N_1N_2 into a two-dimensional DFT of size $N_1 \times N_2$. The same method can be used recursively to define a one to many multidimensional mapping, as is shown explicitly in [792].

In the following a subroutine based on this approach is presented. It is for the case $N = 3072 = N_1N_2 = 1024 \cdot 3$, which is used e. g. for processing the images recorded with the CMOS-arrays of Canon having 3072×2048 pixels. This subroutine contains a 3-point transform in one direction and calls a radix-2 transform as given in the previous subroutine.

```
            SUBROUTINE FFT3072(FR,FI,IFLAG)
            DIMENSION FR(3072),FI(3072)
            DIMENSION GR1(1024),GI1(1024),GR2(1024),GI2(1024),GR3(1024),GI3(1024)
            VA=SQRT(2.)/2.
            S3=SQRT(3.)/2.
            DO 1 I=1,1024
            GR1(I)=FR(3*I-2)
            GI1(I)=FI(3*I-2)
            GR2(I)=FR(MOD(3*I+1021,3072)+1)
            GI2(I)=FI(MOD(3*I+1021,3072)+1)
            GR3(I)=FR(MOD(3*I+2045,3072)+1)
            GI3(I)=FI(MOD(3*I+2045,3072)+1)
1           CONTINUE
            CALL FFT(GR1,GI1,10,IFLAG)
            CALL FFT(GR2,GI2,10,IFLAG)
            CALL FFT(GR3,GI3,10,IFLAG)
            DO 2 I=1,1024
            A=GR1(I)+GR2(I)+GR3(I)
            B=GI1(I)+GI2(I)+GI3(I)
            IF(IFLAG.EQ.1) THEN
            C=GR1(I)-GR2(I)/2.+S3*GI2(I)-GR3(I)/2.-S3*GI3(I)
            D=GI1(I)-S3*GR2(I)-GI2(I)/2.+S3*GR3(I)-GI3(I)/2.
            E=GR1(I)-GR2(I)/2.-3*GI2(I)-GR3(I)/2.+S3*GI3(I)
            F=GI1(I)+S3*GR2(I)-GI2(I)/2.-S3*GR3(I)-GI3(I)/2.
            ELSE
            C=GR1(I)-GR2(I)/2.-S3*GI2(I)-GR3(I)/2.+S3*GI3(I)
            D=GI1(I)+S3*GR2(I)-GI2(I)/2.-S3*GR3(I)-GI3(I)/2.
            E=GR1(I)-GR2(I)/2.+3*GI2(I)-GR3(I)/2.-S3*GI3(I)
            F=GI1(I)-S3*GR2(I)-GI2(I)/2.+S3*GR3(I)-GI3(I)/2.
            ENDIF
            GR1(I)=A/S3
            GI1(I)=B/S3
            GR2(I)=C/S3
            GI2(I)=D/S3
            GR3(I)=E/S3
            GI3(I)=F/S3
2           CONTINUE
            DO 3 I=1,1024
```

```
    FR(MOD(2049*(I-1),3072)+1)=GR1(I)
    FR(MOD(2049*(I-1)+1024,3072)+1)=GR2(I)
    FR(MOD(2049*(I-1)+2048,3072)+1)=GR3(I)
    FI(MOD(2049*(I-1),3072)+1)=GI1(I)
    FI(MOD(2049*(I-1)+1024,3072)+1)=GI2(I)
    FI(MOD(2049*(I-1)+2048,3072)+1)=GI3(I)
3   CONTINUE
    RETURN
    END
```

A.12 Cosine and Hartley Transform

The Fourier transform considered so far has a complex kernel, the complex exponentials. There are transforms with real kernels, which draw advantage from some of the symmetries pointed out in Table A.2.

If $f(x)$ is even, meaning $f(x) = f(-x)$ for all (x,y), or $f(x,y) = f(-x,-y)$ in the two-dimensional case, then its Fourier transform is reduced to the *cosine transform*

$$\mathcal{C}\{f(x)\}(\xi) = \int_0^\infty f(x)\cos(2\pi\xi x)\, dx \tag{A.86}$$

or in two dimensions

$$\mathcal{C}\{f(x,y)\}(\xi,\eta) = \int_0^\infty\int_0^\infty f(x,y)\cos[2\pi(\xi x + \eta y)]\, dx\, dy. \tag{A.87}$$

If the two-dimensional signal also is axially symmetric

$$f(x,y) = f(-x,y) = f(x,-y) = f(-x,-y) \tag{A.88}$$

then its two-dimensional Fourier transform is reduced to the *separable cosine transform* [786]

$$\mathcal{C}\{f(x,y)\}(\xi,\eta) = \int_0^\infty\int_0^\infty f(x,y)\cos(2\pi\xi x)\cos(2\pi\eta y)]\, dx\, dy. \tag{A.89}$$

The inverse cosine transform is identical to the cosine transform.

Another transform with a real-valued kernel like the cosine transform is the *Hartley transform* that is defined in one dimension as

$$\begin{aligned}\mathcal{H}\{f(x)\}(\xi) &= \int_{-\infty}^\infty f(x)[\cos(2\pi\xi x) + \sin(2\pi\xi x)]\, dx \\ &= \frac{\sqrt{2}}{2}\int_{-\infty}^\infty f(x)\cos(2\pi\xi x - \frac{\pi}{4})\, dx.\end{aligned} \tag{A.90}$$

Like the cosine transform, the forward Hartley transform and its inverse are identical. Although the Hartley transform is defined for both real and complex functions $f(x)$, or sequences f_n in the discrete case, its practical value arises from the way in which it takes advantage of the symmetry in the Fourier transform of a real function or sequence [793]. The Fourier spectra and the Hartley spectra of real-valued functions $f(x)$ are related by [102]

$$\begin{aligned}\mathcal{H}\{f(x)\}(\xi) &= \frac{1}{2i}[(i+1)\mathcal{F}\{f(x)\}(\xi) + (i-1)\mathcal{F}\{f(x)\}(-\xi)] \\ &= \mathrm{Re}\mathcal{F}\{f(x)\}(\xi) + \mathrm{Im}\mathcal{F}\{f(x)\}(\xi)\end{aligned} \quad (\text{A.91})$$

and

$$\begin{aligned}\mathcal{F}\{f(x)\}(\xi) =\ & \frac{1}{2}[\mathcal{H}\{f(x)\}(\xi) + \mathcal{H}\{f(x)\}(-\xi)] \\ & - \frac{i}{2}[\mathcal{H}\{f(x)\}(\xi) - \mathcal{H}\{f(x)\}(-\xi)].\end{aligned} \quad (\text{A.92})$$

Thus the real part of the Fourier transform is the even part of the Hartley transform, while the imaginary part of the Fourier transform is the odd part of the Hartley transform.

As in the case of the cosine transform, two possible versions of the two-dimensional Hartley transform exist, namely the *separable Hartley transform* [786]

$$\mathcal{H}_s\{f(x,y)\}(\xi,\eta) = \int_{-\infty}^{\infty}\int_{-\infty}^{\infty} f(x,y)\cos\left(2\pi\xi x - \frac{\pi}{4}\right)\cos\left(2\pi\eta y - \frac{\pi}{4}\right)\,dx\,dy \quad (\text{A.93})$$

and the *inseparable Hartley transform*

$$\mathcal{H}_i\{f(x,y)\}(\xi,\eta) = \int_{-\infty}^{\infty}\int_{-\infty}^{\infty} f(x,y)\cos[2\pi(\xi x + \eta y) - \frac{\pi}{4}]\,dx\,dy. \quad (\text{A.94})$$

The two-dimensional Fourier-Hartley relation with $H_i(\xi,\eta)$ standing for $\mathcal{H}_i\{f(x,y)\}(\xi,\eta)$ and $F(\xi,\eta) = \mathcal{F}\{f(x,y)\}(\xi,\eta)$ is

$$F(\xi,\eta) = \frac{1}{2}[H_i(\xi,\eta) + H_i(-\xi,-\eta)] - \frac{i}{2}[H_i(\xi,\eta) - H_i(-\xi,-\eta)]. \quad (\text{A.95})$$

This relation, which is valid for the inseparable Hartley transform, forms a natural extension of the one-dimensional transform (A.92). But unfortunately due to the lack of separability there is no efficient algorithm for calculating the 2D transform by a cascaded processing of fast 1D transforms along lines and columns.

For the separable Hartley transform on the other hand the Fourier transform is expressed as

$$F(\xi,\eta) = \frac{1}{2}[H_s(\xi,-\eta) + H_s(-\xi,\eta)] - \frac{i}{2}[H_s(\xi,\eta) - H_s(-\xi,-\eta)]. \quad (\text{A.96})$$

Now we have gained separability but paid the price that the simple relation $F = \{$Even part of $H\} - i\{$Odd part of $H\}$, which is valid for the one-dimensional and the inseparable two-dimensional Hartley transform, no longer holds, instead (A.96) must be used [793].

A.13 The Chirp Function and the Fresnel Transform

This allows one to give a lower bound on the variance of $\tilde{f}$ that is independent of f [129]

$$\sigma_{\tilde{f}}^2 \geq \frac{\tau^2}{2\pi}. \tag{A.121}$$

It should be noted that signals can be represented as a linear combination of elementary functions of the form (A.118), which constitutes the so-called *Gabor transform*, another result [794] of the inventor of holography.

The translation of the Fresnel transform to two dimensions is straightforward. The 2D kernel is

$$k_\tau^{(2)}(x,y) = \frac{1}{\tau^2} e^{i\pi \left(\frac{\sqrt{x^2+y^2}}{\tau}\right)^2} \tag{A.122}$$

but since it is separable $k_\tau^{(2)}(x,y) = k_\tau(x)k_\tau(y)$, most mathematical analysis can be performed in one dimension.

The free-space propagation of light as given in (2.73) now can be expressed

$$\begin{aligned} E(\xi, \eta) &= \frac{e^{ikz}}{i\lambda z} \int_{-\infty}^{\infty}\int_{-\infty}^{\infty} U(x,y) e^{\frac{i\pi}{\lambda z}\left[(\xi-x)^2 + (\eta-y)^2\right]} dx\,dy \\ &= -i e^{ikz} \mathcal{FR}_{\sqrt{\lambda z}}^{(2)}\{U(x,y)\}(\xi,\eta). \end{aligned} \tag{A.123}$$

B Computer Aided Tomography

In holographic interferometry of refractive index fields the interference phase is given by the integral over the refractive index distribution along the illuminating ray passing the measurement volume, (4.22). Therefore, except for very simple problems, we have to reconstruct the distribution of the refractive index from a sufficient number of projections, a problem first addressed by Radon [795]. Since the extensive use of this approach in medical diagnosis using X-rays, this so called field of *computer aided tomography* has emerged rapidly: signals from a diversity of sources from the whole electromagnetic spectrum are used [796], and the theoretical background of the computer evaluation algorithms has been refined.

Here only a very basic introduction to the main approaches to computer aided tomography is given to constitute a background for the evaluation techniques used in holographic interferometric measurements at transparent refracting objects, see Section 6.10.

B.1 Mathematical Preliminaries

The measurement of the three-dimensional distribution of a physical quantity like the X-ray attenuation in human tissue or the *refractive index* in a transparent medium is simplified by the treatment of the measurement volume cut into two-dimensional plane slices. After an evaluation many of such slices are stacked to build the three-dimensional result. So here we only have to consider a two-dimensional distribution of a physical quantity $f(x,y)$ in a single plane. Without loss of generality we can assume that $f(x,y)$ is spatially bounded. Then by proper scaling and shifting we can ensure $f(x,y) = 0$ outside the unit circle Ω in the Cartesian coordinate system, Fig. B.1.

For the description of line integrals and projections the use of *polar coordinates* (t, ϕ) is advantageous, Fig. B.1:

$$f(x,y) = f(t\cos\phi, t\sin\phi). \tag{B.1}$$

A line in the plane now can be described by the two parameters s and θ: s is the signed distance of the line to the origin of the coordinate system and θ is the angle between the line and the y-axis, Fig. B.1. The projections of $f(x,y)$ along lines are called the *Radon transforms* $r(s, \theta)$, defined as

$$r(s,\theta) = \int_{-T}^{T} f(s\cos\theta - t\sin\theta, s\sin\theta + t\cos\theta)\, dt \tag{B.2}$$

Handbook of Holographic Interferometry: Optical and Digital Methods. Thomas Kreis
Copyright © 2005 Wiley-VCH Verlag GmbH & Co. KGaA, Weinheim
ISBN: 3-527-40546-1

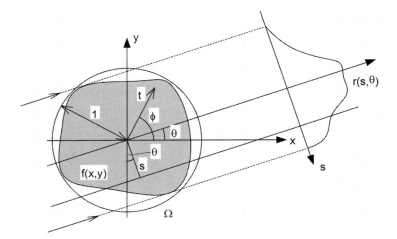

Figure B.1: Projection through a refractive index field.

or employing the delta function

$$r(s,\theta) = \int_{-\infty}^{\infty}\int_{-\infty}^{\infty} f(x,y)\delta(x\,\cos\theta + y\,\sin\theta - s)\,dx\,dy. \tag{B.3}$$

The T of (B.2) can be set to $T(s) = (1-s^2)^{1/2}$ since by assumption $f(s,\theta) = 0$ for $|s| > 1$, which is outside the unit-circle.

(s,θ) and $(-s, \theta + \pi)$ represent the same line in the plane, therefore this is true for the projections

$$r(s,\theta) = r(-s, \theta + \pi). \tag{B.4}$$

In the following some reconstruction methods will be introduced. This outline will be restricted to parallel projections, all rays establishing the projections are parallel. The case of fan beam projection is discussed in the literature [785].

What is measured in practical experiments are the estimated projections $r(s,\theta)$ for discrete values of s and θ. From these projections the distribution $f(x,y)$ has to be reconstructed, which should be a two-dimensional array of numbers, each representing the physical quantity to be measured in an elementary cell.

Algorithms become more easy, if r is uniformly sampled in s and θ. Therefore we assume a set of projections in N angular directions separated by $\Delta\theta$, each of these consisting of M equidistant beams separated by Δs, Fig. B.2.

B.2 The Generalized Projection Theorem

Let $f(x,y)$ be a function as in Section B.1 and $r(s,\theta)$ its Radon transform. Let further $w(s)$ be any function of a single variable such that the following integrals exist. Then for all angles

B.4 Practical Implementation of Filtered Backprojection

In practice the energy contained in the spectra above a certain frequency is negligible, which means for practical purposes the projection may be considered as band-limited. If $\rho_{\max}$ is higher than the highest significant frequency component in each projection, the projections can be sampled error-free at intervals of $\Delta t = 1/(2\rho_{\max})$. We assume further that the projection data are zero for large $|t|$, then the M samples of a projection are

$$r(m\Delta t, \Theta) \quad \text{with} \quad m = -M/2, \ldots, +M/2 - 1. \tag{B.13}$$

Now the FFT algorithm can be used to calculate the Fourier transforms $R(\rho, \Theta)$ of the projections

$$R(\rho, \Theta) \approx R\left(m\frac{2\rho_{\max}}{M}, \Theta\right) = \frac{1}{2\rho_{\max}} \sum_{k=-M/2}^{M/2-1} r\left(\frac{k}{2\rho_{\max}}, \Theta\right) e^{-i2\pi \frac{mk}{M}}. \tag{B.14}$$

This projection now has to be filtered according to (B.11)

$$q_\Theta(t) \approx \frac{2\rho_{\max}}{M} \sum_{m=-M/2}^{M/2-1} R\left(m\frac{2\rho_{\max}}{M}, \Theta\right) \left|m\frac{2\rho_{\max}}{M}\right| e^{i2\pi(2\rho_{\max}/M)t}. \tag{B.15}$$

The multiplication in the frequency domain before taking the inverse transform is equivalent to a convolution in the spatial, the t-domain

$$q_\Theta(t) = \int r(s, \theta) h(t - s)\, ds \tag{B.16}$$

where $h(t)$ is the inverse Fourier transform of the function $|\rho|$ multiplied with a window function in the frequency domain. A simple window function $B(\rho)$ only reflects the band-limitness, so that the convolution kernel is the inverse transform of

$$H(\rho) = |\rho| B(\rho) \tag{B.17}$$

with $B(\rho) = 1$ for $|\rho| < \rho_{\max}$ and zero otherwise. The impulse response of $H(\rho)$ is

$$\begin{aligned} h(t) &= \int_{-\infty}^{\infty} H(\rho) e^{i2\pi\rho t}\, d\rho \\ &= \frac{1}{2\Delta t} \frac{\sin 2\pi t/(2\Delta t)}{2\pi t/(2\Delta t)} - \frac{1}{4(\Delta t)^2} \frac{\sin \pi t/(2\Delta t)}{\pi t/(2\Delta t)}. \end{aligned} \tag{B.18}$$

The projection data are measured with the sampling interval Δt, consequently we only need $h(t)$ at the sampled points. These values are

$$h(m\Delta t) = \begin{cases} \frac{1}{4(\Delta t)^2} & m = 0 \\ 0 & m \text{ even} \\ -\frac{1}{m^2 \pi^2 (\Delta t)^2} & m \text{ odd}. \end{cases} \tag{B.19}$$

This is the well known *Ramachandran-Lakshminarayanan kernel*. Another frequently used kernel is the *Shepp-Logan kernel* [797]

$$h(m\Delta t) = \frac{-2}{\pi^2(\Delta t)^2(4m^2-1)}. \qquad (B.20)$$

Unfortunately the values of $x \cos\theta_i + y \sin\theta_i$ do not always correspond to the sampled points $m\Delta t$. Therefore the filtered projection $q_\Theta(t)$ must be interpolated between the sampling points. Linear interpolation in most applications is adequate.

Altogether a practical implementation of the filtered backprojection method consists of the following steps:

- Choice of parameters: These are the number M of sampling points in each projection, the number N of angular projections, the sampling interval Δt, the angular separation $\Delta\theta = \pi/N$, the cutoff frequency $\rho_{\max} = 1/(2\Delta t)$, the number of pixels $K \times L$ in the reconstructed image, the pixel distances Δx and Δy in the reconstructed image, the interpolation procedure (e. g. piecewise linear), the type of convolution kernel (e. g. Shepp-Logan).

- Projection data input: These are N vectors each having M components.

- Convolution of each projection with the kernel: This is performed in the spatial domain or by multiplication in the frequency domain.

- Interpolation of the filtered projections.

- Backprojection.

- Display of results.

A very simple example is shown in Fig. B.6. The function $f(x,y)$ to be reconstructed from four projections in the directions $\theta = 0°, 45°, 90°,$ and $135°$ is, Fig. B.6a

$$f(x,y) = \begin{cases} 1 & 0 \le x \le 0.25 \text{ and } 0 \le y \le 0.25 \\ 0.5 & 0.25 \le x \le 0.5 \text{ and } 0 \le y \le 0.5 \\ 0.5 & 0.25 \le y \le 0.5 \text{ and } 0 \le x \le 0.25 \\ 0 & \text{elsewhere.} \end{cases} \qquad (B.21)$$

The four projections are shown in Figs. B.6c to f, the Shepp-Logan-kernel to be applied to the projections in Fig. B.6b. The filtered and interpolated projections are given in Figs. B.6g to j. The resulting $f(x,y)$ after backprojection is displayed in pseudo 3D in Fig. B.6k and in gray-values in Fig. B.6l. It must be admitted that this example is far from practical because four projections are never sufficient.

B.5 Algebraic Reconstruction Techniques

The *algebraic reconstruction techniques*, also known as *series expansion reconstruction* methods, principally differ from the transformation based methods as the filtered backprojection

B.5 Algebraic Reconstruction Techniques

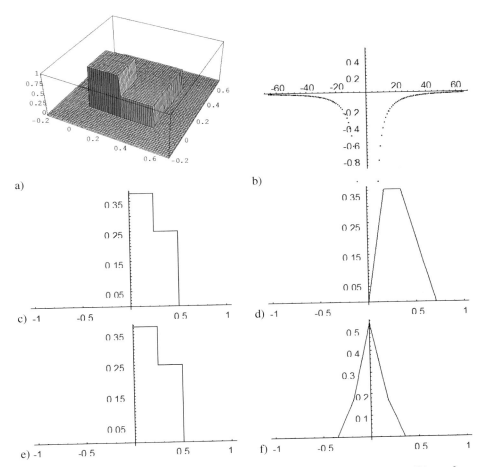

Figure B.6: Numerical example for reconstruction by filtered backprojection: (a) 2D test function $f(x,y)$, (b) Shepp-Logan convolution kernel, (c – f) projections of $f(x,y)$, (c) $\theta = 0°$, (d) $\theta = 45°$, (e) $\theta = 90°$, (f) $\theta = 135°$.

reconstruction. While in the transform methods the problem is treated as a continuous one until it is discretized for computational implementation, the algebraic reconstructions are discretized from the beginning [798–801]. The following short outline again only deals in two dimensions.

The interesting area is partitioned into a Cartesian grid of square cells, the pixels, numbered consecutively from 1 to N, Fig. B.7.

It is assumed that the function $f(x,y)$ to be reconstructed is constant in each cell, f_j being its value in cell j. Let M rays probe the area, one of these rays being indicated in Fig. B.7. The rays now have a finite thickness. The different portions each ray i intercepts with each

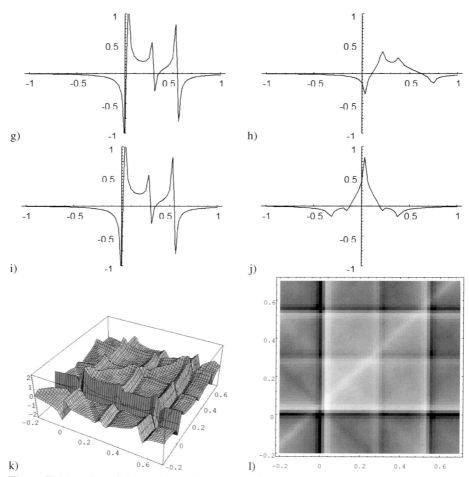

Figure B.6 (continued): (g – j) filtered and interpolated projections, (k, l) reconstructed function, (k) pseudo 3D display, (l) gray-scale display.

cell j are quantified by w_{ij}. After running through the grid the integrated signal p_i of ray i is the sum of the values f_j of all subtended pixels weighted with the area of the cell w_{ij} the ray covers:

$$p_i = \sum_{j=1}^{N} w_{ij} f_j \qquad i = 1, \ldots, M. \tag{B.22}$$

This system of linear equations can be written in matrix form $\boldsymbol{p} = \boldsymbol{W} \cdot \boldsymbol{f}$. Since for each ray only the intercepted pixels yield $w_{ij} \neq 0$, $\boldsymbol{W}$ is a sparsely occupied matrix, but its size frequently is of the order $10^6 \times 10^6$.

A solution by direct matrix inversion is not feasible because of the size of the problem, sometimes the system is underdetermined with less projections than cells, often the system

B.5 Algebraic Reconstruction Techniques

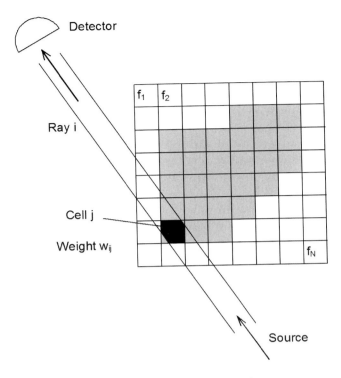

Figure B.7: Geometry for algebraic reconstruction.

is overdetermined and inconsistent, and generally all projections are measured with limited accuracy. Instead one has to look for appropriate iterative solutions.

The primary algebraic reconstruction technique is based on the Kaczmarz algorithm for solving a system of linear equations [798]: Beginning with an arbitrary initial guess, the solutions are iteratively refined by taking into account the measurements along one ray during one iteration step. Convergence of the iteration is forced by introduction of relaxation parameters.

Another approach replaces the equalities (B.22) by inequalities

$$p_i - \varepsilon_i \leq \sum_{j=1}^{N} w_{ij} f_j \leq p_i - \varepsilon_i \qquad i = 1, \ldots, M \qquad (B.23)$$

where the tolerances ε_i reflect the limited measurement accuracy. The iteration looks for a vector in the intersection of all hyperslabs defined by the inequalities (B.23).

Other methods use the concepts of entropy optimization, quadratic optimization, least squares regularization or statistical techniques [785, 798].

This short outline should be finished by the observation that still today most commercial equipment works with transform methods. But due to the versatility and flexibility of the algebraic reconstruction techniques and the enormous increase in computer speed these methods have good prospects for the future.

C Bessel Functions

Bessel functions arise in solving differential equations for systems with cylindrical symmetry. The *Bessel functions* $J_n(z)$ and $Y_n(z)$ are linearly independent solutions to the differential equation

$$z^2 \frac{d^2 y}{dz^2} + z \frac{dy}{dz} + (z^2 - n^2) y = 0. \tag{C.1}$$

$J_n(z)$ is called the *Bessel function of the first kind*, $Y_n(z)$ is referred to as the *Bessel function of the second kind*. For integer n, the $J_n(z)$ are regular at $z = 0$, the $Y_n(z)$ have a logarithmic divergence at $z = 0$.

Alternatively the Bessel function of the first kind can be defined over the integral

$$J_n(z) = \frac{1}{2\pi} \int_0^{2\pi} \cos(z \sin t - nt)\, dt \qquad z \in C, \qquad n = 0, 1, 2, \ldots \tag{C.2}$$

or as the power series

$$\begin{aligned} J_n(z) &= \frac{z^n}{2^n 0! n!} - \frac{z^{n+2}}{2^{n+2} 1!(n+1)!} + \frac{z^{n+4}}{2^{n+4} 2!(n+2)!} - \frac{z^{n+6}}{2^{n+6} 3!(n+3)!} + - \ldots \\ &= \sum_{i=0}^{\infty} \frac{(-1)^i z^{n+2i}}{2^{n+2i} i!(n+i)!} \qquad \text{for} \qquad |z| < \infty. \end{aligned} \tag{C.3}$$

Since the absolute values of the coefficients of this power series decrease very rapidly, this representation is useful for a practical calculation even for large $|z|$.

By proper combination of the power series components and using the Euler formula one obtains the useful formula

$$\sum_{n=-\infty}^{\infty} J_n(z) e^{in\phi} = e^{iz \sin \phi}. \tag{C.4}$$

We see in (C.3) that for real x $J_n(x)$ is also real and an even function for even n and an odd function if n is odd. Figure C.1 shows the first four real Bessel functions $J_0(x), \ldots, J_3(x)$. All real Bessel functions are bounded by the functions $\pm\sqrt{2/(\pi x)}$ which also are shown in Fig. C.1. We recognize the damped oscillation of all curves as well as the distribution of the zeroes becoming more and more regular with increasing x.

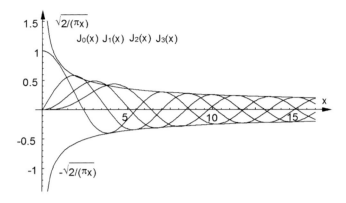

Figure C.1: Real Bessel functions of the first kind.

The most interesting Bessel function in holographic interferometry is the real Bessel function of the first kind and zero order $J_0(x)$ because it describes the intensity distribution resulting from the *time average method* for analyzing harmonically vibrating objects. The zeros b_m, $J_0(b_m) = 0$, $m = 1, 2, \ldots$, can be approximated for sufficiently large m by

$$b_m^* = \left(m - \frac{1}{4}\right)\pi. \tag{C.5}$$

Accordingly an approximating function to $J_0(x)$ for large x is

$$J_0^*(x) = \sqrt{\frac{2}{\pi x}} \cos\left(x - \frac{\pi}{4}\right). \tag{C.6}$$

For higher order n we have

$$J_n^*(x) = \sqrt{\frac{2}{\pi x}} \cos\left(x - \frac{2n+1}{4}\pi\right). \tag{C.7}$$

A table of the first 30 zeros of $J_0(x)$ together with the approximating b_m^* is given with a precision of five decimal places in Table C.1.

At this point it should be mentioned that the n-th order Bessel functions of the first kind also appear in the Zernike description of wavefronts in optical testing: A data spectrum $D(u,v)$ is described by a sum of Zernike terms

$$Z_{m,n}(r, \Theta) = R_{m,n}(r)e^{im\Theta}, \qquad m, n \in N, \quad n > 0, \quad m < n, \quad m+n \text{ even} \tag{C.8}$$

where $R_{m,n}(r)$ are *Zernike polynomials*. The Zernike terms form a complete orthogonal set of functions over the unit circle. The Fourier transforms of $Z_{m,n}(r, \Theta)$ are

$$\mathcal{F}\{Z_{m,n}(r, \Theta)\} = \mathcal{Z}_{m,n}(\rho, \theta) = (-1)^n A_n(\rho)e^{im\theta} \tag{C.9}$$

Table C.1: Zeroes of the real Bessel function of the first kind and zero order.

m	b_m	b_m^*	m	b_m	b_m^*
1	2.40483	2.35619	16	49.48261	49.48008
2	5.52008	5.49779	17	52.62405	52.62168
3	8.65373	8.63938	18	55.76551	55.76327
4	11.79153	11.78097	19	58.90698	58.90486
5	14.93092	14.92257	20	62.04847	62.04645
6	18.07106	18.06416	21	65.18996	65.18805
7	21.21164	21.20575	22	68.33147	68.32964
8	24.35247	24.34734	23	71.47298	71.47123
9	27.49348	27.48894	24	74.61450	74.61283
10	30.63461	30.63053	25	77.75603	77.75442
11	33.77582	33.77212	26	80.89756	80.89601
12	36.91710	36.91371	27	84.03909	84.03760
13	40.05843	40.05531	28	87.18063	87.17920
14	43.19979	43.19690	29	90.32217	90.32078
15	46.34119	46.33849	30	93.46372	93.46238

where

$$A_n(\rho) = \frac{1}{\rho} J_{n+1}(2\pi\rho) \tag{C.10}$$

with (ρ, θ) being the spatial frequency coordinates in polar form [481].

Bibliography

[1] D. Gabor. A new microscopic principle. *Nature*, 161:777–778, 1948. {**3, 37, 169**}

[2] D. Gabor. Microscopy by reconstructed wavefronts. *Proc. Royal Society A*, 197:454–487, 1949. {**3, 37, 169**}

[3] D. Gabor. Microscopy by reconstructed wavefronts: II. *Proc. Royal Society B*, 64:449–469, 1951. {**3, 37**}

[4] G. L. Rogers. Experiments in diffraction microscopy. *Proc. Roy. Soc. Edinb.*, 63A:193–221, 1952. {**3**}

[5] H. M. A. El-Sum and P. Kirkpatrick. Microscopy by reconstructed wavefronts. *Phys. Rev.*, 85:763, 1952. {**3**}

[6] A. W. Lohmann. Optische Einseitenbandübertragung angewandt auf das Gabor-Mikroskop. *Optica Acta*, 3:97–99, 1956. {**3**}

[7] E. N. Leith and J. Upatnieks. Wavefront reconstruction with diffused illumination and three-dimensional objects. *Journ. Opt. Soc. Amer.*, 54:1295–1301, 1964. {**3, 42, 45, 169**}

[8] E. N. Leith and J. Upatnieks. Wavefront reconstruction with continuous-tone objects. *Journ. Opt. Soc. Amer.*, 53:1377–1381, 1963. {**3, 42, 45, 169**}

[9] E. N. Leith and J. Upatnieks. Wavefront reconstruction with diffused illumination and three-dimensional objects. *Journ. Opt. Soc. Amer.*, 54:1295–1301, 1964. {**3, 42, 45**}

[10] Y. N. Denisyuk. Photographic reconstruction of the optical properties of an object in its own scattered field. *Sov. Phys. Dokl.*, 7:543, 1962. {**3**}

[11] A. W. Lohmann and D. P. Paris. Binary Fraunhofer holograms generated by computer. *Appl. Opt.*, 6:1739–1748, 1967. {**4**}

[12] W. H. Lee. Sampled Fourier transform hologram generated by computer. *Appl. Opt.*, 9:639–643, 1970. {**4**}

[13] W. H. Lee. Binary synthetic holograms. *Appl. Opt.*, 13:1677–1682, 1974. {**4**}

[14] J. F. Heanue, M. C. Bashaw, and L. Hesselink. Volume holographic storage and retrieval of digital data. *Science*, 265:749–752, 1994. {**4**}

[15] http://www.aprilisinc.com/. Homepage of Aprilis, Inc. {**4**}

[16] R. L. Powell and K. A. Stetson. Interferometric vibration analysis by wavefront reconstruction. *Journ. Opt. Soc. Amer.*, 55:1593–1598, 1965. {**4, 323, 326**}

[17] K. A. Stetson and R. L. Powell. Interferometric hologram evaluation of real-time vibration analysis of diffuse objects. *Journ. Opt. Soc. Amer.*, 55:1694–1695, 1965. {**4, 332**}

[18] R. J. Collier, E. T. Doherty, and K. S. Pennington. New method for generating depth contours holographically. *Appl. Phys. Lett.*, 7(8):223–225, 1965. {**4**}

[19] R. E. Brooks, L. O. Heflinger, and R. F. Wuerker. Interferometry with a holographically reconstructed comparison beam. *Appl. Phys. Lett.*, 7:248–249, 1965. {**4, 195**}

[20] L. O. Heflinger, R. F. Wuerker, and R. E. Brooks. Holographic interferometry. *Journ. Appl. Phys.*, 37:642–649, 1966. {**4, 195**}

[21] K. A. Haines and B. P. Hildebrand. Surface-deformation measurement using the wavefront reconstruction technique. *Appl. Opt.*, 5:595–602, 1966. {**4, 45, 214**}

[22] K. A. Haines and B. P. Hildebrand. Interferometric measurements on diffuse surfaces by holographic techniques. *IEEE Trans.*, IM 15:149–161, 1966. {**4, 45, 214**}

[23] K. A. Haines and B. P. Hildebrand. Contour generation by wavefront reconstruction. *Phys. Lett.*, 19:10–11, 1965. {**4, 333**}

[24] L. O. Heflinger and R. F. Wuerker. Holographic contouring via multifrequency lasers. *Appl. Phys. Lett.*, 15:28–30, 1969. {**4, 333**}

[25] B. P. Hildebrand and K. A. Haines. The generation of three-dimensional contour maps by wavefront reconstruction. *Phys. Lett.*, 21:422–423, 1966. {**4, 333**}

[26] B. P. Hildebrand and K. A. Haines. Multiple-wavelength and multiple-source holography applied to contour generation. *Journ. Opt. Soc. Amer.*, 57(2):155–162, 1967. {**4, 333**}

[27] T. Tsuruta, N. Shiotake, J. Tsujiuchi, and K. Matsuda. Holographic generation of contour map of diffusely reflecting surface by using immersion method. *Jap. J. Appl. Phys.*, 6:661–662, 1967. {**4, 336**}

[28] J. S. Zelenka and J. R. Varner. New method for generating depth contours holographically. *Appl. Opt.*, 7:2107–2110, 1968. {**4, 333**}

[29] M. H. Horman. An application of wavefront reconstruction to interferometry. *Appl. Opt.*, 4(2):333–336, 1965. {**4**}

[30] D. W. Sweeney and C. M. Vest. Reconstruction of three dimensional refractive index fields from multidirectonal interferometric data. *Appl. Opt.*, 12(11):2649–2663, 1973. {**4, 379–382**}

[31] K. A. Stetson. A rigorous treatment of the fringes of hologram interferometry. *Optik*, 29(4):386–400, 1969. {**4, 215**}

[32] N. E. Molin and K. A. Stetson. Measurement of fringe loci and localization in hologram interferometry for pivot motion, in-plane rotation and in-plane translation. Part I. *Optik*, 31:157–177, 1970. {**4, 215**}

[33] N. E. Molin and K. A. Stetson. Measurement of fringe loci and localization in hologram interferometry for pivot motion, in-plane rotation and in-plane translation. Part II. *Optik*, 31:281–291, 1970. {**4, 215**}

[34] N. E. Molin and K. A. Stetson. Fringe localization in hologram interferometry of mutually independent and dependent rotations around orthogonal, non-intersecting axes. *Optik*, 33:399–422, 1970. {**4, 215**}

[35] K. A. Stetson. The argument of the fringe function in hologram interferometry of general deformations. *Optik*, 31(6):576–591, 1970. {**4, 214, 215, 222, 318**}

[36] E. B. Aleksandrov and A. M. Bonch-Bruevich. Investigation of surface strains by the hologram technique. *Sov. Phys. - Tech. Phys.*, 12(2):258–265, 1967. {**4**}

[37] A. E. Ennos. Measurement of in-plane surface strain by hologram interferometry. *Journ. Phys. E: Scient. Instr.*, 1:731–734, 1968. {**4, 266, 267**}

[38] W. Osten, R. Höfling, and J. Saedler. Two computer-aided methods for data reduction from interferograms. In W. F. Fagan, editor, *Industrial Optoelectronic Measurement Systems using Coherent Light*, volume 863 of *Proc. of Soc. Photo-Opt. Instr. Eng.*, pages 105–113, 1987. {**4, 229**}

[39] P. Carre. Installation et utilisation du comparateur photoelectrique et interferentiel du Bureau International des Poids et Mesures. *Metrologia*, 2(1):13–23, 1966. {**4, 248**}

[40] J. H. Bruning, J. G. Gallagher, D. P. Rosenfeld, D. A. White, D. J. Brangaccio, and D. R. Herriott. Digital wavefront measuring interferometer for testing optical surfaces and lenses. *Appl. Opt.*, 13:2693, 1974. {**4**}

[41] B. Fischer and W. Jüptner. Automatisierte Auswertung holografischer Interferenzmuster mit dem Zeilen-Scan-Verfahren (Automatic evaluation of holographic interference patterns with the line-scan method). In *Spring School of Holographic Interferometry in Technology and Medicine*, 1978. (In German). {**4**}

[42] Th. Kreis, B. Fischer, W. Jüptner, and G. Sepold. Automatisierte Auswertung holografischer Interferenzmuster bei der Untersuchung von Zugproben (Automatic evaluation of holographic interference patterns in the investigation of tensile test specimen). In W. Waidelich, editor, *Laser 81 – Optoelektronik in der Technik*, pages 105–110. Springer-Verlag, 1981. (In German). {**4**}

[43] M. Takeda, H. Ina, and S. Kobayashi. Fourier-transform method of fringe-pattern analysis for computer-based topography and interferometry. *Journ. Opt. Soc. Amer.*, 72(1):156–160, 1982. {**4, 256, 263, 405**}

[44] Th. Kreis. Digital holographic interference-phase measurement using the Fourier-transform method. *Journ. Opt. Soc. Amer. A*, 3(6):847–855, 1986. {**4, 256, 261**}

[45] J. A. Leendertz. Interferometric displacement measuring on scattering surfaces utilizing speckle effect. *J. Phys. E: Sci. Instrum.*, 3:214–218, 1970. {**5**}

[46] J. N. Butters and J. A. Leendertz. Holographic and video techniques applied to engineering measurement. *J. Meas. Control*, 4:349–354, 1971. {**5, 401**}

[47] A. Macovski, D. Ramsey, and L. F. Schaefer. Time lapse interferometry and contouring using television systems. *Appl. Opt.*, 10:2722–2727, 1971. {**5, 401**}

[48] O. Schwomma. Austrian Patent No. 298830, 1972. {**5, 401**}

[49] K. Creath. Phase shifting speckle interferometry. *Appl. Opt.*, 24:3053–3058, 1985. {**5, 404**}

[50] K. A. Stetson and W. R. Brohinsky. Electro-optic holography and its application to hologram interferometry. *Appl. Opt.*, 24(21):3631–3637, 1985. {**5, 404**}

[51] J. W. Goodman and R. W. Lawrence. Digital image formation from electronically detected holograms. *Appl. Phys. Lett.*, 11(3):77–79, 1967. {**5, 6**}

[52] T. S. Huang. Digital holography. *Proc. IEEE*, 59(9):1335–1346, 1971. {**5, 6**}

[53] M. A. Kronrod, N. S. Merzlyakov, and L. P. Yaroslavskii. Reconstruction of holograms with a computer. *Sov. Phys.-Tech. Phys.*, 17:333–334, 1972. {**6**}

[54] T. H. Demetrakopoulos and R. Mittra. Digital and optical reconstruction of images from suboptical diffraction patterns. *Appl. Opt.*, 13:665–670, 1974. {**6**}

[55] L. P. Yaroslavskii and N. S. Merzlyakov. *Methods of Digital Holography*. Consultants Bureau, New York, 1980. {**6**}

[56] U. Schnars. Direct phase determination in hologram interferometry with use of digitally recorded holograms. *Journ. Opt. Soc. Amer. A*, 11(7):2011–2015, 1994. {**6, 269**}

[57] U. Schnars and W. Jüptner. Direct recording of holograms by a CCD target and numerical reconstruction. *Appl. Opt.*, 33(2):179–181, 1994. {**6**}

[58] U. Schnars and W. Jüptner. Digital recording and reconstruction of holograms in hologram interferometry and shearography. *Appl. Opt.*, 33(20):4373–4377, 1994. {**6**}

[59] U. Schnars, Th. Kreis, and W. Jüptner. Digital recording and numerical reconstruction of holograms: Reduction of the spatial frequency spectrum. *Opt. Eng.*, 35(4):977–982, 1995. {**6, 85**}

[60] U. Schnars, Th. Kreis, and W. Jüptner. CCD-recording and numerical reconstruction of holograms and holographic interferograms. In M. Kujawinska, R. J. Pryputniewicz, and M. Takeda, editors, *Interferometry VII: Techniques and Analysis*, volume 2544 of *Proc. of Soc. Photo-Opt. Instr. Eng.*, pages 57–63, 1995. {**6, 85**}

[61] Y. Zou, G. Pedrini, and H. Tiziani. Surface contouring in a video frame by changing the wavelength of a diode laser. *Opt. Eng.*, 35(4):1074–1079, 1996. {**6**}

[62] St. Schedin, G. Pedrini, H. J. Tiziani, and F. Mendoza Santoyo. Simultaneous three-dimensional dynamic deformation measurements with pulsed digital holography. *Appl. Opt.*, 38(34):7056–7062, 1999. {**6, 282, 405**}

[63] G. Pedrini, Y. L. Zou, and H. J. Tiziani. Digital double-pulsed holographic interferometry for vibration analysis. *Journ. Modern Opt.*, 42(2):367–374, 1995. {**6**}

[64] G. Pedrini, Y. L. Zou, and H. J. Tiziani. Comparison of two reflecting surfaces by using digital holographic interferometry. *Opt. Comm.*, 118:186–192, 1995. {**6, 405**}

[65] G. Pedrini, Ph. Fröning, H. Fessler, and H. J. Tiziani. In-line digital holographic interferometry. *Appl. Opt.*, 37(26):6262–6269, 1998. {**6, 332**}

[66] Th. Kreis and U. Schnars. Internal Reports, BMFT-Project HOLOMETEC, 1993+1994. {**6**}

[67] F. Dubois, L. Joannes, and J. C. Legros. Improved three-dimensional imaging with a digital holography microscope with a source of partial spatial coherence. *Appl. Opt.*, 38(34):7085–7094, 1999. {**6**}

[68] F. Dubois, L. Joannes, O. Dupont, J. L. Dewandel, and J. C. Legros. An integrated optical set-up for fluid-physics experiments under microgravity conditions. *Meas. Sci. & Techn.*, 10(10):934–945, 1999. {**6, 385**}

[69] F. Dubois, O. Monnom, C. Yourassowsky, and J.-C. Legros. Border processing in digital holography by extension of the digital hologram and reduction of the higher spatial frequencies. *Appl. Opt.*, 41(14):2621–2626, 2002. {**6, 180**}

[70] F. Dubois, C. Minetti, O. Monnom, C. Yourassowsky, J.-C. Legros, and P. Kischel. Pattern recognition with a digital holographic microscope working in partially coherent illumination. *Appl. Opt.*, 41(20):4108–4119, 2002. {**6**}

[71] A. Albertazzi Jr. and A. Vieira Fantin. Digital complex holography using a shearing interferometer: principles and early results. In K. Creath and J. Schmit, editors, *Interferometry XI: Techniques and Analysis*, volume 4777 of *Proc. of Soc. Photo-Opt. Instr. Eng.*, pages 57–68, 2002. {**6, 127, 128**}

[72] H. J. Kreuzer and R. A. Pawlitzek. Fast implementation of in-line holography for high resolution shape measurement. In Z. Füzessy, W. Jüptner, and W. Osten, editors, *Simulation and Experiment in Laser Metrology*, pages 61–66, 1996. {**6, 175**}

[73] H. J. Kreuzer and R. A. Pawlitzek. Numerical reconstruction for in-line holography in reflection and under glancing incidence. In W. Jüptner and W. Osten, editors, *Fringe '97, Automatic Processing of Fringe Patterns*, pages 364–367, 1997. {**6, 175**}

[74] M. R. A. Shegelski, S. Faltus, Th. Clark, and H. J. Kreuzer. Improvements in the reconstruction of in-line holograms by energy sampling and tomography. *Ultramicroscopy*, 74:169–178, 1998. {**6, 179**}

[75] H. J. Kreuzer, N. Pomerleau, K. Blagrave, and M. H. Jericho. Digital in-line holography with numerical reconstruction. In M. Kujawinska, editor, *Interferometry 99: Techniques and Technology*, volume 3744 of *Proc. of Soc. Photo-Opt. Instr. Eng.*, pages 65–74, 1999. {**6, 175**}

[76] H. J. Kreuzer, M. J. Jericho, I. A. Meinertzhagen, and W. Xu. Digital in-line holography with photons and electrons. *J. Phys.: Condens. Matter*, 13:10729–10741, 2001. {**6, 174**}

[77] W. Xu, M. H. Jericho, I. A. Meinertzhagen, and H. J. Kreuzer. Digital in-line holography for biological applications. *PNAS*, 98(20):11301–11305, 2001. {**6, 169**}

[78] W. Xu, M. H. Jericho, I. A. Meinertzhagen, and H. J. Kreuzer. Digital in-line holography of microspheres. *Appl. Opt.*, 41(25):5367–5375, 2002. {**6, 169**}

[79] W. Xu, M. H. Jericho, H. J. Kreuzer, and I. A. Meinertzhagen. Tracking particles in four dimensions with in-line holographic microscopy. *Opt. Lett.*, 28(3):164–166, 2003. {**6, 169**}

[80] C.-S. Guo, L. Zhang, H.-T. Wang, J. Liao, and Y. Y. Zhu. Phase-shifting error and its elimination in phase-shifting digital holography. *Opt. Lett.*, 27(19):1687–1689, 2002. {**6, 126**}

[81] Ch. Liu, Y. Li, X. Cheng, Zh. Liu, F. Bo, and J. Zhu. Elimination of zero-order diffraction in digital holography. *Opt. Eng.*, 41(10):2434–2437, 2002. {**6, 106**}

[82] H. Sekanina and J. Posposil. Digital holography using a digital photo-camera. *Journ. Modern Opt.*, 49(13):2083–2092, 2002. {**6, 139**}

[83] R. Binet, J. Colineau, and J.-C. Lehureau. Short-range synthetic aperture imaging at 633 nm by digital holography. *Appl. Opt.*, 41(23):4775–4782, 2002. {**6, 137**}

[84] F. Le Clerc, L. Collot, and M. Gross. Numerical heterodyne holography with two-dimensional photodetector arrays. *Opt. Lett.*, 25(10):716–718, 2000. {**6, 127, 242**}

[85] F. Le Clerc, L. Collot, and M. Gross. Selection of ballistic photons in diffusing media by numerical heterodyne holography. In P. Jacquot and J.-M. Fournier, editors, *Interferometry in Speckle Light*, pages 243–248, 2000. {**6, 127**}

[86] F. Le Clerc, L. Collot, and M. Gross. Numerical heterodyne holography. In P. Jacquot and J.-M. Fournier, editors, *Interferometry in Speckle Light*, pages 235–242, 2000. {**6, 127**}

[87] F. Le Clerc, M. Gross, and L. Collot. Synthetic-aperture experiment in the visible with on-axis digital heterodyne holography. *Opt. Lett.*, 26(20):1550–1552, 2001. {**6, 137**}

[88] S. Coetmellec, C. Buraga-Lefebvre, D. Lebrun, and C. Özkul. Application of in-line digital holography to multiple plane velocimetry. *Meas. Sci. & Techn.*, 12:1392–1397, 2001. {**6, 166**}

[89] S. Coetmellec, D. Lebrun, and C. Özkul. Application of the two-dimensional fractional-order Fourier transformation to particle field digital holography. *Journ. Opt. Soc. Amer. A*, 19:1537–1546, 2002. {**6, 162**}

[90] L. Yu and L. Cai. Iterative algorithm with a constraint condition for numerical reconstruction of a three-dimensional object from its hologram. *Journ. Opt. Soc. Amer. A*, 18(5):1033–1045, 2001. {**6, 161**}

[91] L. Yu and L. Cai. Multidimensional data encryption with digital holography. *Opt. Comm.*, 215:271–284, 2003. {**6, 183**}

[92] E. Allaria, S. Brugioni, S. De Nicola, P. Ferraro, S. Grilli, and R. Meucci. Digital holography at 10.6 μm. *Opt. Comm.*, 215:257–262, 2003. {**6, 57, 60, 357**}

[93] S. Grilli, P. Ferraro, S. De Nicola, A. Finizio, G. Pierattini, and R. Meucci. Whole optical wavefields reconstruction by digital holography. *Opt. Expr.*, 9(6):294–302, 2001. {**6, 136**}

[94] S. De Nicola, P. Ferraro, A. Finizio, and G. Pierattini. Compensation of aberrations in Fresnel off-axis digital holography. In W. Osten and W. Jüptner, editors, *Fringe 2001*, pages 407–412, 2001. {**6, 62, 114**}

[95] S. De Nicola, P. Ferraro, A. Finizio, and G. Pierattini. Correct-image reconstruction in the presence of severe anamorphism by means of digital holography. *Opt. Lett.*, 26:974–976, 2001. {**6, 114**}

[96] S. Murata and N. Yasuda. Potential of digital holography in particle measurement. *Opt. Laser Technol.*, 32:567–574, 2000. {**6, 161**}

[97] O. Matoba, T. J. Naughton, Y. Frauel, N. Bertaux, and B. Javidi. Real-time three-dimensional object reconstruction by use of a phase-encoded digital hologram. *Appl. Opt.*, 41(29):6187–6192, 2002. {**6, 127**}

[98] O. Matoba and B. Javidi. Optical retrieval of encrypted digital holograms for secure real-time display. *Opt. Lett.*, 27(5):321–323, 2002. {**6, 183**}

[99] N. Takai, M. Yamada, and T. Idogawa. Holographic interferometry using a reference wave with a sinusoidally modulated amplitude. *Opt. Laser Technol.*, 8:21–23, 1976. {**6, 330**}

[100] Y. Takaki and H. Ohzu. Fast numerical reconstruction technique for high-resolution hybrid holographic microscopy. *Appl. Opt.*, 38(11):2204–2211, 1999. {**6, 170, 171, 173**}

[101] Y. Takaki, H. Kawai, and H. Ohzu. Hybrid holographic microscopy free of conjugate and zero-order images. *Appl. Opt.*, 38(23):4990–4996, 1999. {**6, 105, 108**}

[102] M. Takeda, K. Taniguchi, T. Hirayama, and H. Kohgo. Single-transform Fourier/Hartley fringe analysis for holographic interferometry. In Z. Füzessy, W. Jüptner, and W. Osten, editors, *Simulation and Experiment in Laser Metrology*, pages 67–73, 1996. {**6, 384, 385, 440**}

Bibliography

[103] I. Yamaguchi and T. Zhang. Phase-shifting digital holography. *Opt. Lett.*, 22(16):1268–1270, 1997. {**6, 125, 126**}

[104] T. Zhang and I. Yamaguchi. Three-dimensional microscopy with phase-shifting digital holography. *Opt. Lett.*, 23(15):1221–1223, 1998. {**6, 125, 126**}

[105] T. Zhang and I. Yamaguchi. 3-D microscopy with phase-shifting digital holography. In *Laser Interferometry IX: Applications*, volume 3479 of *Proc. of Soc. Photo-Opt. Instr. Eng.*, pages 152–159, 1998. {**6, 126**}

[106] I. Yamaguchi, J.-I. Kato, S. Ohta, and J. Mizuno. Image formation in phase-shifting digital holography. In G. M. Brown, W. Jüptner, and R. J. Pryputniewicz, editors, *Laser Interferometry X: Applications*, volume 4101 of *Proc. of Soc. Photo-Opt. Instr. Eng.*, pages 330–338, 2000. {**6, 171**}

[107] I. Yamaguchi, S. Ohta, and J.-I. Kato. Surface contouring by phase-shifting digital holography and noise reduction. In P. Jacquot and J.-M. Fournier, editors, *Interferometry in Speckle Light*, pages 249–256, 2000. {**6, 343**}

[108] I. Yamaguchi, J.-I. Kato, S. Ohta, and J. Mizuno. Image formation in phase-shifting digital holography and applications to microscopy. *Appl. Opt.*, 40(34):6177–6186, 2001. {**6, 171**}

[109] I. Yamaguchi, O. Inomoto, and J.-I. Kato. Surface shape measurement by phase-shifting digital holography. In W. Osten and W. Jüptner, editors, *Fringe 2001*, pages 365–372, 2001. {**6, 343**}

[110] I. Yamaguchi, T. Matsumura, and J.-I. Kato. Phase-shifting color digital holography. *Opt. Lett.*, 27(13):1108–1110, 2002. {**6, 129**}

[111] S. Pasko and R. Jozwicki. Novel Fourier approach to digital holography. *Opto-Electron. Rev.*, 10(2):89–95, 2002. {**6, 107**}

[112] Y. Xu, C. M. Vest, and J. D. Murray. Holographic interferometry used to demonstrate a theory of pattern formation in animal coats. *Appl. Opt.*, 22(22):3479–3483, 1983. {**6, 326**}

[113] L. Xu, X. Peng, J. Miao, and A. Asundi. Digital micro-holo-interferometry for microstructure studies. In G. M. Brown, W. Jüptner, and R. J. Pryputniewicz, editors, *Laser Interferometry X: Applications*, volume 4101 of *Proc. of Soc. Photo-Opt. Instr. Eng.*, pages 543–548, 2000. {**6, 179, 311**}

[114] L. Xu, X. Peng, J. Miao, and A. Asundi. Studies of digital microscopic holography with applications to microstructure testing. *Appl. Opt.*, 40(28):5046–5051, 2001. {**6, 179, 311**}

[115] B. Nilsson and T. E. Carlsson. Simultaneous measurement of shape and deformation using digital light-in-flight recording by holography. *Opt. Eng.*, 39(1):244–253, 2000. {**6, 346, 347**}

[116] B. Nilsson and T. E. Carlsson. Direct three-dimensional shape measurement by digital light-in-flight holography. *Appl. Opt.*, 37(34):7954–7959, 1998. {**6, 346**}

[117] T. E. Carlsson. Measurement of three-dimensional shapes using light-in-flight recording by holography. *Opt. Eng.*, 32(10):2587–2592, 1993. {**6, 339**}

[118] B. Nilsson and T. E. Carlsson. Resolution improvements of the digital light-in-flight recording by holography method. In G. M. Brown, W. Jüptner, and R. J. Pryputniewicz, editors, *Laser Interferometry X: Applications*, volume 4101 of *Proc. of Soc. Photo-Opt. Instr. Eng.*, pages 369–377, 2000. {**6, 346, 348**}

[119] B. Skarman, J. Becker, and K. Wozniak. Simultaneous 3D-PIV and temperature measurements using a new CCD-based holographic interferometer. *Flow Meas. Instrum.*, 7(1):1–6, 1996. {**6, 127**}

[120] M. Jacquot, P. Sandoz, and G. Tribillon. Digital holography with improved resolution by spatial sampling of holograms. In P. Jacquot and J.-M. Fournier, editors, *Interferometry in Speckle Light*, pages 219–226, 2000. {**6, 84, 137**}

[121] M. Jacquot, P. Sandoz, and G. Tribillon. High resolution digital holography. *Opt. Comm.*, 190:87–94, 2001. {**6, 84, 137**}

[122] T. Colomb, P. Dahlgren, E. Cuche, D. Beghuin, and Chr. Depeursinge. A method for polarisation imaging using digital holography. In P. Jacquot and J.-M. Fournier, editors, *Interferometry in Speckle Light*, pages 227–234, 2000. {**6, 314**}

[123] T. Colomb, P. Dahlgren, D. Beghuin, E. Cuche, P. Marquet, and Chr. Depeursinge. Polarization imaging by use of digital holography. *Appl. Opt.*, 41(1):27–37, 2002. {**6, 314**}

[124] D. E. Cuche. Determination of the Poisson's ratio by the holographic moire technique. In W. Jüptner, editor, *Holography Techniques and Applications*, volume 1026 of *Proc. of Soc. Photo-Opt. Instr. Eng.*, pages 165–170, 1988. {**6, 176, 178, 179, 318, 359**}

[125] D. Beghuin, E. Cuche, P. Dahlgren, C. Depeursinge, G. Delacretaz, and R. P. Salathe. Single acquisition polarisation imaging with digital holography. *Electron. Lett.*, 35(23):2053–2055, 1999. {**6, 314**}

[126] D. Cuche and W. Schumann. Fringe modification with amplification in holographic interferometry and application of this to determine strain and rotation. In W. F. Fagan, editor, *Industrial Applications of Laser Technology*, volume 398 of *Proc. of Soc. Photo-Opt. Instr. Eng.*, pages 35–45, 1983. {**6, 317**}

[127] E. Cuche, P. Marquet, and C. Depeursinge. Spatial filtering for zero-order and twin-image elimination in digital off-axis holography. *Appl. Opt.*, 39(23):4070–4075, 2000. {**6, 107**}

[128] E. Cuche, P. Marquet, P. Dahlgren, and Chr. Depeursinge. Digital holographic microscopy, a new method for simultaneous amplitude- and quantitative phase-contrast imaging. In P. Jacquot and J.-M. Fournier, editors, *Interferometry in Speckle Light*, pages 213–218, 2000. {**6, 176**}

[129] M. Liebling, Th. Blu, and M. Unser. Fresnelets: New multiresolution wavelet bases for digital holography. *IEEE Trans. on Image Proc.*, 12(1):29–43, 2003. {**6, 133, 134, 443–445**}

[130] L. Onural and P. D. Scott. Digital decoding of in-line holograms. *Opt. Eng.*, 26(11):1124–1132, 1987. {**6, 105, 108**}

[131] L. Onural and M. T. Özgen. Extraction of three-dimensional object-location information directly from in-line holograms using Wigner analysis. *Journ. Opt. Soc. Amer. A*, 9(2):252–260, 1992. {**6, 108**}

[132] L. Onural. Sampling of the diffraction field. *Appl. Opt.*, 39(32):5929–5935, 2000. {**6, 146**}

[133] Y. Frauel and B. Javidi. Digital three-dimensional image correlation by use of computer-reconstructed integral imaging. *Appl. Opt.*, 41(26):5488–5496, 2002. {**6**}

[134] E. Tajahuerce and B. Javidi. Encrypting three-dimensional information with digital holography. *Appl. Opt.*, 39(35):6595–6601, 2000. {**6, 180, 182**}

[135] E. Tajahuerce, O. Matoba, S. C. Verrall, and B. Javidi. Optoelectronic information encryption with phase-shifting interferometry. *Appl. Opt.*, 39(14):2313–2320, 2000. {**6, 180, 183**}

[136] Y. Y. Hung and C. Y. Liang. Image-shearing camera for direct measurement of surface strain. *Appl. Opt.*, 18(7):1046–1051, 1979. {**6, 405**}

[137] G. Indebetouw and P. Klysubun. Imaging through scattering media with depth resolution by use of low-coherence gating in spatiotemporal digital holography. *Opt. Lett.*, 25(4):212–214, 2000. {**6, 179, 180**}

[138] G. Indebetouw and P. Klysubun. Space-time digital holography: A three-dimensional microscopic imaging scheme with an arbitrary degree of spatial coherence. *Adv. Phys. Lett.*, 75(14):2017–2019, 1999. {**6, 179**}

[139] T.-C. Poon, T. Kim, G. Indebetouw, B. W. Schilling, M. H. Wu, K. Shinoda, and Y. Suzuki. Twin-image elimination experiments for three-dimensional images in optical scanning holography. *Opt. Lett.*, 25(4):215–217, 2000. {**6, 127**}

[140] M. K. Kim. Wavelength-scanning digital interference holography for optical section imaging. *Opt. Lett.*, 24(23):1693–1695, 1999. {**6, 137**}

[141] M. K. Kim. Tomographic three-dimensional imaging of a biological specimen using wavelength-scanning digital interference holography. *Optics Expr.*, 7(9):305–310, 2000. {**6, 137**}

[142] S. Lai and M. A. Neifeld. Digital wavefront reconstruction and its application to image encryption. *Opt. Comm.*, 178(4-6):283–289, 2000. {**6, 183, 263**}

[143] G. Lai and T. Yatagai. Dual-reference holographic interferometry with a double pulsed laser. In F.-P. Chiang, editor, *International Conference on Photomechanics and Speckle Metrology*, volume 814 of *Proc. of Soc. Photo-Opt. Instr. Eng.*, pages 346–351, 1987. {**6, 47, 202**}

[144] Z. Liu, M. Centurion, G. Panotopoulos, J. Hong, and D. Psaltis. Holographic recording of fast events on a CCD camera. *Opt. Lett.*, 27(1):22–24, 2002. {**6, 286**}

[145] J. H. Milgram and W. Li. Computational reconstruction of images from holograms. *Appl. Opt.*, 41(5):853–864, 2002. {**6, 169**}

[146] R. B. Owen and A. A. Zozulya. In-line digital holographic sensor for monitoring and characterizing marine particulates. *Opt. Eng.*, 39(8):2187–2197, 2000. {**6, 385**}

[147] R. B. Owen, A. A. Zozulya, M. R. Benoit, and D. M. Klaus. Microgravity materials and life sciences research applications of digital holography. *Appl. Opt.*, 41(19):3927–3935, 2002. {**6, 385**}

[148] R. B. Owen and A. A. Zozulya. Comparative study with double-exposure digital holographic interferometry and a shack-hartmann sensor to characterize transparent materials. *Appl. Opt.*, 41(28):5891–5895, 2002. {**6, 385**}

[149] M.R. Benoit, D. M. Klaus, and R. B. Owen. Quantifying extracellular mass transport using digital holography. *ASGSB 2001 Abstracts*, 2001. {**6**}

[150] G. Pan and H. Meng. Digital in-line holographic PIV for 3D particulate flow diagnostics. In *4th International Symposium on Particle Image Velocimetry, Göttingen*, 2001. {**6, 160**}

[151] X. Xiao and I. K. Puri. Digital recording and numerical reconstruction of holograms: an optical diagnostic for combustion. *Appl. Opt.*, 41(19):3890–3899, 2002. {**6, 384**}

[152] L. Yaroslavsky. Digital holography: 30 years later. In *Practical Holography XVI*, volume 4659 of *Proc. of Soc. Photo-Opt. Instr. Eng.*, pages 1–11, 2002. {**6**}

[153] L. Yaroslavsky. *Digital Holography and Digital Image Processing*. Kluwer Academic Publishers, 2003. {**6**}

[154] P. Hariharan. Laser interferometry: current trends and future prospects. In R. J. Pryputniewicz, editor, *Laser Interferometry IV: Computer-Aided Interferometry*, volume 1553 of *Proc. of Soc. Photo-Opt. Instr. Eng.*, pages 2–11, 1991. {**7**}

[155] H. Rottenkolber and W. Jüptner. Holographic interferometry in the next decade. In R. J. Pryputniewicz, editor, *Laser Interferometry: Quantitative Analysis of Interferograms*, volume 1162 of *Proc. of Soc. Photo-Opt. Instr. Eng.*, pages 2–15, 1989. {**7**}

[156] H. J. Tiziani. Optical methods for precision measurements. *Opt. Quantum Electron.*, 21:253–282, 1989. {**7**}

[157] J. Durnin. Exact solutions for nondiffracting beams. I. The scalar theory. *Journ. Opt. Soc. Amer. A*, 4:651–654, 1987. {**12**}

[158] J. W. Goodman. *Introduction to Fourier Optics*. McGraw-Hill, 1968. {**17, 19**}

[159] W. Lauterborn, T. Kurz, and M. Wiesenfeldt. *Coherent Optics*. Springer, 1993. (In German). {**17–19, 52, 61**}

[160] J. W. Goodman. *Introduction to Fourier Optics*. McGraw-Hill, New York, second edition, 1996. {**21–23, 25, 27–29, 101, 140, 141, 143, 171, 178, 411, 427**}

[161] K. Iizuka. *Engineering Optics*. Springer Verlag, Berlin, Heidelberg, New York, second edition, 1987. {**28**}

[162] B. E. A. Saleh and M. C. Teich. *Fundamentals of Photonics*. John Wiley and Sons, 1991. {**29**}

[163] J. W. Goodman. Statistical properties of laser speckle patterns. In J. C. Dainty, editor, *Laser Speckle and Related Phenomena*, volume 9 of *Springer Series Topics in Applied Physics*, pages 9–75, 1975. {**30, 33, 35**}

[164] K. J. Gasvik. *Optical Metrology*. John Wiley and Sons, 1987. {**36, 306, 400**}

[165] E. B. Champagne. Non-paraxial imaging, magnification and aberration properties in holography. *Journ. Opt. Soc. Amer.*, 57:51–55, 1967. {**44, 47**}

[166] R. W. Meier. Magnification and third-order aberrations in holography. *Journ. Opt. Soc. Amer.*, 55:987–992, 1965. {**44, 45, 47**}

[167] J. F. Miles. Imaging and magnification properties in holography. *Optica Acta*, 19:165–186, 1972. {**44**}

[168] D. B. Neumann. Geometrical relationships between the original object and the two images of a hologram reconstruction. *Journ. Opt. Soc. Amer.*, 56:858–861, 1966. {**44**}

[169] W. Schumann and M. Dubas. On the motion of holographic images caused by movements of the reconstruction light source, with the aim of application to deformation analysis. *Optik*, 46(4):377–392, 1976. {**44**}

[170] C. M. Vest. *Holographic Interferometry*. John Wiley and Sons, 1979. {**45, 185, 195, 204, 206, 211, 214, 224, 311, 313–315, 318, 329, 330, 367, 370, 372, 374, 376, 383**}

[171] E. N. Leith and J. Upatnieks. Reconstructed wavefronts and communication theory. *Journ. Opt. Soc. Amer.*, 52:1123–1130, 1962. {**45**}

[172] P. Hariharan. *Optical Holography: Principles, Techniques and Applications*. Cambridge University Press, 1984. {**45–47, 50, 58, 59, 160, 329**}

[173] P. Hariharan. Basic principles. In P. K. Rastogi, editor, *Holographic Interferometry*, volume 68 of *Springer Series in Optical Sciences*, pages 7–32, 1994. {**46**}

[174] D. Paoletti, S. Amadesi, and A. D'Altorio. A fringe control method for real time HNDT. *Opt. Comm.*, 49(2):98–102, 1984. {**47**}

[175] J. N. Latta. Computer-based analysis of hologram imagery and aberrations. I. Hologram types and their nonchromatic aberrations. *Appl. Opt.*, 10:599–608, 1971. {**47**}

[176] J. N. Latta. Computer-based analysis of hologram imagery and aberrations. II. Aberrations induced by a wavelength shift. *Appl. Opt.*, 10:609–618, 1971. {**47**}

[177] J. N. Latta. Computer-based analysis of holography using ray tracing. *Appl. Opt.*, 10:2698–2710, 1971. {**47**}

[178] J. F. Miles. Evaluation of the wavefront aberration in holography. *Optica Acta*, 20:19–31, 1973. {**47**}

[179] G. W. Stroke and A. E. Labeyrie. White light reconstruction of holographic images using the Lippmann-Bragg diffraction effect. *Phys. Lett.*, 20(4):368–370, 1966. {**51**}

[180] Yu. I. Ostrovsky, M. M. Butusov, and G. V. Ostrovskaya. *Interferometry by Holography*. Springer-Verlag, 1980. {**52, 386**}

[181] R. T. Pitlak and R. Page. Pulsed lasers for holographic interferometry. *Opt. Eng.*, 24(4):639–644, 1985. {**56**}

[182] A. J. Decker. Holographic interferometry with an injection seeded Nd:YAG laser and two reference beams. *Appl. Opt.*, 29(18):2697–2700, 1990. {**57**}

[183] G. Hüttmann, W. H. Lauterborn, and E. Schmitz. Holography with a frequency-doubled Nd:YAG laser. In W. Jüptner, editor, *Holography Techniques and Applications*, volume 1026 of *Proc. of Soc. Photo-Opt. Instr. Eng.*, pages 14–21, 1988. {**57**}

[184] L. Crawforth, C.-K. Lee, and A. C. Munce. Application of pulsed laser holographic interferometry to the study of magnetic disc drive component motions. In K. Stetson and R. Pryputniewicz, editors, *International Conference on Hologram Interferometry and Speckle Metrology*, pages 404–412, 1990. {**57**}

[185] P. Hariharan. High-precision, digital, phase-stepping interferometry with laser diode. In R. J. Pryputniewicz, editor, *Laser Interferometry: Quantitative Analysis of Interferograms*, volume 1162 of *Proc. of Soc. Photo-Opt. Instr. Eng.*, pages 86–91, 1989. {**57**}

[186] Y. Ishii, J. Chen, and K. Murata. Digital phase-measuring interferometry with a tunable laser diode. *Opt. Lett.*, 12(4):233–235, 1987. {**57**}

[187] R. J. Parker. A quarter century of thermoplastic holography. In K. Stetson and R. Pryputniewicz, editors, *International Conference on Hologram Interferometry and Speckle Metrology*, pages 217–224, 1990. {**59**}

[188] M. Georges, C. Thizy, V. Scauflaire, S. Ryhon, G. Pauliat, P. Lemaire, and G. Roosen. Dynamic holographic interferometry with photorefractive crystals: review of applications and advanced techniques. In K. Gastinger, O. Løkberg, and S. Winther, editors, *Speckle Metrology 2003*, volume 4933 of *Proc. of Soc. Photo-Opt. Instr. Eng.*, pages 250–255, 2003. {**60**}

[189] J. P. Herriau, A. Delboulbe, and J. P. Huignard. Non-destructive testing using real time holographic interferometry in BSO crystals. In W. F. Fagan, editor, *Industrial Applications of Laser Technology*, volume 398 of *Proc. of Soc. Photo-Opt. Instr. Eng.*, pages 123–129, 1983. {**60**}

[190] H. J. Tiziani. Real-time metrology with BSO crystals. *Optica Acta*, 29(4):463–470, 1982. {**60**}

[191] X. Wang, R. Magnusson, and A. Haji-Sheikh. Real-time interferometry with photorefractive reference holograms. *Appl. Opt.*, 32(11):1983–1986, 1993. {**60**}

[192] D. H. Barnhart, N. Hampp, N. A. Halliwell, and J. M. Coupland. Digital holographic velocimetry with bacteriorhodopsin (BR) for real-time recording and numeric reconstruction. In *11th International Symposium on Applications of Laser Techniques to Fluid Mechanics, Lisbon*, 2002. {**61**}

[193] J. P. Bentley. *Principles of Measurement Systems*, chapter 15: Optical measurement systems, pages 364–405. Longman Scientific and Technical, 2nd edition, 1988. {**61**}

[194] G. O. Reynolds, J. B. DeVelis, G. B. Parrent, and B. J. Thompson. *The New Physical Optics Notebook: Tutorials in Fourier Optics*. SPIE Optical Engineering Press, 1989. {**61**}

[195] J. A. Gilbert, D. R. Matthys, and Ch. M. Hendren. Displacement analysis of the interior walls of a pipe using panoramic holo-interferometry. In F. P. Chiang, editor, *Second International Conference on Photomechanics and Speckle Metrology*, volume 1554 of *Proc. of Soc. Photo-Opt. Instr. Eng.*, pages 128–134, 1991. {**61**}

[196] N. Abramson. *The Making and Evaluation of Holograms*. Academic Press, 1981. {**61, 337, 365**}

[197] B. E. Jones. Optical fibre sensors and systems for industry. *Journ. Phys. E: Sci. Instr.*, 18:770–782, 1985. {**61**}

[198] F. Solitro, L. Gatti, F. Bedarida, and L. Michetti. Multidirectional holographic interferometer (MHOI) with fiber optics for study of crystal growth in microgravity. In R. J. Pryputniewicz, editor, *Laser Interferometry: Quantitative Analysis of Interferograms*, volume 1162 of *Proc. of Soc. Photo-Opt. Instr. Eng.*, pages 62–65, 1989. {**61, 377**}

[199] J. A. Gilbert, T. D. Dudderar, M. E. Schultz, and A. J. Boehnlein. The monomode fiber – a new tool for holographic interferometry. *Exp. Mech.*, 23(2):190–195, 1983. {**62**}

[200] J. A. Gilbert, T. D. Dudderar, and A. Nose. Remote deformation field measurement through different media using fiber optics. *Opt. Eng.*, 24(4):628–631, 1985. {**62**}

Bibliography

[201] S. Pflüger, R. Noll, J. Hertzberg, V. Sturm, and W. Meesters. Pulsed holography with elongated q-switch pulse. In VDI-Technology Centre, editor, *Holographic-Interferometric Metrology*, Laser-Research and Laser-Technology, pages 131–145, 1995. (In German). {62}

[202] K. Creath. Phase-shifting holographic interferometry. In P. K. Rastogi, editor, *Holographic Interferometry*, volume 68 of *Springer Series in Optical Sciences*, pages 109–150, 1994. {62, 242, 404}

[203] G. E. Sommargren. Double exposure holographic interferometry using commonpath reference waves. *Appl. Opt.*, 16(6):1736–1741, 1977. {63, 245}

[204] M. Kujawinska and D. W. Robinson. Automatic fringe pattern analysis for holographic measurement of transient event. In W. Jüptner, editor, *Holography Techniques and Applications*, volume 1026 of *Proc. of Soc. Photo-Opt. Instr. Eng.*, pages 93–103, 1988. {65}

[205] M. Kujawinska and D. W. Robinson. Multichannel phase-stepped holographic interferometry. *Appl. Opt.*, 27:312–320, 1988. {65, 243}

[206] C. R. Mercer and G. Beheim. Fiber optic phase stepping system for interferometry. *Appl. Opt.*, 30(7):729–734, 1991. {65}

[207] M. Takeda and M. Kitoh. Spatio-temporal frequency-multiplex heterodyne interferometry. In R. J. Pryputniewicz, editor, *Laser Interferometry IV: Computer-Aided Interferometry*, volume 1553 of *Proc. of Soc. Photo-Opt. Instr. Eng.*, 1991. {65, 265}

[208] Th. V. Higgins. The technology of image capture. *Laser Focus World*, 30(12):53–60, 1994. {66}

[209] W. S. Boyle and G. E. Smith. Charge coupled semiconductor devices. *Bell Systems Tech. Journ.*, 49:587–593, 1970. {66}

[210] G. C. Holst. *CCD Arrays, Cameras, and Displays*. SPIE Optical Engineering Press, 1996. {66, 68, 71, 72, 74, 76}

[211] R. Spooren. Standard charge-coupled device cameras for video speckle interferometry. *Opt. Eng.*, 33(3):889–896, 1994. {69, 73}

[212] Th. Baumbach, W. Osten, C. Falldorf, and W. Jüptner. Remote interferometry by digital holography for shape control. In W. Osten, editor, *Interferometry XI*, volume 4778 of *Proc. of Soc. Photo-Opt. Instr. Eng.*, pages 338–349, 2002. {73, 354, 356}

[213] http://www.extremetech.com/article2/0,3973,474239,00.asp. {73}

[214] J. Adams, K. Parulski, and K. Spaulding. Color processing in digital cameras. *IEEE Micro*, 18(6):20–29, 1998. {77}

[215] D. Merrill. The next generation digital camera. *Optics and Photonics News*, Jan.:26–33, 2003. {77}

[216] D. Alleysson, S. Süsstrunk, and J. Herault. Color demosaicing by estimating luminance and opponent chromatic signals in the Fourier domain. In *Proc. IS&T/SID 10th Color Imaging Conference*, pages 331–336, 2002. {77, 79}

[217] R. Ramanath, W. E. Snyder, G. L. Bilbro, and W. A. Sander III. Demosaicking methods for Bayer color arrays. *Journ. Electronic Imaging*, 11(3):306–315, 2002. {79}

[218] P. Longere, X. Zhang, P. DelaHunt, and D. Brainard. Perceptual assessment of demosaicing algorithm performance. *Proc. IEEE*, 90(1):123–132, 2002. {80}

[219] J. P. Allebach, N. C. Gallagher, and B. Liu. Aliasing error in digital holography. *Appl. Opt.*, 15(9):2183–2188, 1976. {**81**}

[220] M. Young. *Optics and Lasers: An Engineering Physics Approach*. Springer-Verlag, 1977. {**86, 371**}

[221] P. Rastogi and A. Sharma. Systematic approach to image formation in digital holography. *Opt. Eng.*, 42(5):1208–1214, 2003. {**88, 102, 141**}

[222] G. Pedrini, S. Schedin, and H. J. Tiziani. Lensless digital-holographic interferometry for the measurement of large objects. *Opt. Comm.*, 171:29–36, 1999. {**88, 102**}

[223] A. F. Doval, C. Trillo, O. Lopez, D. Cernadas, C. Lopez, B. V. Dorrio, J. L. Fernandez, and M. Perez-Amor. Lensless zooming Fourier-transform digital holography. In K. Gastinger, O. Løkberg, and S. Winther, editors, *Speckle Metrology 2003*, volume 4933 of *Proc. of Soc. Photo-Opt. Instr. Eng.*, pages 111–116, 2003. {**89, 102**}

[224] S. De Nicola, P. Ferraro, A. Finizio, S. Grilli, and G. Pierattini. Experimental demonstration of the longitudinal image shift in digital holography. *Opt. Eng.*, 42(6):1625–1630, 2003. {**91**}

[225] M. Y. Y. Hung, H. M. Shang, and L. Yang. Unified approach for holography and shearography in surface deformation measurement and nondestructive testing. *Opt. Eng.*, 42(5):1197–1207, 2003. {**92, 400**}

[226] E. Torres and J. Guerrero Bermudez. Digital Fourier holography: recording and optical reconstruction. *Revista Colombiana de Fisica*, 35(1):168–171, 2003. {**93**}

[227] Th. Kreis, P. Aswendt, and R. Höfling. Hologram reconstruction using a digital micromirror device. *Opt. Eng.*, 40(6):926–933, 2001. {**93, 146**}

[228] E. Cuche, F. Bevilacqua, and C. Depeursinge. Digital holography for quantitative phase-contrast imaging. *Opt. Lett.*, 24(5):291–293, 1999. {**96**}

[229] C. Wagner, S. Seebacher, W. Osten, and W. Jüptner. Digital recording and numerical reconstruction of lensless Fourier holograms in optical metrology. *Appl. Opt.*, 38(22):4812–4820, 1999. {**101**}

[230] N. Demoli, J. Mestrovic, and I. Sovic. Subtraction digital holography. *Appl. Opt.*, 42:798–804, 2003. {**105, 108, 385**}

[231] Th. Kreis and W. Jüptner. Suppression of the dc term in digital holography. *Opt. Eng.*, 36(8):2357–2360, 1997. {**105**}

[232] Ø . Skotheim. Holovision – a software package for reconstruction and analysis of digitally sampled holograms. In K. Gastinger, O. Løkberg, and S. Winther, editors, *Speckle Metrology 2003*, volume 4933 of *Proc. of Soc. Photo-Opt. Instr. Eng.*, pages 311–316, 2003. {**106, 124**}

[233] S. Pasko and R. Jozwicki. Improvement methods of reconstruction process in digital holography. *Opto-Electron. Rev.*, 11(3):203–209, 2003. {**107**}

[234] G. Liu. Object reconstruction from noisy holograms. *Opt. Eng.*, 29(1):19–24, 1990. {**107**}

[235] G. Pedrini, P. Fröning, H. Fessler, and H. J. Tiziani. In-line digital holographic interferometry. *Appl. Opt.*, 37(26):6262–6269, 1998. {**108**}

[236] A. Stadelmeier and J. H. Massing. Compensation of lens aberrations in digital holography. *Opt. Lett.*, 25:1630–1632, 2000. {**114**}

[237] Th. Kreis. *Holographic Interferometry: Principles and Methods*. Wiley-VCH, 1996. {**126, 332, 342**}

[238] S.-G. Kim, B. Lee, and E.-S. Kim. Removal of bias and the conjugate image in incoherent on-axis triangular holography and real-time reconstruction of the complex hologram. *Appl. Opt.*, 36(20):4784–4791, 1997. {**127**}

[239] C. Furlong and R. J. Pryputniewicz. Optoelectronic characterization of shape and deformation of MEMS accelerometers used in transportation applications. *Opt. Eng.*, 42(5):1223–1231, 2003. {**127, 311, 405**}

[240] A. Albertazzi Jr. and A. Vieira Fantin. Digital complex holography using a shearing interferometer: improvements and recent results. In K. Gastinger, O. Løkberg, and S. Winther, editors, *Speckle Metrology 2003*, volume 4933 of *Proc. of Soc. Photo-Opt. Instr. Eng.*, pages 42–47, 2003. {**127, 128**}

[241] M. Liebling, Th. Blu, E. Cuche, P. Marquet, Chr. Depeursinge, and M. Unser. Local amplitude and phase retrieval method for digital holography applied to microscopy. In *Conference on Biomedical Optics 2003*, volume 5143 of *Proc. of Soc. Photo-Opt. Instr. Eng.*, 2003. {**130, 132**}

[242] M. Liebling, Th. Blu, and M. Unser. Complex-wave retrieval from a single off-axis hologram. *Journ. Opt. Soc. Amer. A*, 21(3):367–377, 2004. {**130**}

[243] M. Liebling, Th. Blu, E. Cuche, P. Marquet, Chr. Depeursinge, and M. Unser. A novel non-diffractive reconstruction method for digital holographic microscopy. In *Proc. of the First 2002 IEEE International Symposium on Biomedical Imaging: Macro to Nano*, pages 625–628, 2002. {**130**}

[244] Th. Kreis. *Auswertung holografischer Interferenzmuster mit Methoden der Ortsfrequenzanalyse (Evaluation of holographic interference patterns using methods of spatial frequency analysis)*. Fortschritt-Berichte VDI, Reihe 8, Nr. 108. VDI-Verlag, 1986. (In German). {**132, 390**}

[245] M. Liebling, Th. Blu, and M. Unser. Non-linear Fresnelet approximation for interference term suppression in digital holography. In *Wavelets: Applications in Signal and Image Processing X*, Proc. of Soc. Photo-Opt. Instr. Eng., 2003. {**134**}

[246] S. Belaid, D. Lebrun, and C. Özkul. Application of two-dimensional wavelet transform to hologram analysis: visualization of glass fibers in a turbulent flame. *Opt. Eng.*, 36(7):1947–1951, 1997. {**134**}

[247] M. Malek, S. Coetmellec, D. Allano, and D. Lebrun. Formulation of in-line holography process by a linear shift invariant system: application to the measurement of fiber diameter. *Opt. Comm.*, 223:263–271, 2003. {**134**}

[248] D. Lebrun, S. Belaid, and C. Özkul. Hologram reconstruction by use of optical wavelet transform. *Appl. Opt.*, 38(17):3730–3734, 1999. {**134**}

[249] S. Soontaranon, J. Widjaja, and T. Asakura. Direct analysis of in-line particle holograms by using wavelet transform and envelope reconstruction method. *Optik*, 113(11):489–494, 2002. {**134**}

[250] L. Onural and M. Kocatepe. Family of scaling chirp functions, diffraction, and holography. *IEEE Trans. Signal Proc.*, 43(71):1568–1578, 1995. {**134**}

[251] L. Onural. Diffraction from a wavelet point of view. *Opt. Lett.*, 18(11):846–848, 1993. {**134**}

[252] Th. Kreis, M. Adams, and W. Jüptner. Methods of digital holography: A comparison. In *European Symposium on Lasers and Optics in Manufacturing*, volume 3098 of *Proc. of Soc. Photo-Opt. Instr. Eng.*, pages 224–233, 1997. {**134**}

[253] Th. Kreis and W. Jüptner. Principles of digital holography. In W. Jüptner and W. Osten, editors, *Fringe '97*, pages 353–363, 1997. {**136, 162**}

[254] J. W. Goodman. Statistical properties of laser speckle patterns. In J. C. Dainty, editor, *Laser Speckle and Related Phenomena*, volume 9 of *Topics in Applied Physics*, pages 9–75, 1975. {**136**}

[255] Th. Kreis, M. Adams, and W. Jüptner. Aperture synthesis in digital holography. In K. Creath and J. Schmitt, editors, *Laser Interferometry XI: Techniques and Analysis*, volume 4777 of *Proc. of Soc. Photo-Opt. Instr. Eng.*, pages 69–76, 2002. {**137**}

[256] Th. Kreis. Frequency analysis of digital holography. *Opt. Eng.*, 41(4):771–778, 2002. {**141**}

[257] Th. Kreis. Frequency analysis of digital holography with reconstruction by convolution. *Opt. Eng.*, 41(8):1829–1839, 2002. {**141**}

[258] L. Xu, J. Miao, and A. Asundi. Properties of digital holography based on in-line configuration. *Opt. Eng.*, 39(12):3214–3219, 2000. {**146, 179**}

[259] M. L. Huebschman, B. Munjuluri, and H. R. Garner. Dynamic holographic 3-D image projection. *Opt. Expr.*, 11(5):437–445, 2003. {**146**}

[260] E. Tajahuerce, O. Matoba, and B. Javidi. Shift-invariant three-dimensional object recognition by means of digital holography. *Appl. Opt.*, 40(23):3877–3886, 2001. {**159**}

[261] Th. Kreis, M. Adams, and W. Jüptner. Digital in-line holography in particle measurement. In M. Kujawinska, editor, *Interferometry 99: Techniques and Technology*, volume 3744 of *Proc. of Soc. Photo-Opt. Instr. Eng.*, pages 54–64, 1999. {**160, 165**}

[262] Th. Kreis. Digital holography for metrologic applications. In P. Jacquot and J.-M. Fournier, editors, *Interferometry in Speckle Light*, pages 205–212, 2000. {**160, 333, 348**}

[263] G. Pan and H. Meng. Digital holography of particle fields: reconstruction by use of complex amplitude. *Appl. Opt.*, 42(5):827–833, 2003. {**160**}

[264] C. Fournier, C. Ducottet, and Th. Fournel. Digital in-line holography: influence of the reconstruction function on the axialprofile of a reconstructed particle image. *Meas. Sci. & Techn.*, 15:686–693, 2004. {**160**}

[265] M. Adams, Th. Kreis, and W. Jüptner. Particle size and position measurement with digital holography. In *European Symposium on Lasers and Optics in Manufacturing*, volume 3098 of *Proc. of Soc. Photo-Opt. Instr. Eng.*, pages 234–240, 1997. {**160, 162, 165**}

[266] Th. Kreis, W. Jüptner, and J. Geldmacher. Digital holography: Methods and applications. In *International Conference on Applied Optical Metrology*, volume 3407 of *Proc. of Soc. Photo-Opt. Instr. Eng.*, pages 169–177, 1998. {**162**}

[267] J. Lobera, N. Andres, and M. P. Arroyo. Digital image plane holography as a three-dimensional flow velocimetry technique. In K. Gastinger, O. Løkberg, and S. Winther, editors, *Speckle Metrology 2003*, volume 4933 of *Proc. of Soc. Photo-Opt. Instr. Eng.*, pages 279–284, 2003. {**166, 403**}

[268] S. Lai, B. King, and M. A. Neifeld. Wavefront reconstruction by means of phase-shifting digital in-line holography. *Opt. Comm.*, 173:155–160, 2000. {**167**}

[269] S. Lai, B. Kemper, and G. von Bally. Off-axis reconstruction of in-line holograms for twin-image elimination. *Opt. Comm.*, 169:37–43, 1999. {**167**}

[270] H. Sun, H. Dong, M. A. Player, J. Watson, D. M. Paterson, and R. Perkins. In-line digital video holography for the study of erosion processes in sediments. *Meas. Sci. & Techn.*, 13:L7–L12, 2002. {**168**}

[271] Y. Takaki and H. Ohzu. Hybrid holographic microscopy: visualization of three-dimensional object information by use of viewing angles. *Appl. Opt.*, 39(29):5302–5308, 2000. {**174**}

[272] L. Yu, Y. An, and L. Cai. Numerical reconstruction of digital holograms with variable viewing angles. *Opt. Expr.*, 10(22):1250–1257, 2002. {**174**}

[273] D. Lebrun, A. M. Benkouider, S. Coetmellec, and M. Malek. Particle field digital holographic reconstruction in arbitrary tilted planes. *Opt. Expr.*, 11(3):224–229, 2003. {**174**}

[274] E. Cuche, P. Marquet, and Chr. Depeursinge. Simultaneous amplitude-contrast and quantitative phase-contrast microscopy by numerical reconstruction of Fresnel off-axis holograms. *Appl. Opt.*, 38(34):6994–7001, 1999. {**178**}

[275] P. Ferraro, S. De Nicola, A. Finizio, G. Coppola, S. Grilli, C. Magro, and G. Pierattini. Compensation of the inherent wave front curvature in digital holographic coherent microscopy for quantitative phase-contrast imaging. *Appl. Opt.*, 42:1938–1946, 2003. {**178**}

[276] L. Xu, X. Peng, A. Asundi, and J. Miao. Digital microholointerferometer: development and validation. *Opt. Eng.*, 42(8):2218–2224, 2003. {**179**}

[277] G. Pedrini and H. J. Tiziani. Short-coherence digital microscopy by use of a lensless holographic imaging system. *Appl. Opt.*, 41(22):4489–4496, 2002. {**180**}

[278] G. Pedrini, S. Schedin, and H. J. Tiziani. Aberration compensation in digital holographic reconstruction of microscopic objects. *Journ. Modern Opt.*, 48(6):1035–1041, 2001. {**180**}

[279] G. Pedrini, S. Schedin, and H. J. Tiziani. Spatial filtering in digital holographic microscopy. *Journ. Modern Opt.*, 47(8):1447–1454, 2000. {**180**}

[280] Th. J. Naughton, J. B. McDonald, and B. Javidi. Efficient compression of Fresnel fields for internet transmission of three-dimensional images. *Appl. Opt.*, 42:4758–4764, 2003. {**180**}

[281] B. Javidi and T. Nomura. Securing information by use of digital holography. *Opt. Lett.*, 25(1):28–30, 2000. {**180**}

[282] N. Takai and Y. Mifune. Digital watermarking by a holographic technique. *Appl. Opt.*, 41(5):865–873, 2002. {**183**}

[283] S. Kishk and B. Javidi. Watermarking of three-dimensional objects by digital holography. *Opt. Lett.*, 28(3):167–169, 2003. {**183**}

[284] R. W. Larson, J. S. Zelenka, and E. L. Johansen. Microwave radar imagery. In *Proceedings of Symposium on Engineering Applications of Holography*, Proc. of Soc. Photo-Opt. Instr. Eng., page 14, 1972. {**185**}

[285] K. Suzuki and B. P. Hildebrand. Holographic interferometry with acoustic waves. In N. Booth, editor, *Acoustical Holography*, volume 6, pages 577–595. Plenum Press, New York, 1975. {**185**}

[286] J. E. Sollid and J. B. Swint. A determination of the optimum beam ratio to produce maximum contrast photographic reconstructions from double-exposure holographic interferograms. *Appl. Opt.*, 9(12):2717–2719, 1970. {**186**}

[287] J. E. Sollid. Holographic interferometry applied to measurements of small static displacements of diffusely reflecting surfaces. *Appl. Opt.*, 8(8):1587–1595, 1969. {**191, 266, 301**}

[288] R. Pawluczyk and Z. Kraska. Diffuse illumination in holographic double-aperture interferometry. *Appl. Opt.*, 24(18):3072–3078, 1985. {**194**}

[289] J. D. Trolinger. The holography of phase objects. In N. A. Massie, editor, *Interferometric Metrology, Critical Reviews*, volume 816 of *Proc. of Soc. Photo-Opt. Instr. Eng.*, pages 128–139, 1987. {**194**}

[290] N. L. Hecht, J. E. Minardi, D. Lewis, and R. L. Fusek. Quantitative theory for predicting fringe pattern formation in holographic interferometry. *Appl. Opt.*, 12(11):2665–2676, 1973. {**196**}

[291] J. Janta and M. Miler. Model interferogram as an aid for holographic interferometry. *J. Optics*, 8(5):301–307, 1977. {**196**}

[292] R. Höfling and W. Osten. Displacement measurement by image processed speckle patterns. *Journ. Modern Opt.*, 34:607–617, 1987. {**196**}

[293] Th. Kreis, W. Jüptner, and R. Biedermann. Neural network approach to holographic nondestructive testing. *Appl. Opt.*, 34(8):1407–1415, 1995. {**197, 387, 391, 395**}

[294] D. I. Farrant and J. N. Petzing. Sensitivity errors in interferometric deformation metrology. *Appl. Opt.*, 42:5634–5641, 2003. {**198, 200**}

[295] G. S. Rightley, L. K. Matthews, and G. P. Mulholland. Holographic analysis and experimental error. In K. Stetson and R. Pryputniewicz, editors, *International Conference on Hologram Interferometry and Speckle Metrology*, pages 343–350, 1990. {**200**}

[296] W. Jüptner, K. Ringer, and H. Welling. The evaluation of interference fringes for holographic strain and translation measurement (in German). *Optik*, 38:437–448, 1973. {**200**}

[297] N. Abramson. The holo-diagram: A practical device for making and evaluating holograms. *Appl. Opt.*, 8:1235–1240, 1969. {**200**}

[298] N. Abramson. The holo-diagram II: A practical device for information retrieval in hologram interferometry. *Appl. Opt.*, 9:97–101, 1970. {**200**}

[299] N. Abramson. The holo-diagram III: A practical device for predicting fringe patterns in hologram interferometry. *Appl. Opt.*, 9:2311–2320, 1970. {**200**}

[300] N. Abramson. The holo-diagram IV: A practical device for simulating fringe patterns in hologram interferometry. *Appl. Opt.*, 10:2155–2161, 1971. {**200**}

[301] N. Abramson. The holo-diagram V: A device for practical interpreting of hologram fringes. *Appl. Opt.*, 11:1143–1147, 1972. {**200**}

[302] S. Toyooka. Holographic interferometry with increased sensitivity for diffusely reflecting objects. *Appl. Opt.*, 16(4):1054–1057, 1977. {**201**}

[364] A. Choudry, H. Dekker, and D. Enard. Automated interferometric evaluation of optical components at the European Southern Observatory (ESO). In W. F. Fagan, editor, *Industrial Applications of Laser Technology*, volume 398 of *Proc. of Soc. Photo-Opt. Instr. Eng.*, pages 66–72, 1983. {**221**}

[365] A. C. Gillies. Image processing approach to fringe patterns. *Opt. Eng.*, 27(10):861–866, 1988. {**221**}

[366] F. Ginesu and F. Bertolino. Numerical analysis of fringe patterns for structural engineering problems. In R. J. Pryputniewicz, editor, *Laser Interferometry IV: Computer-Aided Interferometry*, volume 1553 of *Proc. of Soc. Photo-Opt. Instr. Eng.*, pages 313–324, 1991. {**221**}

[367] J. M. Huntley. Fringe analysis today and tomorrow. In K. Gastinger, O. Løkberg, and S. Winther, editors, *Speckle Metrology 2003*, volume 4933 of *Proc. of Soc. Photo-Opt. Instr. Eng.*, pages 167–174, 2003. {**221**}

[368] Th. Kreis. Computer aided evaluation of fringe patterns. *Opt. Lasers in Eng.*, 19:221–240, 1993. {**221**}

[369] M. Kujawinska. The architecture of a multipurpose fringe pattern analysis system. *Opt. Lasers in Eng.*, 19:261–268, 1993. {**221**}

[370] G. E. Maddux. Video/computer techniques for static and dynamic experimental mechanics. In R. J. Pryputniewicz, editor, *Industrial Laser Interferometry*, volume 746 of *Proc. of Soc. Photo-Opt. Instr. Eng.*, pages 52–57, 1987. {**221**}

[371] R. J. Pryputniewicz. Review of methods for automatic analysis for fringes in hologram interferometry. In N. A. Massie, editor, *Interferometric Metrology, Critical Reviews*, volume 816 of *Proc. of Soc. Photo-Opt. Instr. Eng.*, pages 140–148, 1987. {**221**}

[372] G. T. Reid. Automatic fringe pattern analysis: A review. *Opt. Lasers in Eng.*, 7:37–68, 1986. {**221**}

[373] D. W. Robinson and G. T. Reid, editors. *Interferogram Analysis: Digital Fringe Pattern Measurement Techniques*. Institute of Physics, Bristol and Philadelphia, 1993. {**221**}

[374] C. A. Sciammarella and G. Bhat. Computer assisted techniques to evaluate fringe patterns. In R. J. Pryputniewicz, editor, *Laser Interferometry IV: Computer-Aided Interferometry*, volume 1553 of *Proc. of Soc. Photo-Opt. Instr. Eng.*, pages 252–262, 1991. {**221**}

[375] J. S. Sirkis, Y.-M. Chen, H. Singh, and A. Y. Cheng. Computerized optical fringe pattern analysis in photomechanics: a review. *Opt. Eng.*, 31(2):304–314, 1992. {**221**}

[376] R. J. Pryputniewicz. Quantitative determination of displacements and strains from holograms. In P. K. Rastogi, editor, *Holographic Interferometry*, volume 68 of *Springer Series in Optical Sciences*, pages 33–74, 1994. {**222, 318, 329, 404**}

[377] N. Abramson. The rose of error or the importance of sign. In *Technical Digest of Topical Meeting On Hologram Interferometry and Speckle Metrology*. Opt. Soc. Amer., 1980. {**222**}

[378] Th. Kreis. Computer-aided evaluation of holographic interferograms. In P. K. Rastogi, editor, *Holographic Interferometry*, volume 68 of *Springer Series in Optical Sciences*, pages 151–212, 1994. {**222**}

[379] D. R. Matthys, J. A. Gilbert, T. D. Dudderar, and K. W. Koenig. A windowing technique for the automated analysis of holo-interferograms. *Opt. Lasers in Eng.*, 8:123–136, 1988. {**224**}

[380] P. D. Plotkowski, M. Y. Y. Hung, J. D. Hovanesian, and G. Gerhardt. Improved fringe carrier techniques for unambiguous determination of holographically recorded displacements. *Opt. Eng.*, 24(5):754–756, 1985. {**224**}

[381] J. Petkovsek and K. Rankel. Measurement of three-dimensional displacement by four small holograms. In P. Meyrueis and M. Grosmann, editors, *2nd European Congress on Optics Applied to Metrology*, volume 210 of *Proc. of Soc. Photo-Opt. Instr. Eng.*, pages 173–177, 1979. {**224, 299**}

[382] N. Eichhorn and W. Osten. An algorithm for the fast derivation of line structures from interferograms. *Journ. Modern Opt.*, 35(10):1717–1725, 1988. {**225, 229, 231**}

[383] W. Osten. *Digital Processing and Evaluation of Interference Images*. Akademie-Verlag, Berlin, 1991. (In German). {**226**}

[384] W. Luth. *Isolation of Moving Objects in Digital Image Sequences*. Academy of Sciences of the GDR, 1989. (In German). {**226**}

[385] K. Creath. Submicron linewidth measurement using an interferometric optical profiler. In W. H. Arnold, editor, *Integrated Circuit Metrology, Inspection, and Process Control*, volume 1464 of *Proc. of Soc. Photo-Opt. Instr. Eng.*, pages 474–483, 1991. {**228**}

[386] K. Creath. Holographic contour and deformation measurement using a 1.4 million element detector array. *Appl. Opt.*, 28(11):2170–2175, 1989. {**228, 333**}

[387] K. Creath. Phase-measurement interferometry: Beware these errors. In R. J. Pryputniewicz, editor, *Laser Interferometry IV: Computer-Aided Interferometry*, volume 1553 of *Proc. of Soc. Photo-Opt. Instr. Eng.*, pages 213–220, 1991. {**228, 255**}

[388] K. Freischlad and C. L. Koliopoulos. Fourier description of digital phase-measuring interferometry. *Journ. Opt. Soc. Amer. A*, 7(4):542–551, 1990. {**228, 255**}

[389] C. P. Brophy. Effect of intensity error correlation on the computed phase of phase shifting interferometry. *Journ. Opt. Soc. Amer. A*, 7(4):537–541, 1990. {**228, 255**}

[390] O. A. Skydan, F. Lilley, M. J. Lalor, and D. R. Burton. Quantization error of CCD cameras and their influence on phase calculation in fringe pattern analysis. *Appl. Opt.*, 42:5302–5307, 2003. {**228**}

[391] W. R. J. Funnell. Image processing applied to the interactive analysis of interferometric fringes. *Appl. Opt.*, 20(18):3245–3250, 1981. {**229, 231, 233, 235, 374**}

[392] W. Osten, J. Saedler, and H. Rottenkolber. Interpretation of interferometric fringe patterns using digital image processing (in German). *Technisches Messen tm*, 54:285–290, 1988. {**229**}

[393] T. Yatagai. Intensity based analysis methods. In D. W. Robinson and G. T. Reid, editors, *Interferogram Analysis: Digital Fringe Pattern Measurement Techniques*, pages 72–93. Institute of Physics, Bristol and Philadelphia, 1993. {**229, 231, 235**}

[394] Th. Kreis and H. Kreitlow. Quantitative evaluation of holographic interference patterns under image processing aspects. In P. Meyrueis and M. Grosmann, editors, *2nd European Congress on Optics Applied to Metrology*, volume 210 of *Proc. of Soc. Photo-Opt. Instr. Eng.*, pages 196–202, 1979. {**230, 231, 287**}

[395] A. K. Jain and C. R. Christensen. Digital processing of images in speckle noise. In W. H. Carter, editor, *Applications of Speckle Phenomena*, volume 243 of *Proc. of Soc. Photo-Opt. Instr. Eng.*, pages 46–50, 1980. {**230**}

[396] J. S. Lim and H. Nawab. Techniques for speckle noise removal. *Opt. Eng.*, 20(3):472–480, 1981. {**230, 409**}

[397] F. A. Sadjadi. Perspectives on techniques for enhancing speckled imagery. *Opt. Eng.*, 29(1):25–30, 1990. {**230**}

[398] T. R. Crimmins. Geometric filter for speckle reduction. *Appl. Opt.*, 24(10):1438–1443, 1985. {**230**}

[399] T. R. Crimmins. Geometric filter for reducing speckle. *Opt. Eng.*, 25(5):651–654, 1986. {**230**}

[400] E. Bieber and W. Osten. Improvement of speckled fringe patterns by Wiener filtration. In Z. Jaroszewicz and M. Pluta, editors, *Interferometry 89*, volume 1121 of *Proc. of Soc. Photo-Opt. Instr. Eng.*, pages 393–399, 1989. {**230**}

[401] Y. Katzir, I. Glaser, A. A. Friesem, and B. Sharon. On-line acquisition and analysis for holographic nondestructive evaluation. *Opt. Eng.*, 21:1016, 1982. {**231**}

[402] F. Becker, G. E. A. Meier, and H. Wegner. Automatic evaluation of interferograms. In A. G. Tescher, editor, *Applications of Digital Image Processing*, volume 359 of *Proc. of Soc. Photo-Opt. Instr. Eng.*, pages 386–393, 1982. {**231**}

[403] F. Becker and Y. H. Yu. Digital fringe reduction technique applied to the measurement of three-dimensional transonic flow fields. *Opt. Eng.*, 24(3):429–434, 1985. {**231, 234, 235**}

[404] J. Budzinski. SNOP: a method for skeletonization of a fringe pattern along a fringe direction. *Appl. Opt.*, 31(16):3109–3113, 1992. {**231, 233**}

[405] S. Nakadate, N. Magome, T. Honda, and J. Tsujiuchi. Hybrid holographic interferometer for measuring three-dimensional deformations. *Opt. Eng.*, 20(2):246–252, 1981. {**231, 235**}

[406] T. Yatagai, S. Nakadate, M. Idesawa, and H. Saito. Automatic fringe analysis using digital image processing techniques. *Opt. Eng.*, 21(3):432–435, 1982. {**231, 235**}

[407] E. A. Mnatsakanyan and S. V. Nefyodov. Skeletonizing of interferometric images for finding maximum and minimum centres of fringes. In K. Stetson and R. Pryputniewicz, editors, *International Conference on Hologram Interferometry and Speckle Metrology*, pages 351–355, 1990. {**231**}

[408] H. Winter, S. Unger, and W. Osten. The application of adaptive and anisotropic filtering for the extraction of fringe pattern skeletons. In W. Osten, R. J. Pryputniewicz, G. T. Reid, and H. Rottenkolber, editors, *Fringe '89, Automatic Processing of Fringe Patterns*, Physical Research, pages 158–166. Akademie-Verlag, Berlin, 1989. {**231**}

[409] J. A. Aparicio, J. L. Molpeceres, A. M. de Frutos, C. de Castro, S. Caceres, and F. A. Frechoso. Improved algorithm for the analysis of holographic interferograms. *Opt. Eng.*, 32(5):963–969, 1993. {**233**}

[410] F. Bertolino and F. Ginesu. Semiautomatic approach of grating techniques. *Opt. Lasers in Eng.*, 19:313–323, 1993. {**233**}

[411] A. E. Ennos, D. W. Robinson, and D. C. Williams. Automatic fringe analysis in holographic interferometry. *Optica Acta*, 32(2):135–145, 1985. {**233**}

[412] S. Krishnaswamy. Algorithm for computer tracing of interference fringes. *Appl. Opt.*, 30(13):1624–1628, 1991. {**233**}

[413] R. Nübel. Computer-aided evaluation method for interferograms. *Exp. Fluids*, 12:166–172, 1992. {**233**}

[414] G. W. Johnson, D. C. Leiner, and D. T. Moore. Phase-locked interferometry. *Opt. Eng.*, 18(1):46–52, 1979. {**233**}

[415] G. A. Mastin and D. C. Ghiglia. Digital extraction of interference fringe contours. *Appl. Opt.*, 24(12):1727–1728, 1985. {**234**}

[416] V. Srinivasan, S.-T. Yeo, and P. Chaturvedi. Fringe processing and analysis with a neural network. *Opt. Eng.*, 33(4):1166–1171, 1994. {**234**}

[417] K. H. Womack, K. L. Underwood, and D. Forbes. Microprocessor-based video interferogram analysis system. In *Minicomputers and Microprocessors in Optical Systems*, volume 230 of *Proc. of Soc. Photo-Opt. Instr. Eng.*, pages 168–179, 1980. {**235**}

[418] A. Colin and W. Osten. Automatic support for consistent labelling of skeletonized fringe patterns. *Journ. Modern Opt.*, 42:945–954, 1995. {**235**}

[419] J. B. Schemm and C. M. Vest. Fringe pattern recognition and interpolation using nonlinear regression analysis. *Appl. Opt.*, 22(18):2850–2853, 1983. {**235**}

[420] U. Mieth and W. Osten. Three methods for the interpolation of phase values between fringe pattern skeletons. In W. Osten, R. J. Pryputniewicz, G. T. Reid, and H. Rottenkolber, editors, *Fringe '89, Automatic Processing of Fringe Patterns*, Physical Research, pages 118–123. Akademie-Verlag, Berlin, 1989. {**235**}

[421] R. Dändliker, B. Ineichen, and F. M. Mottier. High resolution hologram interferometry by electronic phase measurement. *Opt. Comm.*, 9:412–416, 1973. {**235**}

[422] P. V. Farrell, G. S. Springer, and C. M. Vest. Heterodyne holographic interferometry: concentration and temperature measurements in gas mixtures. *Appl. Opt.*, 21(9):1624–1627, 1982. {**235**}

[423] R. J. Pryputniewicz. Heterodyne holography, applications in studies of small components. *Opt. Eng.*, 24(5):849–854, 1985. {**235, 387**}

[424] R. Thalmann and R. Dändliker. Strain measurement by heterodyne holographic interferometry. *Appl. Opt.*, 26(10):1964–1971, 1987. {**235**}

[425] Th. Kreis, J. Geldmacher, and R. Biedermann. Theoretical and experimental investigations of the accuracy of diverse methods for evaluating interference patterns. In *Report on DFG-Project Kr953/2-1*, 1993. (In German). {**239**}

[426] Th. Kreis, J. Geldmacher, and W. Jüptner. A comparison of interference phase determination methods with respect to achievable accuracy. In W. Jüptner and W. Osten, editors, *Fringe '93*, pages 51–59, 1993. {**241**}

[427] B. Breuckmann and W. Thieme. Computer-aided analysis of holographic interferograms using the phase-shift method. *Appl. Opt.*, 24(14):2145–2149, 1985. {**242**}

[428] R. Dändliker and R. Thalmann. Heterodyne and quasi-heterodyne holographic interferometry. *Opt. Eng.*, 24(5):824–831, 1985. {**242**}

[494] G. Schönebeck. New holographic means to exactly determine coefficients of elasticity. In W. F. Fagan, editor, *Industrial Applications of Laser Technology*, volume 398 of *Proc. of Soc. Photo-Opt. Instr. Eng.*, pages 130–136, 1983. {**280, 317**}

[495] G. Schönebeck. Holography and torsional problems. In W. F. Fagan, editor, *Industrial Optoelectronic Measurement Systems using Coherent Light*, volume 863 of *Proc. of Soc. Photo-Opt. Instr. Eng.*, pages 173–177, 1987. {**280, 317**}

[496] P. Picart, E. Moisson, and D. Mounier. Mechanical measurement using multiplexing/demultiplexing of digital Fresnel holograms. In K. Gastinger, O. Løkberg, and S. Winther, editors, *Speckle Metrology 2003*, volume 4933 of *Proc. of Soc. Photo-Opt. Instr. Eng.*, pages 346–354, 2003. {**282**}

[497] P. Picart, E. Moisson, and D. Mounier. Twin sensitivity measurement by spatial multiplexing of digitally recorded holograms. *Appl. Opt.*, 42:1947–1957, 2003. {**282**}

[498] G. Pedrini and H. J. Tiziani. Quantitative evaluation of two-dimensional dynamic deformations using digital holography. *Opt. Laser Technol.*, 29(5):249–256, 1997. {**283**}

[499] J. M. Huntley and J. R. Buckland. Characterization of sources of 2π phase discontinuity in speckle interferograms. *Journ. Opt. Soc. Amer. A*, 12(9):1990–1996, 1995. {**287**}

[500] W. Osten and R. Höfling. The inverse modulo process in automatic fringe analysis – problems and approaches. In K. Stetson and R. Pryputniewicz, editors, *International Conference on Hologram Interferometry and Speckle Metrology*, pages 301–309, 1990. {**287**}

[501] J. Munoz, G. Paez, and M. Strojnik. Two-dimensional phase unwrapping of subsampled phase-shifted interferograms. *Journ. Modern Opt.*, 51(1):49–63, 2004. {**287, 292**}

[502] J. Munoz, M. Strojnik, and G. Paez. Phase recovery from a single undersampled interferograms. *Appl. Opt.*, 42(34):6846–6852, 2003. {**287**}

[503] D. R. Burton and M. J. Lalor. Multichannel Fourier fringe analysis as an aid to automatic phase unwrapping. *Appl. Opt.*, 33(14):2939–2948, 1994. {**288**}

[504] D. J. Bone. Fourier fringe analysis: the two-dimensional phase unwrapping problem. *Appl. Opt.*, 30(25):3627–3632, 1991. {**288, 292**}

[505] K. Itoh. Analysis of the phase unwrapping algorithm. *Appl. Opt.*, 21(14):2470, 1982. {**289**}

[506] J. M. Tribolet. A new phase unwrapping algorithm. *IEEE Trans. on Acoust., Speech, and Signal Proc.*, 25(2):170–177, 1977. {**289**}

[507] D. W. Robinson. Phase unwrapping methods. In D. W. Robinson and G. T. Reid, editors, *Interferogram Analysis: Digital Fringe Pattern Measurement Techniques*, pages 194–229. Institute of Physics, Bristol and Philadelphia, 1993. {**289, 292**}

[508] M. Takeda. Fringe formula for projection type moire topography. *Opt. Lasers in Eng.*, 1(3):45–52, 1982. {**289**}

[509] D. W. Robinson and D. C. Williams. Digital phase stepping speckle interferometry. *Opt. Comm.*, 57(1):26–30, 1986. {**289**}

[510] H. A. Vrooman and A. A. M. Maas. Image processing in digital speckle interferometry. In H. Halliwell, editor, *Fringe Analysis 89*, FASIG, 1989. {**289**}

[511] K. Andresen and Q. Yu. Robust phase unwrapping by spin filtering using a phase direction map. In W. Jüptner and W. Osten, editors, *Fringe '93*, pages 154–156, 1993. {**289**}

[512] P. J. Bryanston-Cross and C. Quan. Examples of automatic phase unwrapping applied to interferometric and photoelastic images. In W. Jüptner and W. Osten, editors, *Fringe '93*, pages 121–135, 1993. {**289**}

[513] D. G. Ghiglia and L. A. Romero. Robust two-dimensional weighted and unweighted phase unwrapping that uses fast transforms and iterative methods. *Journ. Opt. Soc. Amer. A*, 11(1):107–117, 1994. {**289, 295**}

[514] T. R. Judge, Ch. Quan, and P. J. Bryanston-Cross. Holographic deformation measurements by Fourier transform technique with automatic phase unwrapping. *Opt. Eng.*, 31(3):533–543, 1992. {**289**}

[515] J. A. Quiroga and E. Bernabeu. Phase-unwrapping algorithm for noisy phase-map processing. *Appl. Opt.*, 33(29):6725–6731, 1994. {**289**}

[516] G. Schirripa Spagnolo, D. Ambrosini, D. Paoletti, and R. Borghi. High-speed digital processing of electro-optic holography images for a quantitative analysis. *J. Opt.*, 28:118–124, 1997. {**289, 400**}

[517] J. Schörner, A. Ettemeyer, U. Neupert, H. Rottenkolber, C. Winter, and P. Obermeier. New approaches in interpreting holographic images. *Opt. Lasers in Eng.*, 14:283–291, 1991. {**289**}

[518] R. Cusack, J. M. Huntley, and H. T. Goldrein. Improved noise-immune phase-unwrapping algorithm. *Appl. Opt.*, 34(5):781–789, 1995. {**290**}

[519] J. M. Huntley. Noise-immune phase unwrapping algorithm. *Appl. Opt.*, 28(15):3268–3270, 1989. {**290, 295**}

[520] J. J. Gierloff. Phase unwrapping by regions. In R. E. Fischer and W. J. Smith, editors, *Current Developments in Optical Engineering II*, volume 818 of *Proc. of Soc. Photo-Opt. Instr. Eng.*, pages 2–9, 1987. {**292**}

[521] O. Y. Kwon. Advanced wavefront sensing at Lockheed. In N. A. Massie, editor, *Interferometric Metrology*, volume 816 of *Proc. of Soc. Photo-Opt. Instr. Eng.*, pages 196–211, 1987. {**292**}

[522] D. Winter, R. Ritter, and H. Sadewasser. Evaluation of modulo 2π phase-images with regional discontinuities: area based unwrapping. In W. Jüptner and W. Osten, editors, *Fringe '93*, pages 157–159, 1993. {**292**}

[523] P. Stephenson, D. R. Burton, and M. J. Lalor. Data validation techniques in a tiled phase unwrapping algorithm. *Opt. Eng.*, 33(11):3703–3708, 1994. {**292**}

[524] D. P. Towers, T. Judge, and P. J. Bryanston-Cross. A quasi heterodyne holographic technique and automatic algorithms for phase unwrapping. In G. T. Reid, editor, *Fringe Pattern Analysis*, volume 1163 of *Proc. of Soc. Photo-Opt. Instr. Eng.*, pages 95–119, 1989. {**292**}

[525] D. P. Towers, T. R. Judge, and P. J. Bryanston-Cross. Automatic interferogram analysis techniques applied to quasi-heterodyne holography and ESPI. *Opt. Lasers in Eng.*, 14:239–281, 1991. {**292**}

[526] A. Baldi. Phase unwrapping by region growing. *Appl. Opt.*, 42(14):2498–2505, 2003. {**292**}

[527] D. G. Ghiglia, G. A. Mastin, and L. A. Romero. Cellular-automata method for phase unwrapping. *Journ. Opt. Soc. Amer. A*, 4(1):267–280, 1987. {**292, 294**}

Bibliography

[528] A. Spik and D. W. Robinson. Investigation of the cellular automata method for phase unwrapping and its implementation on an array processor. *Opt. Lasers in Eng.*, 14:25–37, 1991. {**292, 294**}

[529] R. J. Green and J. G. Walker. Phase unwrapping using a priori knowledge about the band limits of a function. In D. W. Braggins, editor, *Industrial Inspection*, volume 1010 of *Proc. of Soc. Photo-Opt. Instr. Eng.*, pages 36–43, 1989. {**294**}

[530] R. J. Green, J. G. Walker, and D. W. Robinson. Investigation of the Fourier-transform method of fringe pattern analysis. *Opt. Lasers in Eng.*, 8:29–44, 1988. {**294**}

[531] Th. Kreis, R. Biedermann, and W. Jüptner. Evaluation of holographic interference patterns by artificial neural networks. In M. Kujawinska, R. J. Pryputniewicz, and M. Takeda, editors, *Interferometry VII: Techniques and Analysis*, volume 2544 of *Proc. of Soc. Photo-Opt. Instr. Eng.*, pages 11–24, 1995. {**294, 395**}

[532] M. Takeda, K. Nagatome, and Y. Watanabe. Phase unwrapping by neural network. In W. Jüptner and W. Osten, editors, *Fringe '93*, pages 136–141, 1993. {**294**}

[533] J. M. Huntley, R. Cusack, and H. Saldner. New phase unwrapping algorithms. In W. Jüptner and W. Osten, editors, *Fringe '93*, pages 148–153, 1993. {**295**}

[534] J. M. Huntley and H. Saldner. Temporal phase-unwrapping algorithm for automated interferogram analysis. *Appl. Opt.*, 32(17):3047–3052, 1993. {**295**}

[535] G. Pedrini, I. Alexeenko, W. Osten, and H. J. Tiziani. Temporal phase unwrapping of digital hologram sequences. *Appl. Opt.*, 42:5846–5854, 2003. {**295**}

[536] J. S. Lim. *Two-Dimensional Signal and Image Processing*. Prentice-Hall Intern., 1990. {**295**}

[537] J. Arines. Least-squares modal estimation of wrapped phases: application to phase unwrapping. *Appl. Opt.*, 42:3373–3378, 2003. {**295**}

[538] J. Ebbeni. Measurement of mechanical deformations from holographic interferograms. In P. Smigielski, editor, *Deuxieme Colloque Franco-Allemand sur les Applications de l'Holographie*, pages 1–18, 1989. {**298**}

[539] Z. Füzessy and N. Abramson. Measurement of 3-D displacement: sandwich holography and regulated path length interferometry. *Appl. Opt.*, 21(2):260–264, 1982. {**298**}

[540] E. Müller, V. Hrdliczka, and D. E. Cuche. Computer-based evaluation of holographic interferograms. In W. F. Fagan, editor, *Industrial Applications of Laser Technology*, volume 398 of *Proc. of Soc. Photo-Opt. Instr. Eng.*, pages 46–52, 1983. {**298**}

[541] R. J. Pryputniewicz and W. W. Bowley. Techniques of holographic displacement measurement: an experimental comparison. *Appl. Opt.*, 17(11):1748–1756, 1978. {**299**}

[542] K. A. Stetson. Use of sensivity vector variations to determine absolute displacements in double exposure hologram interferometry. *Appl. Opt.*, 29(4):502–504, 1990. {**300**}

[543] Z. Füzessy. Application of double-pulse holography for the investigations of machines and systems. In G. Frankowski, N. Abramson, and Z. Füzessy, editors, *Application of Metrological Laser Methods in Machines and Systems*, volume 15 of *Physical Research*, pages 75–107. Akademie Verlag, Berlin, 1991. {**301**}

[544] Z. Füzessy and F. Gyimesi. Difference holographic interferometry: displacement measurement. *Opt. Eng.*, 23(6):780–783, 1984. {**301, 350, 351**}

[545] Z. Füzessy and P. Wesolowski. Simplified static holographic evaluation method omitting the zero-order fringe. *Opt. Eng.*, 24(6):1023–1025, 1985. {**301**}

[546] W. Jüptner, J. Geldmacher, Th. Bischof, and Th. Kreis. Measurement of the deformation of a pressure vessel above a weld point. In R. J. Pryputniewicz, G. M. Brown, and W. Jüptner, editors, *Interferometry: Applications*, volume 1756 of *Proc. of Soc. Photo-Opt. Instr. Eng.*, pages 98–105, 1992. {**301, 305**}

[547] Th. Kreis and W. Jüptner. Determination of defects by combination of holographic interferometry and finite-element-method. *VDI-Berichte*, 631:139–151, 1987. (In German). {**301, 321, 322**}

[548] M. Medhat. Computer analysis of interferograms by fitting quadratic forms. *Journ. Modern Opt.*, 38(1):121–128, 1991. {**303**}

[549] J. D. Trolinger, D. C. Weber, G. C. Pardoen, G. T. Gunnarsson, and W. F. Fagan. Application of long-range holography in earthquake engineering. *Opt. Eng.*, 30(9):1315–1319, 1991. {**304**}

[550] W. Jüptner and H. Kreitlow. Holographic recording of non-vibration protected objects (in German). In W. Waidelich, editor, *Laser 77 Opto-Electronics*, pages 420–425. ipc science and technology press, 1977. {**304**}

[551] D. B. Neumann and H. W. Rose. Improvement of recorded holographic fringes by feedback control. *Appl. Opt.*, 6(6):1097–1104, 1967. {**304**}

[552] V. Kebbel, H.-J. Hartmann, W. Jüptner, U. Schnars, L. Gatti, and J. Becker. Detection and compensation of misalignment for interferometric diagnostic tools applied in space-borne facilities. In G. M. Brown, W. Jüptner, and R. Pryputniewicz, editors, *Interferometry X: Applications*, volume 4101 of *Proc. of Soc. Photo-Opt. Instr. Eng.*, pages 468–476, 2000. {**304, 385**}

[553] H. W. Rose and H. D. Pruett. Stabilization of holographic fringes by FM feedback. *Appl. Opt.*, 7(1):87–89, 1968. {**304, 305**}

[554] H. Kreitlow, Th. Kreis, and W. Jüptner. Holographic interferometry with reference beams modulated by the object motion. *Appl. Opt.*, 26(19):4256–4262, 1987. {**305, 331**}

[555] F. M. Mottier. Holography of randomly moving objects. *Appl. Phys. Lett.*, 15(2):44–45, 1969. {**305**}

[556] J. P. Waters. Object motion compensation by speckle reference beam interferometry. *Appl. Opt.*, 11:630–636, 1972. {**305, 331**}

[557] D. B. Neumann and R. C. Penn. Object motion compensation using reflection holography. *Journ. Opt. Soc. Amer.*, 62:1373, 1972. {**305**}

[558] N. Abramson. Sandwich hologram interferometry: A new dimension in holographic comparison. *Appl. Opt.*, 13:2019–2025, 1974. {**305**}

[559] N. Abramson. Sandwich hologram interferometry 2: Some practical calculations. *Appl. Opt.*, 14:981–984, 1975. {**305**}

[560] N. Abramson. Sandwich hologram interferometry 4: Holographic studies of two milling machines. *Appl. Opt.*, 16:2521–2531, 1977. {**305**}

[561] N. Abramson and H. Bjelkhagen. Industrial holographic measurements. *Appl. Opt.*, 12:2792–2796, 1973. {**305**}

[562] N. Abramson and H. Bjelkhagen. Pulsed sandwich holography 2: Practical application. *Appl. Opt.*, 17:187–191, 1978. {**305**}

[563] N. Abramson and H. Bjelkhagen. Sandwich hologram interferometry 5: Measurement of in-plane displacement and compensation for rigid body motion. *Appl. Opt.*, 18:2870–2880, 1979. {**305**}

[564] N. Abramson, H. Bjelkhagen, and P. Skande. Sandwich holography for storing information interferometrically with a high degree of security. *Appl. Opt.*, 18:2017–2021, 1979. {**305**}

[565] H. Bjelkhagen. Pulsed sandwich holography. *Appl. Opt.*, 16:1727–1731, 1977. {**305**}

[566] G. Lai and T. Yatagai. Dual-reference holographic interferometry with a double pulsed laser. *Appl. Opt.*, 27(18):3855–3858, 1988. {**305**}

[567] A. Stimpfling and P. Smigielski. New method for compensating and measuring any motion of three-dimensional objects in holographic interferometry. *Opt. Eng.*, 24(5):821–823, 1985. {**305**}

[568] P. Smigielski. New possibilities of holographic interferometry. In W. Jüptner, editor, *Holography Techniques and Applications*, volume 1026 of *Proc. of Soc. Photo-Opt. Instr. Eng.*, pages 90–92, 1988. {**305**}

[569] B. Lutz and W. Schumann. Approach to extend the domain of visibility of recovered modified fringes when holographic interferometry is applied to large deformations. *Opt. Eng.*, 34(7):1879–1886, 1995. {**305**}

[570] W. Schumann. Holographic interferometry applied to the case of large deformations. *Journ. Opt. Soc. Amer. A*, 6(11):1738–1747, 1989. {**305**}

[571] S. Toyooka. Holographic interferometer, proof against external vibrations. *Optica Acta*, 29(6):861–865, 1982. {**305**}

[572] P. Cielo. *Optical Techniques for Industrial Inspection*. Academic Press, 1988. {**306**}

[573] R. J. Pryputniewicz. Determination of the sensitivity vectors directly from holograms. *Journ. Opt. Soc. Amer.*, 67(10):1351–1353, 1977. {**306**}

[574] R. J. Pryputniewicz and K. A. Stetson. Determination of sensitivity vectors in hologram interferometry from two known rotations of the object. *Appl. Opt.*, 19(13):2201–2205, 1980. {**307**}

[575] D. Vogel, V. Grosser, W. Osten, J. Vogel, and R. Höfling. Holographic 3d-measurement technique based on a digital image processing system. In W. Jüptner and W. Osten, editors, *Fringe '89*, pages 33–42, 1989. {**307**}

[576] R. Dändliker and R. Thalmann. Determiation of 3-D displacement and strain by holographic interferometry for non-plane objects. In W. F. Fagan, editor, *Industrial Applications of Laser Technology*, volume 398 of *Proc. of Soc. Photo-Opt. Instr. Eng.*, pages 11–16, 1983. {**308**}

[577] S. K. Dhir and J. P. Sikora. An improved method for obtaining the general displacement field from a holographic interferogram. *Exp. Mech.*, 12:323–327, 1972. {**311**}

[578] P. W. King III. Holographic interferometry technique utilizing two plates and relative fringe orders for measuring micro-displacements. *Appl. Opt.*, 13:231–233, 1974. {**311**}

[579] C. A. Sciammarella and T. Y. Chang. Holographic interferometry applied to the solution of a shell problem. *Exp. Mech.*, 14:217–224, 1974. {**311**}

[580] C. A. Sciammarella and J. A. Gilbert. Strain analysis of a disk subjected to diametral compression by means of holographic interferometry. *Appl. Opt.*, 12:1951–1956, 1973. {**311**}

[581] G. Coppola, S. De Nicola, P. Ferraro, A. Finizio, S. Grilli, M. Iodice, C. Magro, and G. Pierattini. Evaluation of residual stress in MEMS structures by digital holography. In K. Gastinger, O. Løkberg, and S. Winther, editors, *Speckle Metrology 2003*, volume 4933 of *Proc. of Soc. Photo-Opt. Instr. Eng.*, pages 226–231, 2003. {**311**}

[582] C. A. Sciammarella and F. M. Sciammarella. Measurement of mechanical properties of materials in the micrometer range using electronic holographic moire. *Opt. Eng.*, 42(5):1215–1222, 2003. {**311, 403**}

[583] W. Schumann, J.-P. Zürcher, and D. Cuche. *Holography and Deformation Analysis*. Springer-Verlag, 1985. {**311**}

[584] M. Y. Y. Hung, L. Lin, and H. M. Shang. Simple method for direct determination of bending strains by use of digital holography. *Appl. Opt.*, 40(25):4514–4518, 2001. {**317**}

[585] C. A. Sciammarella and R. Narayanan. The determination of the components of the strain tensor in holographic interferometry. *Exp. Mech.*, 24:257–264, 1984. {**317**}

[586] K. A. Stetson. The relationship between strain and derivatives of observed displacement in coherent optical metrology. *Exp. Mech.*, 7:273–275, 1981. {**317**}

[587] L. H. Taylor and G. B. Brandt. An error analysis of holographic strains determined by cubic splines. *Exp. Mech.*, 12:543–548, 1972. {**317**}

[588] R. J. Pryputniewicz and K. A. Stetson. Holographic strain analysis: extension of fringe vector method to include perspective. *Appl. Opt.*, 15(3):725–728, 1976. {**318**}

[589] R. J. Pryputniewicz. Holographic strain analysis: an experimental implementation of the fringe-vector theory. *Appl. Opt.*, 17(22):3613–3618, 1978. {**318**}

[590] K. A. Stetson. Homogeneous deformations: determination by fringe vectors in hologram interferometry. *Appl. Opt.*, 14:2256–2259, 1975. {**318**}

[591] F. Chen, G. M. Brown, M. M. Marchi, and M. Dale. Recent advances in brake noise and vibration engineering using laser metrology. *Opt. Eng.*, 42(5):1359–1369, 2003. {**321, 323**}

[592] G. M. Brown. Fringe analysis for automotive applications. *Opt. Lasers in Eng.*, 19:203–220, 1993. {**321**}

[593] W. Jüptner and Th. Kreis, editors. *An External Interface for Processing 3-D Holographic and X-Ray Images*. Research Reports ESPRIT. Springer-Verlag, 1989. {**321**}

[594] A. S. Kobayashi. Hybrid experimental-numerical stress analysis. *Exp. Mech.*, 23(3):338–347, 1983. {**321**}

[595] M. M. Ratnam and W. T. Evans. Comparison of measurement of piston deformation using holographic interferometry and finite elements. *Exp. Mech.*, 12:336–342, 1993. {**321**}

[596] J. M. Weathers, W. A. Foster, W. F. Swinson, and J. L. Turner. Integration of laser-speckle and finite-element techniques of stress analysis. *Exp. Mech.*, 3:60–65, 1985. {**321**}

[597] K. Oh and R. J. Pryputniewicz. Application of electro-optic holography in the study of cantilever plate vibration with concentrated masses. In K. Stetson and R. Pryputniewicz, editors, *International Conference on Hologram Interferometry and Speckle Metrology*, pages 245–253, 1990. {**321**}

[598] H. Borner, M. Schulz, J. Villain, and H. Steinbichler. Application of holographic interferometry supported by FEM-calculations during the development of a new assembly technique. In W. Jüptner, editor, *Holography Techniques and Applications*, volume 1026 of *Proc. of Soc. Photo-Opt. Instr. Eng.*, pages 171–175, 1988. {**321**}

[599] M. A. Caponero, A. De Angelis, V. R. Filetti, and S. Gammella. Structural analysis of an aircraft turbine blade prototype by use of holographic interferometry. In R. J. Pryputniewicz, G. M. Brown, and W. Jüptner, editors, *Interferometry VI: Applications*, volume 2004 of *Proc. of Soc. Photo-Opt. Instr. Eng.*, pages 150–161, 1993. {**321**}

[600] X. Bohineust, V. Linet, and F. Dupuy. Development and application of holographic modal decomposition techniques to acoustic analysis of vehicle structures. In R. J. Pryputniewicz, G. M. Brown, and W. Jüptner, editors, *Interferometry VI: Applications*, volume 2004 of *Proc. of Soc. Photo-Opt. Instr. Eng.*, pages 118–129, 1993. {**321**}

[601] R. J. Pryputniewicz. Holographic and finite element studies of vibrating beams. In W. F. Fagan, editor, *Optics in Engineering Measurement*, volume 599 of *Proc. of Soc. Photo-Opt. Instr. Eng.*, pages 54–62, 1985. {**321**}

[602] Th. Bischof and W. Jüptner. Investigation of the stress distribution in intact bonds by holographic interferometry and finite element method. In R. J. Pryputniewicz, editor, *Laser Interferometry IV: Computer-Aided Interferometry*, volume 1553 of *Proc. of Soc. Photo-Opt. Instr. Eng.*, pages 326–331, 1991. {**321**}

[603] G. C. Brown and R. J. Pryputniewicz. Holographic microscope for measuring displacements of vibrating microbeams using time-averaged, electro-optic holography. *Opt. Eng.*, 37(5):1398–1405, 1998. {**321**}

[604] W. Jüptner, Th. Kreis, J. Geldmacher, and Th. Bischof. Detection of defects in adhesion bonds using holographic interferometry (in German). *Qualität und Zuverlässigkeit*, 36(7):417–423, 1991. {**321**}

[605] J. Balas, J. Sladek, and M. Drzik. Stress analysis by combination of holographic interferometry and boundary-integral method. *Exp. Mech.*, 83:196–202, 1983. {**322**}

[606] C. A. Sciammarella. Contributions of interferometry to the field of fracture mechanics. In R. J. Pryputniewicz, G. M. Brown, and W. Jüptner, editors, *Interferometry VI: Applications*, volume 2004 of *Proc. of Soc. Photo-Opt. Instr. Eng.*, pages 204–214, 1993. {**323**}

[607] J. M. Huntley and L. R. Benckert. Measurement of dynamic crack tip displacement field by speckle photography and interferometry. *Opt. Lasers in Eng.*, 19:299–312, 1993. {**323**}

[608] L. W. Meyer, W. Jüptner, and H.-D. Steffens. Fracture toughness investigations using holographic interferometry. In W. Waidelich, editor, *Laser 75 Opto-Electronics*, pages 203–205. IPC Science and Technology Press, 1975. {**323**}

[609] A. J. Moore and J. R. Tyrer. The evaluation of fracture mechanics parameters from electronic speckle pattern interferometric fringe patterns. *Opt. Lasers in Eng.*, 19:325–336, 1993. {**323**}

[610] W. Jüptner, K. Grünewald, R. Zirn, and H. Kreitlow. Measurement of the stress intensity factor k_i in large specimens by means of holographic interferometry. In D. Vukicevic, editor, *Holographic Data Nondestructive Testing*, volume 370 of *Proc. of Soc. Photo-Opt. Instr. Eng.*, pages 62–65, 1982. {**323**}

[611] P. Will, W. Totzauer, and B. Michel. Generalized J-integral of fracture mechanics from holographic data. *Phys. Stat. Sol. (a)*, 95:K113–K116, 1986. {**323**}

[612] K. A. Stetson and I. R. Harrison. Computer-aided holographic vibration analysis for vectorial displacements of bladed disks. *Appl. Opt.*, 17(11):1733–1738, 1978. {**323**}

[613] K. A. Stetson. Holographic vibration analysis. In R. K. Erf, editor, *Holographic Non-Destructive Testing*, pages 181–220. Academic Press, 1974. {**323**}

[614] J. Janta and M. Miler. Time average holographic interferometry of damped oscillations. *Optik*, 36:185–195, 1972. {**324**}

[615] P. C. Gupta and K. Singh. Time-average hologram interferometry of periodic, non-cosinusoidal vibrations. *Appl. Phys.*, 6:233–240, 1975. {**324**}

[616] P. C. Gupta and K. Singh. Characteristic fringe function for time-average holography of periodic nonsinusoidal vibrations. *Appl. Opt.*, 14:129–133, 1975. {**324**}

[617] P. C. Gupta and K. Singh. Hologram interferometry of vibrations represented by the square of a Jacobian elliptic function. *Nouv. Rev. Opt.*, 7:95–100, 1976. {**324**}

[618] P. Hariharan, B. F. Oreb, and C. H. Freund. Stroboscopic holographic interferometry: measurements of vector components of a vibration. *Appl. Opt.*, 26(18):3899–3903, 1987. {**325**}

[619] P. Sajenko and C. D. Johnson. Stroboscopic holographic interferometry. *Appl. Phys. Lett.*, 13:44–46, 1968. {**326**}

[620] S. Nakadate, H. Saito, and T. Nakajima. Vibration measurement using phase-shifting stroboscopic holographic interferometry. *Optica Acta*, 33(10):1295–1309, 1986. {**326**}

[621] B. Ineichen and J. Mastner. Vibration analysis by stroboscopic two-reference-beam heterodyne holographic interferometry. In P. Meyrueis and M. Grosmann, editors, *2nd European Congress on Optics Applied to Metrology*, volume 210 of *Proc. of Soc. Photo-Opt. Instr. Eng.*, pages 207–212, 1979. {**326, 331**}

[622] K. A. Stetson. Method of vibration measurements in heterodyne interferometry. *Opt. Lett.*, 7:233–234, 1982. {**326, 331**}

[623] P. Hariharan. Application of holographic subtraction to time-average hologram interferometry of vibrating objects. *Appl. Opt.*, 12:143–146, 1973. {**326**}

[624] R. J. Pryputniewicz. Time-average holography in vibration analysis. *Opt. Eng.*, 24:843–848, 1985. {**326**}

[625] R. Tonin and D. A. Bies. Analysis of 3-D vibrations from time-averaged holograms. *Appl. Opt.*, 17(23):3713–3721, 1978. {**326**}

[626] R. Tonin and D. A. Bies. General theory of time-averaged holography for the study of three-dimensional vibrations at a single frequency. *Journ. Opt. Soc. Amer.*, 68(7):924–931, 1978. {**326**}

[627] K. Antropius. Fundamentals of time-average holographic interferometry and its applications to vibration measurements. In G. Frankowski, N. Abramson, and Z. Füzessy, editors, *Application of Metrological Laser Methods in Machines and Systems*, volume 15 of *Physical Research*, pages 167–194. Akademie Verlag, Berlin, 1991. {**327**}

[735] C. M. Vest. Tomography for properties of materials that bend rays: a tutorial. *Appl. Opt.*, 24(23):4089–4094, 1985. {**383, 384**}

[736] R. Snyder and L. Hesselink. Optical tomography for flow visualization of the density around a revolving helicopter rotor blade. *Appl. Opt.*, 23(20):3650–3656, 1984. {**383**}

[737] S. J. Norton and M. Linzer. Correcting for ray refraction in velocity and attenuation tomography. *Ultrason. Imaging*, 4:201, 1982. {**383**}

[738] K. Iwata and R. Nagata. Calculation of refractive index distribution from interferograms using the Born and Rytov's approximation. *Jpn. J. Appl. Phys.*, 14(1):379, 1975. {**383**}

[739] S. K. Kenue and J. F. Greenleaf. Limited angle multifrequency diffraction tomography. *IEEE Trans. Son. Ultrason.*, 29:213, 1982. {**383**}

[740] S. S. Cha and H. Sun. Tomography for reconstructing continuous fields from ill-posed multidirectional interferometric data. *Appl. Opt.*, 29(2):251–258, 1990. {**383**}

[741] I. Prikryl and C. M. Vest. Holographic interferometry of transparent media using light scattered by embedded test objects. *Appl. Opt.*, 21(14):2554–2557, 1982. {**383**}

[742] S. E. Reichenbach, J. C. Burton, and K. W. Miller. Comparison of algorithms for computing the two-dimensional discrete Hartley transform. *Journ. Opt. Soc. Amer. A*, 6:818–821, 1989. {**385**}

[743] V. Kebbel, H. J. Hartmann, and W. Jüptner. Characterization of micro-optics using digital holography. In G. M. Brown, W. Jüptner, and R. Pryputniewicz, editors, *Laser Interferometry X: Applications*, volume 4101 of *Proc. of Soc. Photo-Opt. Instr. Eng.*, pages 477–487, 2000. {**385**}

[744] V. Kebbel, M. Adams, H. J. Hartmann, and W. Jüptner. Digital holography as a versatile optical diagnostic method for microgravity experiments. *Meas. Sci. & Techn.*, 10(10):893–899, 1999. {**385**}

[745] N. Demoli, D. Vukicevic, and M. Torzynski. Dynamic digital holographic interferometry with three wavelengths. *Opt. Expr.*, 11(7):767–774, 2003. {**385**}

[746] J. D. Trolinger and J. C. Hsu. Flowfield diagnostics by holographic interferometry and tomography. In W. Jüptner and W. Osten, editors, *Fringe '93*, pages 423–439, 1993. {**386**}

[747] B. A. Tozer, R. Glanville, A. L. Gordon, M. J. Little, J. M. Webster, and D. G. Wright. Holography applied to inspection and measurement in an industrial environment. *Opt. Eng.*, 24(5):746–753, 1985. {**387**}

[748] H. Hong, D. B. Sheffer, and C. W. Loughry. Detection of breast lesions by holographic interferometry. *J. Biomed. Opt.*, 4(03):368–375, 1999. {**387**}

[749] J Woisetschläger, D. B. Sheffer, C. W. Loughry, K. Somasundaram, S. K. Chawla, and P. J. Wesolowski. Phase-shifting holographic interferometry for breast cancer detection. *Appl. Opt.*, 33:5011–5015, 1994. {**387**}

[750] M. A. Schulze, M. A. Hunt, E. Voelkl, J. D. Hickson, W. Usry, R. G. Smith, R. Bryant, and C. E. Thomas Jr. Semiconductor wafer defect detection using digital holography. In *Process and Materials Characterization and Diagnostics in IC Manufacturing II*, Proc. of Soc. Photo-Opt. Instr. Eng., 2003. {**387**}

[751] U. Mieth, W. Osten, and W. Jüptner. Knowledge assisted fault detection based on line features of skeletons. In W. Jüptner and W. Osten, editors, *Fringe '93*, Physical Research, pages 367–373. Akademie-Verlag, Berlin, 1993. {**388**}

[752] D. A. Tichenor and V. P. Madsen. Computer analysis of holographic interferograms for nondestructive testing. *Opt. Eng.*, 18(5):469–472, 1979. {**390**}

[753] D. W. Robinson. Automatic fringe analysis with a computer image-processing system. *Appl. Opt.*, 22(14):2169–2176, 1983. {**390**}

[754] J. Kornis and G. Vasarhelyi. Application of artificial neural network in holographic and speckle interferometry. In K. Gastinger, O. Løkberg, and S. Winther, editors, *Speckle Metrology 2003*, volume 4933 of *Proc. of Soc. Photo-Opt. Instr. Eng.*, pages 212–217, 2003. {**395**}

[755] Y. Frauel and B. Javidi. Neural network for three-dimensional object recognition based on digital holography. *Opt. Lett.*, 26(19):1478–1480, 2001. {**398**}

[756] Y. Frauel, E. Tajahuerce, M.-A. Castro, and B. Javidi. Distortion-tolerant three-dimensional object recognition with digital holography. *Appl. Opt.*, 40(23):3887–3893, 2001. {**398**}

[757] N. A. Halliwell and C. J. Pickering. Analysis methods in laser speckle photography and particle image velocimetry. In D. W. Robinson and G. T. Reid, editors, *Interferogram Analysis: Digital Fringe Pattern Measurement Techniques*, pages 230–261. Institute of Physics, Bristol and Philadelphia, 1993. {**400**}

[758] B. Ineichen, P. Eglin, and R. Dändliker. Hybrid optical and electronic image processing for strain measurements by speckle photography. *Appl. Opt.*, 19:2191–2195, 1980. {**400**}

[759] H. Kreitlow and Th. Kreis. Automatic evaluation of Young's fringes related to the study of in-plane-deformations by speckle techniques. In P. Meyrueis and M. Grosmann, editors, *2nd European Congress on Optics Applied to Metrology*, volume 210 of *Proc. of Soc. Photo-Opt. Instr. Eng.*, pages 18–24, 1979. {**400**}

[760] J. M. Huntley. Speckle photography fringe analysis by the Walsh transform. *Appl. Opt.*, 25:382–386, 1986. {**400**}

[761] H. Kreitlow, Th. Kreis, and W. Jüptner. Optimierung der automatisierten Auswertung von Specklegrammen beim Einsatz eines schnellen Fouriertransformators (Optimization of automatic evaluation of specklegrams using a fast Fourier-transformer). In W. Waidelich, editor, *Laser 83 - Optoelektronik in der Technik*, pages 159–164. Springer-Verlag, 1983. (In German). {**400**}

[762] K. A. Stetson. K100 electronic holography system and HG7000 interferometric computer: instruments for automotive component analysis. *Opt. Eng.*, 42(5):1348–1353, 2003. {**400, 404**}

[763] R. Mattsson, P. Gren, and A. Wåhlin. Laser ignition of pre-mixed gases studies by pulse TV holography. In K. Gastinger, O. Løkberg, and S. Winther, editors, *Speckle Metrology 2003*, volume 4933 of *Proc. of Soc. Photo-Opt. Instr. Eng.*, pages 285–290, 2003. {**400**}

[764] P. Gren. Bending wave propagation in rotating objects measured by pulsed TV holography. *Appl. Opt.*, 41(34):7237–7240, 2002. {**400**}

[765] P. Gren. Pulsed TV holography combined with digital speckle photography restores lost interference phase. *Appl. Opt.*, 40(14):2304–2309, 2001. {**400**}

[766] G. Schirripa Spagnolo, D. Ambrosini, and G. Guattari. Electro-optic holography system and digital image processing for in situ analysis of microclimate variation on artworks. *J. Opt.*, 28:99–106, 1997. {**400**}

[767] G. Schirripa Spagnolo, G. Guattari, E. Grinzato, P. G. Bison, D. Paoletti, and D. Ambrosini. Frescoes diagnostics by electro-optic holography and infrared thermography. *NDT.net*, 5(1), 2000. http://www.ndt.net/article/v05n01/schirrip/schirrip.htm. {**400**}

[768] S. Schedin, G. Pedrini, and H. J. Tiziani. Pulsed digital holography for deformation measurements on biological tissues. *Appl. Opt.*, 39(16):2853–2857, 2000. {**400, 403**}

[769] G. Pedrini, H. J. Tiziani, and M. E. Gusev. Pulsed digital holographic interferometry with 694- and 347-nm wavelengths. *Appl. Opt.*, 39(2):246–249, 2000. {**400, 403**}

[770] S. Schedin, G. Pedrini, H. J. Tiziani, A. K. Aggarwal, and M. E. Gusev. Highly sensitive pulsed digital holography for built-in defect analysis with a laser excitation. *Appl. Opt.*, 40:100–117, 2001. {**400, 403**}

[771] E. Marquardt and J. Richter. Digital image holography. *Opt. Eng.*, 37(5):1514–1519, 1998. {**400**}

[772] P. Aswendt, R. Höfling, and W. Totzauer. Digital speckle pattern interferometry applied to thermal strain measurements of metal-ceramic compounds. *Opt. Laser Technol.*, 22(4):278–282, 1990. {**403**}

[773] A. Shulev, I. Russev, and V. Sainov. New automatic FFT filtration method for phase maps and its application in speckle interferometry. In K. Gastinger, O. Løkberg, and S. Winther, editors, *Speckle Metrology 2003*, volume 4933 of *Proc. of Soc. Photo-Opt. Instr. Eng.*, pages 323–327, 2003. {**403**}

[774] C. Trillo, A. F. Doval, D. Cernadas, O. Lopez, C. Lopez, B. V. Dorrio, J. L. Fernandez, and M. Perez-Amor. Measurement of the mechanical amplitude and phase of transient surface acoustic waves using double pulsed TV holography and the spatial Fourier transform method. In K. Gastinger, O. Løkberg, and S. Winther, editors, *Speckle Metrology 2003*, volume 4933 of *Proc. of Soc. Photo-Opt. Instr. Eng.*, pages 66–71, 2003. {**403**}

[775] R. Krupka, Th. Walz, and A. Ettemeyer. Fast and full-field measurement of brake squeal using pulsed ESPI technique. *Opt. Eng.*, 42(5):1354–1358, 2003. {**403**}

[776] P. Fröning, G. Pedrini, H. J. Tiziani, and F. Mendoza Santoyo. Vibration mode separation of transient phenomena using multi-pulse digital holography. *Opt. Eng.*, 38(12):2062–2068, 1999. {**403**}

[777] F. Mendoza Santoyo, G. Pedrini, S. Schedin, and H. J. Tiziani. 3D displacement measurements of vibrating objects with multi-pulse digital holography. *Meas. Sci. & Technol.*, 10(12):1305–1308, 1999. {**403**}

[778] S. Leidenbach. The direct phase measurement – a new method for determination of a phase image from a single intensity image (in German). In W. Waidelich, editor, *Laser 91 - Optoelektronik Mikrowellen*, pages 68–72, 1991. {**405**}

[779] H. O. Saldner, N.-E. Molin, and K. A. Stetson. Fourier transform evaluation of phase data in spatially phase-biased TV holograms. *Appl. Opt.*, 35(2):332–336, 1996. {**405**}

[780] Y. Y. Hung. Electronic shearography versus ESPI for nondestructive evaluation. In F.-P. Chiang, editor, *Moire Techniques, Holographic Interferometry, Optical NDT, and Applications to Fluid Mechanics*, volume 1554B of *Proc. of Soc. Photo-Opt. Instr. Eng.*, pages 692–700, 1991. {**405**}

[781] A. R. Ganesan, D. K. Sharma, and M. P. Kothiyal. Universal digital speckle shearing interferometer. *Appl. Opt.*, 27(22):4731–4734, 1988. {**407**}

[782] C. Falldorf, S. G. Hanson, W. Osten, and W. Jüptner. Fringe compensation in multiband speckle shearography using a wedge prism. In K. Gastinger, O. Løkberg, and S. Winther, editors, *Speckle Metrology 2003*, volume 4933 of *Proc. of Soc. Photo-Opt. Instr. Eng.*, pages 82–89, 2003. {**408**}

[783] D. C. Champeney. *A Handbook of Fourier Theorems*. Cambridge University Press, 1987. {**411**}

[784] A. Papoulis. *Signal Analysis*. McGraw-Hill International Book Company, 1977. {**412, 442**}

[785] A. C. Kak and M. Slaney. *Principles of Computerized Tomographic Imaging*. IEEE Press, 1988. {**423, 448, 450, 457**}

[786] L. P. Yaroslavsky. Efficient algorithm for discrete sinc interpolation. *Appl. Opt.*, 36(2):460–463, 1997. {**428, 439, 440**}

[787] L. Yaroslavsky. Boundary effect free and adaptive discrete signal sinc-interpolation algorithms for signal and image resampling. *Appl. Opt.*, 42(20+32):4166–4175+6495, 2003. {**428**}

[788] L. Yaroslavsky and M. Eden. *Fundamentals of Digital Optics*. Birkhäuser, 1996. {**428**}

[789] A. Rosenfeld and A. C. Kak. *Digital Picture Processing*. Academic Press, 2nd edition, 1982. {**431**}

[790] E. O. Brigham. *The Fast Fourier Transform*. Prentice-Hall, 1974. {**437**}

[791] S. D. Stearns. *Digital Signal Analysis*. Hayden Book Comp., 1975. {**437**}

[792] H. J. Nussbaumer. *Fast Fourier Transform and Convolution Algorithms*. Springer Series in Information Sciences, Vol. 2. 1982. {**437, 438**}

[793] A. B. Watson and A. Poirson. Separable two-dimensional discrete Hartley transform. *Journ. Opt. Soc. Amer. A*, 3(12):2001–2004, 1986. {**440**}

[794] D. Gabor. Theory of communication. *J. Inst. Electr. Eng. (London)*, 93:429–457, 1946. {**445**}

[795] J. Radon. Über die Bestimmung von Funktionen durch ihre Integralwerte längs gewisser Mannigfaltigkeiten. *Ber. der Sächs. Akad. der Wissensch., Math.-Phys. Klasse*, 69:262–277, 1917. {**447**}

[796] R. H. T. Bates, K. L. Garden, and T. M. Peters. Overview of computerized tomography with emphasis on future developments. *Proc. IEEE*, 71(3):356–372, 1983. {**447**}

[797] R. M. Lewitt. Reconstruction algorithms: transform methods. *Proc. IEEE*, 71(3):390–408, 1983. {**454**}

[798] Y. Censor. Finite series-expansion reconstruction methods. *Proc. IEEE*, 71(3):409–419, 1983. {**455, 457**}

[799] A. L. Collins, M. W. Collins, and J. Hunter. The application of tomographic reconstruction techniques to the extraction of data from holographic interferograms. In R. J. Pryputniewicz, G. M. Brown, and W. Jüptner, editors, *Interferometry VI: Applications*, volume 2004 of *Proc. of Soc. Photo-Opt. Instr. Eng.*, pages 234–243, 1993. {**455**}

[800] H. Tan and D. Modarress. Algebraic reconstruction technique code for tomographic interferometry. *Opt. Eng.*, 24:435–440, 1985. {**455**}

[801] D. D. Verhoeven. MART-type CT algorithms for the reconstruction of multidirectional interferometric data. In R. J. Pryputniewicz, editor, *Laser Interferometry IV: Computer-Aided Interferometry*, volume 1553 of *Proc. of Soc. Photo-Opt. Instr. Eng.*, pages 376–387, 1991. {**455**}

Author Index

Abe, A., see Watanabe, M. [348]
Abendroth, H., see Lu, B. [352]
Abramson, N. [196, 297–301, 377, 558–564, 649, 654–660]
—, see Carlsson, T. E. [706]
—, see Füzessy, Z. [539]
Adams, J. [214]
Adams, M. [265]
—, see Kebbel, V. [744]
—, see Kreis, Th. [252, 255, 261]
Aggarwal, A. K., see Schedin, S. [718, 770]
Albertazzi Jr., A. [71, 240]
Albrecht, D., see Zanetta, P. [459]
Aleksandrov, E. B. [36, 484]
Aleksoff, C. C. [630, 633]
Alexeenko, I., see Pedrini, G. [535, 640, 720]
Allano, D., see Malek, M. [247]
Allaria, E. [92]
Allebach, J. P. [219]
Alleysson, D. [216]
Amadesi, S. [324]
—, see Paoletti, D. [174]
Ambrosini, D., see Spagnolo, G. Schirripa [516, 766, 767]
An, Y., see Yu, L. [272]
Andres, N., see Lobera, J. [267]
Andresen, K. [511]
Angelis, A. De, see Caponero, M. A. [599]
Antropius, K. [627]
Aparicio, J. A. [409]
Arines, J. [537]
Arroyo, M. P., see Lobera, J. [267]
Asakura, T., see Soontaranon, S. [249]
Asundi, A., see Xu, L. [113, 114, 258, 276]
Aswendt, P. [772]
—, see Kreis, Th. [227]
Awatsuji, Y., see Kubota, T. [661]

Bachor, H.-A., see Bone, D. J. [461]
Bakker, P. G., see Lanen, T. A. W. M. [340]
Balas, J. [605]
Baldi, A. [526]
Banyasz, I., see Füzessy, Z. [674]
Barakat, R. [727]
Barillot, M., see Rastogi, P. K. [684]
Barnhart, D. H. [192]
Bashaw, M. C., see Heanue, J. F. [14]
Bates, R. H. T. [796]
Baumbach, T., see Osten, W. [686, 687]
Baumbach, Th. [212]
Becker, F. [402, 403]
Becker, J., see Kebbel, V. [552]
—, see Skarman, B. [119]
Bedarida, F., see Solitro, F. [198]
Beeck, M.-A. [702, 703]
Beghuin, D. [125]
—, see Colomb, T. [122, 123]
Beheim, G., see Mercer, C. R. [206]
Belaid, S. [246]
—, see Lebrun, D. [248]
Benckert, L. R., see Huntley, J. M. [607]
Benkouider, A. M., see Lebrun, D. [273]
Benoit, M. R., see Owen, R. B. [147]
Benoit, M.R. [149]
Bentley, J. P. [193]
Bermudez, J. Guerrero, see Torres, E. [226]
Bernabeu, E., see Quiroga, J. A. [515]
Berry, M. V. [733]
Bertaux, N., see Matoba, O. [97]
Bertolino, F. [410]
—, see Ginesu, F. [366]
Bevilacqua, F., see Cuche, E. [228]
Bhat, G., see Sciammarella, C. A. [374]
Bieber, E. [400]
Biedermann, K., see Ek, L. [485, 487]

Biedermann, R., see Kreis, Th. [293, 425, 531]
Bies, D. A., see Tonin, R. [625, 626]
Bilbro, G. L., see Ramanath, R. [217]
Binet, R. [83]
Birnbaum, G. [323]
Bischof, Th. [602]
—, see Jüptner, W. [546, 604]
Bison, P. G., see Spagnolo, G. Schirripa [767]
Bjelkhagen, H. [565]
—, see Abramson, N. [561–564]
Blagrave, K., see Kreuzer, H. J. [75]
Blanco-Garcia, J. [311]
Blu, Th., see Liebling, M. [129, 241–243, 245]
Bo, F., see Liu, Ch. [81]
Boehnlein, A. J., see Gilbert, J. A. [199]
Bohineust, X. [600]
Böhmer, M., see Hinrichs, H. [663]
Bonch-Bruevich, A. M., see Aleksandrov, E. B. [36, 484]
Bone, D. J. [461, 504]
Boone, P. M. [315, 721]
Borghi, R., see Spagnolo, G. Schirripa [516]
Borner, H. [598]
Bowley, W. W., see Pryputniewicz, R. J. [541]
Boyle, W. S. [209]
Brainard, D., see Longere, P. [218]
Brandt, G. B., see Taylor, L. H. [587]
Brangaccio, D. J., see Bruning, J. H. [40]
Breuckmann, B. [427]
Brigham, E. O. [790]
Brock, N., see Trolinger, J. D. [342]
Brohinsky, W. R., see Stetson, K. A. [50]
Brooks, R. E. [19]
—, see Heflinger, L. O. [20]
Brophy, C. P. [389]
Brown, G. C. [603]
Brown, G. M. [325, 592]
—, see Chen, F. [591]
—, see Neumann, D. B. [632]
Brown, N., see Hariharan, P. [430]
Brugioni, S., see Allaria, E. [92]
Bruning, J. H. [40]
Bryanston-Cross, P. J. [462, 512]
—, see Holden, C. M. E. [338]
—, see Judge, T. R. [514]
—, see Lanen, T. A. W. M. [340]
Bryanston-Cross, P. J., see Towers, D. P. [525]
Bryant, R., see Schulze, M. A. [750]
Buckberry, C. H., see Tatam, R. P. [646]

Buckland, J. R., see Huntley, J. M. [499]
Budzinski, J. [404]
Buraga-Lefebvre, C., see Coetmellec, S. [88]
Burow, R., see Schwider, J. [443]
Burton, D. R. [449, 463, 503]
—, see Malcolm, A. A. [453]
—, see Skydan, O. A. [390]
—, see Stephenson, P. [523]
Burton, J. C., see Reichenbach, S. E. [742]
Butters, J. N. [46]
Butusov, M. M., see Ostrovsky, Yu. I. [180]

Caceres, S., see Aparicio, J. A. [409]
Cai, L., see Yu, L. [90, 91, 272]
Caponero, M. A. [599]
Carelli, P. [651]
Carlsson, T. E. [117, 669, 706]
—, see Nilsson, B. [115, 116, 118]
Carre, P. [39]
Casey, R. T., see Watanabe, M. [348]
Castro, M.-A., see Frauel, Y. [756]
Cavaccini, G. [326]
Censor, Y. [798]
Centurion, M., see Liu, Z. [144]
Cernadas, D., see Doval, A. F. [223]
—, see Trillo, C. [774]
Cha, S. S. [722, 740]
Champagne, E. B. [165]
Champeney, D. C. [783]
Chang, M. [440]
Chang, T. Y., see Sciammarella, C. A. [579]
Chaturvedi, P., see Srinivasan, V. [416]
Chawla, S. K. [701]
—, see Sciammarella, C. A. [700]
—, see Woisetschläger, J [749]
Chen, F. [591]
—, see Gu, J. [465]
Chen, J., see Ishii, Y. [186]
Chen, Y.-M., see Sirkis, J. S. [375]
Cheng, A. Y., see Sirkis, J. S. [375]
Cheng, X., see Liu, Ch. [81]
Cheng, Y. Y. [442]
Choudry, A. [362–364]
Christensen, C. R., see Jain, A. K. [395]
Cielo, P. [572]
Ciliberto, A., see Cavaccini, G. [326]
Clark, Th., see Shegelski, M. R. A. [74]
Clerc, F. Le [84]
Cline, H. E. [642, 643]

Author Index

Coetmellec, S. [88, 89]
—, see Lebrun, D. [273]
—, see Malek, M. [247]
Colin, A. [418]
Colineau, J., see Binet, R. [83]
Collier, R. J. [18]
Collins, A. L. [799]
Collins, D. J., see Matulka, R. D. [344]
Collins, M. W., see Collins, A. L. [799]
Collot, L., see Clerc, F. Le [84]
—, see Le Clerc, F. [85–87]
Colomb, T. [122, 123]
Coppola, G. [581]
—, see Ferraro, P. [275]
Cornejo-Rodriguez, A., see Ohyama, N. [460]
Coupland, J. M., see Barnhart, D. H. [192]
Crawforth, L. [184]
Creath, K. [49, 202, 385–387]
Crimmins, T. R. [398, 399]
Crostack, H.-A. [357]
—, see Pohl, K.-J. [358]
Cuche, D. [126]
Cuche, D. E. [124]
—, see Müller, E. [540]
Cuche, D., see Schumann, W. [583]
Cuche, E. [127, 128, 228, 274]
—, see Beghuin, D. [125]
—, see Colomb, T. [122, 123]
—, see Liebling, M. [241, 243]
Cusack, R. [518]
—, see Huntley, J. M. [533]

Dahlgren, P., see Beghuin, D. [125]
—, see Colomb, T. [122, 123]
—, see Cuche, E. [128]
Dale, M., see Chen, F. [591]
D'Altorio, A., see Amadesi, S. [324]
D'Altorio, A., see Carelli, P. [651]
D'Altorio, A., see Paoletti, D. [174]
Dändliker, R. [303–305, 421, 428, 576, 690]
—, see Ineichen, B. [758]
—, see Thalmann, R. [424, 431]
D'Antonio, L., see Cavaccini, G. [326]
Davies, J. C., see Tatam, R. P. [646]
de Castro, C., see Aparicio, J. A. [409]
de Frutos, A. M., see Aparicio, J. A. [409]
de Groot, P. J. [446, 447]
Deason, V. A. [438]
DeBarber, P., see Trolinger, J. D. [342]

Decker, A. J. [182]
Dekker, H., see Choudry, A. [364]
Delacretaz, G., see Beghuin, D. [125]
DelaHunt, P., see Longere, P. [218]
Delboulbe, A., see Herriau, J. P. [189]
DeMattia, P. [652]
Demetrakopoulos, T. H. [54]
Demoli, N. [230, 745]
—, see Lovric, D. [456]
Denisyuk, Y. N. [10]
Depeursinge, C., see Beghuin, D. [125]
—, see Cuche, E. [127, 228]
Depeursinge, Chr., see Colomb, T. [122, 123]
—, see Cuche, E. [128, 274]
—, see Liebling, M. [241, 243]
DeVelis, J. B., see Reynolds, G. O. [194, 474]
Dewandel, J. L., see Dubois, F. [68]
Dhir, S. K. [577]
Dirr, B., see Vogt, E. [708]
Doherty, E. T., see Collier, R. J. [18]
Dong, H., see Sun, H. [270]
Dorrio, B. V., see Doval, A. F. [223]
—, see Trillo, C. [774]
Doval, A. F. [223]
—, see Trillo, C. [774]
Drzik, M., see Balas, J. [605]
Dubas, M., see Schumann, W. [169]
Dubois, F. [67–70]
Ducottet, C., see Fournier, C. [264]
Dudderar, T. D., see Gilbert, J. A. [199, 200]
—, see Matthys, D. R. [379, 475]
Dupont, O., see Dubois, F. [68]
Dupuy, F., see Bohineust, X. [600]
Durnin, J. [157]

Ebbeni, J. [538]
Eden, M., see Yaroslavsky, L. [788]
Eggers, H., see Lu, B. [352]
Eglin, P., see Ineichen, B. [758]
Ehrlich, M. J. [345]
—, see Steckenrider, J. S. [351]
Eichhorn, N. [382]
Ek, L. [485, 487]
El-Sum, H. M. A. [5]
Elssner, K.-E., see Schwider, J. [443]
Enard, D., see Choudry, A. [364]
Engelsberger, J., see Steinbichler, H. [322]
Ennos, A. E. [37, 411]
Ettemeyer, A., see Krupka, R. [775]

—, see Schörner, J. [517]
Evans, W. T., see Ratnam, M. M. [595]

Fagan, W. F., see Trolinger, J. D. [549]
Falldorf, C. [782]
—, see Baumbach, Th. [212]
Faltus, S., see Shegelski, M. R. A. [74]
Fantin, A. Vieira, see Albertazzi Jr., A. [71, 240]
Farrant, D. I. [294]
Farrell, P. V. [422]
Fernandez, J. L., see Blanco-Garcia, J. [311]
—, see Doval, A. F. [223]
—, see Trillo, C. [774]
Ferraro, P. [275]
—, see Allaria, E. [92]
—, see Cavaccini, G. [326]
—, see Coppola, G. [581]
—, see Grilli, S. [93]
—, see Nicola, S. De [94, 95, 224]
Fessler, H., see Pedrini, G. [65, 235, 641]
Filetti, V. R., see Caponero, M. A. [599]
Finizio, A., see Coppola, G. [581]
—, see Ferraro, P. [275]
—, see Grilli, S. [93]
—, see Nicola, S. De [94, 95, 224]
Fischer, B. [41]
—, see Kreis, Th. [42]
Forbes, D., see Womack, K. H. [417]
Fossati Bellani, V. [486]
Fossati-Bellani, V., see DeMattia, P. [652]
Foster, W. A., see Weathers, J. M. [596]
Fournel, Th., see Fournier, C. [264]
Fourney, W. L., see Holloway, D. C. [694]
Fournier, C. [264]
Fraile, D. [360]
Frankena, H. J., see van Wingerden, J. [444]
Frauel, Y. [133, 755, 756]
—, see Matoba, O. [97]
Frechoso, F. A., see Aparicio, J. A. [409]
Freischlad, K. [388]
Freund, C. H., see Hariharan, P. [618]
Friesem, A. A. [645]
—, see Katzir, Y. [401]
Fröning, P. [776]
—, see Pedrini, G. [235, 641, 644]
Fröning, Ph., see Pedrini, G. [65]
Funnell, W. R. J. [391]
Furlong, C. [239]
Fusek, R. L., see Hecht, N. L. [290]

Füzessy, Z. [539, 543–545, 673, 674, 676, 677, 679]
—, see Gombkötö, B. [493]
—, see Gyimesi, F. [675, 678]

Gabor, D. [1–3, 794]
Gallagher, J. G., see Bruning, J. H. [40]
Gallagher, N. C., see Allebach, J. P. [219]
Gammella, S., see Caponero, M. A. [599]
Ganesan, A. R. [781]
Garden, K. L., see Bates, R. H. T. [796]
Garner, H. R., see Huebschman, M. L. [259]
Gascon, F., see Fraile, D. [360]
Gaskill, J. D., see Lam, P. [691]
Gasvik, K. J. [164]
Gatti, L., see Kebbel, V. [552]
—, see Solitro, F. [198]
Geldmacher, J. [707]
—, see Jüptner, W. [546, 604, 693]
—, see Kreis, Th. [266, 425, 426, 491]
—, see Vogt, E. [708]
Georges, M. [188]
Georges, M. P., see Ninane, N. [692]
Gerhardt, G., see Plotkowski, P. D. [380]
Ghiglia, D. C., see Mastin, G. A. [415]
Ghiglia, D. G. [513, 527]
Gibbs, D. F., see Berry, M. V. [733]
Gierloff, J. J. [520]
Gilbert, J. A. [195, 199, 200]
—, see Matthys, D. R. [379, 475]
—, see Sciammarella, C. A. [580, 697, 698]
Gillies, A. C. [365]
Ginesu, F. [366]
—, see Bertolino, F. [410]
Gladic, J., see Lovric, D. [456]
Glanville, R., see Tozer, B. A. [747]
Glaser, I., see Katzir, Y. [401]
Goldrein, H. T., see Cusack, R. [518]
Gombkötö, B. [493]
Goodman, J. W. [51, 158, 160, 163, 254, 631]
Gordon, A. L., see Tozer, B. A. [747]
Gorecki, C. [464]
Green, R. J. [529, 530]
Greenleaf, J. F., see Kenue, S. K. [739]
Greivenkamp, J. E. [433]
Gren, P. [359, 764, 765]
—, see Mattsson, R. [763]
Grigull, U., see Hauf, W. [723]
Grilli, S. [93]

Author Index

—, see Allaria, E. [92]
—, see Coppola, G. [581]
—, see Ferraro, P. [275]
—, see Nicola, S. De [224]
Grinzato, E., see Spagnolo, G. Schirripa [767]
Gross, M., see Clerc, F. Le [84]
—, see Le Clerc, F. [85–87]
Gross, T. S., see Watt, D. W. [361]
Grosser, V., see Vogel, D. [575]
Grünewald, K., see Jüptner, W. [610]
Gryzagoridis, J. [327]
Grzanna, J., see Schwider, J. [443]
Gu, J. [465]
Guattari, G., see Spagnolo, G. Schirripa [766, 767]
Guerrero, J. A., see Lopez, C. Perez [715]
Gunnarsson, G. T., see Trolinger, J. D. [549]
Guo, C.-S. [80]
Gupta, P. C. [615–617]
Gusev, M. E., see Pedrini, G. [644, 769]
—, see Schedin, S. [718, 770]
Gustafsson, J., see Carlsson, T. E. [669, 706]
Gyimesi, F. [675, 678]
—, see Füzessy, Z. [544, 673, 674, 676, 679]

Haines, K. A. [21–23]
—, see Hildebrand, B. P. [25, 26]
Haji-Sheikh, A., see Wang, X. [191]
Halliwell, N. A. [757]
—, see Barnhart, D. H. [192]
Hampp, N., see Barnhart, D. H. [192]
Hanson, S. G., see Falldorf, C. [782]
Hariharan, P. [154, 172, 173, 185, 429, 430, 618, 623, 637]
Harrison, I. R., see Stetson, K. A. [612]
Hartmann, H. J., see Kebbel, V. [743, 744]
Hartmann, H.-J., see Pomarico, J. [668]
Hauf, W. [723]
Haupt, U. [713]
Heanue, J. F. [14]
Hecht, N. L. [290]
Heflinger, L. O. [20, 24]
—, see Brooks, R. E. [19]
Hendren, Ch. M., see Gilbert, J. A. [195]
Hening, S. D., see Watt, D. W. [361]
Herault, J., see Alleysson, D. [216]
Herriau, J. P. [189]
Herriott, D. R., see Bruning, J. H. [40]
Hertzberg, J., see Pflüger, S. [201]

Hesselink, L., see Heanue, J. F. [14]
—, see Snyder, R. [736]
Hickson, J. D., see Schulze, M. A. [750]
Higgins, Th. V. [208]
Hildebrand, B. P. [25, 26]
—, see Haines, K. A. [21–23]
—, see Suzuki, K. [285]
Hilton, P. D., see Reynolds, G. O. [474]
Hinrichs, H. [663]
Hinsch, K. D. [337]
—, see Hinrichs, H. [663]
Hirayama, T., see Takeda, M. [102]
Höfling, R. [292]
—, see Aswendt, P. [772]
—, see Kreis, Th. [227]
—, see Osten, W. [38, 500]
—, see Vogel, D. [575]
Holden, C. M. E. [338]
Holik, A. S., see Cline, H. E. [642, 643]
Holloway, D. C. [694]
Holst, G. C. [210]
Honda, T., see Nakadate, S. [405]
—, see Ohyama, N. [460]
—, see Ru, Q.-S. [454, 455]
Hong, H. [748]
Hong, J., see Liu, Z. [144]
Horman, M. H. [29]
Hovanesian, J. D., see Plotkowski, P. D. [380]
Hrdliczka, V., see Müller, E. [540]
Hsu, D., see Long, P. [478]
Hsu, J. C., see Trolinger, J. D. [746]
Hsu, J., see Trolinger, J. D. [342]
Hu, Ch.-P., see Chang, M. [440]
Huang, T. S. [52]
Huebschman, M. L. [259]
Huignard, J. P., see Herriau, J. P. [189]
Hung, M. Y. Y. [225, 584]
—, see Plotkowski, P. D. [380]
Hung, Y. Y. [136, 328, 780]
Hunt, M. A., see Schulze, M. A. [750]
Hunter, J., see Collins, A. L. [799]
Huntley, J. M. [367, 499, 519, 533, 534, 607, 760]
—, see Cusack, R. [518]
Hüttmann, G. [183]

Idesawa, M., see Yatagai, T. [406]
Idogawa, T., see Takai, N. [99]
III, P. W. King [578]
Iizuka, K. [161]

Ina, H., see Takeda, M. [43]
Indebetouw, G. [137, 138]
—, see Poon, T.-C. [139]
Ineichen, B. [621, 758]
—, see Dändliker, R. [421]
Inomoto, O., see Yamaguchi, I. [109]
Iodice, M., see Coppola, G. [581]
Ishii, Y. [186]
Itoh, K. [505]
Itoh, Y., see Tsuruta, T. [317, 705]
Iwata, K. [738]

Jacobsen, C. F., see Neumann, D. B. [632]
Jacquot, M. [120, 121]
Jacquot, P., see Boone, P. M. [721]
Jäger, H., see Vukicevic, D. [724]
Jain, A. K. [395]
Janta, J. [291, 614]
Javidi, B. [281]
—, see Frauel, Y. [133, 755, 756]
—, see Kishk, S. [283]
—, see Matoba, O. [97, 98]
—, see Naughton, Th. J. [280]
—, see Tajahuerce, E. [134, 135, 260]
Jericho, M. H., see Kreuzer, H. J. [75]
—, see Xu, W. [77–79]
Jericho, M. J., see Kreuzer, H. J. [76]
Joannes, L., see Dubois, F. [67, 68]
Johansen, E. L., see Larson, R. W. [284]
Johnson, C. D., see Sajenko, P. [619]
Johnson, G. W. [414]
Jones, B. E. [197]
Jones, J. D. C., see Tatam, R. P. [646]
Jones, R. [310]
Jozwicki, R., see Pasko, S. [111, 233]
Jr., C. E. Thomas, see Schulze, M. A. [750]
Judge, T. R. [514]
—, see Bryanston-Cross, P. J. [462]
—, see Towers, D. P. [525]
Judge, T., see Towers, D. P. [524]
Junginger, H. G. [734]
Jüptner, W. [296, 436, 546, 550, 604, 610, 693]
—, see Adams, M. [265]
—, see Baumbach, Th. [212]
—, see Bischof, Th. [602]
—, see Falldorf, C. [782]
—, see Fischer, B. [41]
—, see Kebbel, V. [552, 743, 744]
—, see Kolenovic, E. [717]

—, see Kreis, Th. [42, 231, 252, 253, 255, 261, 266, 293, 426, 451, 452, 491, 531, 547]
—, see Kreitlow, H. [554, 761]
—, see Meyer, L. W. [608]
—, see Mieth, U. [751]
—, see Osten, W. [686, 687]
—, see Pomarico, J. [668]
—, see Rottenkolber, H. [155]
—, see Schnars, U. [57–60]
—, see Seebacher, S. [664]
—, see Wagner, C. [229]

Kak, A. C. [785]
—, see Rosenfeld, A. [789]
Kalal, M. [729]
Kasprzak, H. [329]
Kato, J.-I., see Yamaguchi, I. [106–110, 665]
Katzir, Y. [401]
Kaufmann, G. H. [683]
—, see Rastogi, P. K. [684]
Kawai, H., see Takaki, Y. [101]
Kebbel, V. [552, 743, 744]
Kemper, B., see Lai, S. [269]
Kenue, S. K. [739]
Kickstein, J., see Hinrichs, H. [663]
Kim, E.-S., see Kim, S.-G. [238]
Kim, M. K. [140, 141]
Kim, S.-G. [238]
Kim, T., see Poon, T.-C. [139]
King, B., see Lai, S. [268]
Kinnstaetter, K. [445]
Kinoshita, S., see Ohyama, N. [460]
Kirkpatrick, P., see El-Sum, H. M. A. [5]
Kischel, P., see Dubois, F. [70]
Kishk, S. [283]
Kiss, M., see Gombkötö, B. [493]
Kitoh, M., see Takeda, M. [207]
Klattenhoff, R., see Kolenovic, E. [717]
Klaus, D. M., see Benoit, M.R. [149]
—, see Owen, R. B. [147]
Klumpp, P. A., see Schnack, E. [321]
Klysubun, P., see Indebetouw, G. [137, 138]
Kobayashi, A. S. [594]
Kobayashi, S., see Takeda, M. [43]
Kocatepe, M., see Onural, L. [250]
Koenig, K. W., see Matthys, D. R. [379]
Kohgo, H., see Takeda, M. [102]
Kokal, J. V., see Ransom, P. L. [480]
Kolenovic, E. [717]

Koliopoulos, C. L., see Freischlad, K. [388]
Kornis, J. [754]
—, see Füzessy, Z. [676]
—, see Gombkötö, B. [493]
Kothiyal, M. P., see Ganesan, A. R. [781]
Kovacs, P., see Gombkötö, B. [493]
Kraska, Z., see Pawluczyk, R. [288]
Kreis, Th. [42, 44, 66, 227, 231, 237, 244, 252, 253, 255–257, 261, 262, 266, 293, 368, 378, 394, 425, 426, 441, 450–452, 457, 490–492, 531, 547]
—, see Adams, M. [265]
—, see Jüptner, W. [436, 546, 604, 693]
—, see Kreitlow, H. [554, 759, 761]
—, see Schnars, U. [59, 60]
Kreitlow, H. [554, 759, 761]
—, see Geldmacher, J. [707]
—, see Jüptner, W. [436, 550, 610]
—, see Kreis, Th. [394]
—, see Vogt, E. [708]
Kreuzer, H. J. [72, 73, 75, 76]
—, see Shegelski, M. R. A. [74]
—, see Xu, W. [77–79]
Krishnaswamy, S. [412]
Kronrod, M. A. [53]
Krupka, R. [775]
Kubota, T. [661]
Kujawinska, M. [204, 205, 369]
Kurz, T., see Lauterborn, W. [159]
Kwon, O. Y. [432, 521]

Labeyrie, A. E., see Stroke, G. W. [179]
Ladenburg, R. [726]
Lai, G. [143, 566]
Lai, S. [142, 268, 269]
—, see Kolenovic, E. [717]
Lalor, M. J., see Burton, D. R. [449, 463, 503]
—, see Skydan, O. A. [390]
—, see Stephenson, P. [523]
Lam, P. [691]
—, see Chang, M. [440]
Landry, M. J. [483]
Lanen, T. A. W. M. [339–341]
Larkin, K. G. [448]
Larson, R. W. [284]
Lassahn, G. D., see Deason, V. A. [438]
Latta, J. N. [175–177]
Lauterborn, W. [159]
Lauterborn, W. H., see Hüttmann, G. [183]

Lawrence, R. W., see Goodman, J. W. [51]
Le Clerc, F. [85–87]
Lebrun, D. [248, 273]
—, see Belaid, S. [246]
—, see Coetmellec, S. [88, 89]
—, see Malek, M. [247]
Lee, B., see Kim, S.-G. [238]
Lee, C.-K., see Crawforth, L. [184]
Lee, W. H. [12, 13]
Leendertz, J. A. [45]
—, see Butters, J. N. [46]
Legros, J.-C., see Dubois, F. [69, 70]
Lehureau, J.-C., see Binet, R. [83]
Leidenbach, S. [778]
—, see Steinbichler, H. [322]
Leiner, D. C., see Johnson, G. W. [414]
Leis, H. G. [330]
Leith, E. N. [7–9, 171]
Lemaire, P., see Georges, M. [188]
Levitt, J. A. [634]
Levy, U., see Friesem, A. A. [645]
Lewis, D., see Hecht, N. L. [290]
Lewitt, R. M. [797]
Li, W., see Milgram, J. H. [145]
Li, Y., see Liu, Ch. [81]
Liang, C. Y., see Hung, Y. Y. [136]
Liao, J., see Guo, C.-S. [80]
Liebling, M. [129, 241–243, 245]
Lilley, F., see Skydan, O. A. [390]
Lim, J. S. [396, 536]
Lin, L., see Hung, M. Y. Y. [584]
Linet, V., see Bohineust, X. [600]
Linzer, M., see Norton, S. J. [737]
Lira, I. H. [731]
Little, M. J., see Tozer, B. A. [747]
Liu, B., see Allebach, J. P. [219]
Liu, Ch. [81]
Liu, G. [234]
Liu, Z. [144]
Liu, Zh., see Liu, Ch. [81]
Lobera, J. [267]
Lohmann, A. W. [6, 11]
—, see Kinnstaetter, K. [445]
Long, P. [478]
Longere, P. [218]
Lopez, C. Perez [714, 715]
Lopez, C., see Doval, A. F. [223]
—, see Trillo, C. [774]
Lopez, O., see Doval, A. F. [223]

—, see Trillo, C. [774]
Lorensen, W. E., see Cline, H. E. [642, 643]
Loughry, C. W., see Hong, H. [748]
—, see Woisetschläger, J [749]
Lovric, D. [456]
Ltd., Lasermet [712]
Lu, B. [352]
Lucia, A. C., see Zanetta, P. [335]
Luth, W. [384]
Lutz, B. [569]

Maas, A. A. M., see Vrooman, H. A. [510]
Machado Gama, M. A. [316]
Machida, H., see Yonemura, M. [716]
Macovski, A. [47]
Macy Jr., W. W. [479]
Maddux, G. E. [370]
Mader, D. L. [355]
Madsen, V. P., see Tichenor, D. A. [752]
Magnusson, R., see Wang, X. [191]
Magome, N., see Nakadate, S. [405]
Magro, C., see Coppola, G. [581]
—, see Ferraro, P. [275]
Malcolm, A. A. [453]
Malek, M. [247]
—, see Lebrun, D. [273]
Marchi, M. M., see Chen, F. [591]
Marom, E., see Dändliker, R. [303]
Marquardt, E. [771]
Marquet, P., see Colomb, T. [123]
—, see Cuche, E. [127, 128, 274]
—, see Liebling, M. [241, 243]
Massing, J. H., see Stadelmeier, A. [236]
Mastin, G. A. [415]
—, see Ghiglia, D. G. [527]
Mastner, J., see Ineichen, B. [621]
Matoba, O. [97, 98]
—, see Tajahuerce, E. [135, 260]
Matsuda, K., see Tsuruta, T. [27]
Matsumura, T., see Yamaguchi, I. [110]
Matsuzaki, H., see Yamaguchi, I. [665]
Matthews, L. K., see Rightley, G. S. [295]
Matthys, D. R. [379, 475]
—, see Gilbert, J. A. [195]
Mattsson, R. [763]
Matulka, R. D. [344]
Mayville, R. A., see Reynolds, G. O. [474]
McDonald, J. B., see Naughton, Th. J. [280]
McKelvie, J., see Perry Jr., K. E. [458]

Medhat, M. [548]
Meesters, W., see Pflüger, S. [201]
Meier, G. E. A., see Becker, F. [402]
Meier, R. W. [166]
Meinertzhagen, I. A., see Kreuzer, H. J. [76]
—, see Xu, W. [77–79]
Mendenhall, F. T., see Sikora, J. P. [704]
Meng, H., see Pan, G. [150, 263]
Mercer, C. R. [206]
Merkel, K., see Schwider, J. [443]
Merrill, D. [215]
Merzkirch, W. [353]
Merzlyakov, N. S., see Kronrod, M. A. [53]
—, see Yaroslavskii, L. P. [55]
Mestrovic, J., see Demoli, N. [230]
Meucci, R., see Allaria, E. [92]
—, see Grilli, S. [93]
Meyer, E. H., see Crostack, H.-A. [357]
Meyer, L. W. [608]
Miao, J., see Xu, L. [113, 114, 258, 276]
Michel, B., see Will, P. [611]
Michetti, L., see Solitro, F. [198]
Mieth, U. [420, 751]
Mifune, Y., see Takai, N. [282]
Miida, S., see Ueda, M. [629]
Milas, M, see Lovric, D. [456]
Miler, M., see Janta, J. [291, 614]
Miles, J. F. [167, 178]
Milgram, J. H. [145]
Miller, K. W., see Reichenbach, S. E. [742]
Millerd, J., see Trolinger, J. D. [342]
Minardi, J. E., see Hecht, N. L. [290]
Minetti, C., see Dubois, F. [70]
Mitrovic, S., see Lovric, D. [456]
Mittra, R., see Demetrakopoulos, T. H. [54]
Mizuno, J., see Yamaguchi, I. [106, 108]
Mnatsakanyan, E. A. [407]
Modarress, D., see Tan, H. [800]
Moisson, E., see Picart, P. [496, 497]
Molin, N. E. [32–34]
Molin, N.-E., see Saldner, H. O. [779]
Molpeceres, J. L., see Aparicio, J. A. [409]
Monneret, J., see Spajer, M [699]
Monnom, O., see Dubois, F. [69, 70]
Moore, A. J. [609]
Moore, D. T., see Johnson, G. W. [414]
Mottier, F. M. [555]
—, see Dändliker, R. [303, 421]
Mounier, D., see Picart, P. [496, 497]

Mulholland, G. P., see Rightley, G. S. [295]
Müller, E. [540]
Munce, A. C., see Crawforth, L. [184]
Munjuluri, B., see Huebschman, M. L. [259]
Munoz, J. [501, 502]
Murata, K., see Ishii, Y. [186]
Murata, S. [96]
Murray, J. D., see Xu, Y. [112]

Nagata, R., see Iwata, K. [738]
Nagatome, K., see Takeda, M. [532]
Nakadate, S. [405, 620, 638]
—, see Yatagai, T. [406]
Nakajima, T., see Nakadate, S. [620]
Narayanan, R., see Sciammarella, C. A. [585]
Naughton, T. J., see Matoba, O. [97]
Naughton, Th. J. [280]
Nawab, H., see Lim, J. S. [396]
Nebbeling, C., see Lanen, T. A. W. M. [341]
Nefyodov, S. V., see Mnatsakanyan, E. A. [407]
Neger, T., see Vukicevic, D. [724]
Neifeld, M. A., see Lai, S. [142, 268]
Neumann, D. B. [168, 551, 557, 632, 671, 672]
Neupert, U., see Schörner, J. [517]
Newman, J. W. [331]
Nicola, S. De [94, 95, 224]
—, see Allaria, E. [92]
—, see Coppola, G. [581]
—, see Ferraro, P. [275]
—, see Grilli, S. [93]
Nilsson, B. [115, 116, 118]
—, see Carlsson, T. E. [669, 706]
Ninane, N. [692]
Nishida, H., see Toyooka, S. [473]
Nishisaka, T., see Yonemura, M. [716]
Noll, R., see Pflüger, S. [201]
Nomura, T., see Javidi, B. [281]
Norton, S. J. [737]
Nose, A., see Gilbert, J. A. [200]
Nösekabel, E.-H., see Steinbichler, H. [322]
Novak, J. [437]
Nübel, R. [413]
Nugent, K. A. [466]
—, see Kalal, M. [729]
Nussbaumer, H. J. [792]

Obermeier, P., see Schörner, J. [517]
Oh, K. [597]
Ohta, S., see Yamaguchi, I. [106–108]

Ohyama, N. [460]
—, see Ru, Q.-S. [454, 455]
Ohzu, H., see Takaki, Y. [100, 101, 271]
Olsson, R., see Gren, P. [359]
Onural, L. [130–132, 250, 251]
Oreb, B. F., see Hariharan, P. [430, 618, 637]
—, see Larkin, K. G. [448]
Ortona, A., see Cavaccini, G. [326]
Osten, W. [38, 383, 392, 500, 667, 686, 687]
—, see Baumbach, Th. [212]
—, see Bieber, E. [400]
—, see Colin, A. [418]
—, see Eichhorn, N. [382]
—, see Falldorf, C. [782]
—, see Höfling, R. [292]
—, see Kolenovic, E. [717]
—, see Mieth, U. [420, 751]
—, see Pedrini, G. [535]
—, see Seebacher, S. [664]
—, see Vogel, D. [575]
—, see Wagner, C. [229, 666]
—, see Winter, H. [408]
Ostrovskaya, G. V., see Ostrovsky, Yu. I. [180]
Ostrovsky, Yu. I. [180]
Ovryn, B. [332]
Owen, R. B. [146–148]
—, see Benoit, M.R. [149]
Özgen, M. T., see Onural, L. [131]
Özkul, C., see Belaid, S. [246]
—, see Coetmellec, S. [88, 89]
—, see Lebrun, D. [248]

Paez, G., see Munoz, J. [501, 502]
Page, R., see Pitlak, R. T. [181]
Pan, G. [150, 263]
Panotopoulos, G., see Liu, Z. [144]
Paoletti, D. [174, 467]
—, see Amadesi, S. [324]
—, see Carelli, P. [651]
—, see Spagnolo, G. Schirripa [516, 767]
—, see Zanetta, P. [459]
Papoulis, A. [784]
Pardoen, G. C., see Trolinger, J. D. [549]
Paris, D. P., see Lohmann, A. W. [11]
Parker, R. J. [187, 346, 347]
Parker, S. C. J., see Holden, C. M. E. [338]
Parrent, G. B., see Reynolds, G. O. [194]
Parulski, K., see Adams, J. [214]
Pasko, S. [111, 233]

Patacca, A. M., see Holloway, D. C. [694]
Paterson, D. M., see Sun, H. [270]
Paulet, P. [333]
Pauliat, G., see Georges, M. [188]
Pawlitzek, R. A., see Kreuzer, H. J. [72, 73]
Pawluczyk, R. [288]
Pedrini, G. [63–65, 222, 235, 277–279, 498, 535, 640, 641, 644, 720, 769]
—, see Fröning, P. [776]
—, see Lopez, C. Perez [714]
—, see Santoyo, F. Mendoza [777]
—, see Schedin, S. [306, 718, 719, 768, 770]
—, see Schedin, St. [62]
—, see Zou, Y. [61]
Peirce, D. C., see Reynolds, G. O. [474]
Peng, X., see Xu, L. [113, 114, 276]
Penn, R. C., see Neumann, D. B. [557]
Pennington, K. S., see Collier, R. J. [18]
Peralta-Fabi, R. [488]
Perez-Amor, M., see Blanco-Garcia, J. [311]
Perez-Amor, M., see Doval, A. F. [223]
—, see Trillo, C. [774]
Perkins, R., see Sun, H. [270]
Perry Jr., K. E. [458]
Peters, T. M., see Bates, R. H. T. [796]
Petkovsek, J. [381]
Petzing, J. N., see Farrant, D. I. [294]
Pflug, L., see Rastogi, P. K. [650]
Pflüger, S. [201]
Philipp, H., see Vukicevic, D. [724]
Picart, P. [496, 497]
Pickering, C. J., see Halliwell, N. A. [757]
Pierattini, G., see Coppola, G. [581]
—, see Ferraro, P. [275]
—, see Grilli, S. [93]
—, see Nicola, S. De [94, 95, 224]
Pitlak, R. T. [181]
Player, M. A., see Sun, H. [270]
Plotkowski, P. D. [380]
Podbielska, H., see Kasprzak, H. [329]
Pohl, K.-J. [358]
—, see Crostack, H.-A. [357]
Poirson, A., see Watson, A. B. [793]
Polhemus, C. [689]
Politch, J. [636]
Pomarico, J. [668]
Pomerleau, N., see Kreuzer, H. J. [75]
Poon, T.-C. [139]
Posposil, J., see Sekanina, H. [82]

Powell, R. L. [16]
—, see Stetson, K. A. [17]
Preater, R. [468]
Prikryl, I. [741]
Prongue, D., see Dändliker, R. [690]
Pruett, H. D., see Rose, H. W. [553]
Pryputniewicz, R. J. [371, 376, 423, 541, 573, 574, 588, 589, 601, 624]
—, see Brown, G. C. [603]
—, see Furlong, C. [239]
—, see Oh, K. [597]
Psaltis, D., see Liu, Z. [144]
Puri, I. K., see Xiao, X. [151]

Quan, C., see Bryanston-Cross, P. J. [462, 512]
Quan, Ch., see Judge, T. R. [514]
Quiroga, J. A. [515]

Radon, J. [795]
Ramanath, R. [217]
Ramos-Izquierdo, L., see Reynolds, G. O. [474]
Ramsey, D., see Macovski, A. [47]
Rankel, K., see Petkovsek, J. [381]
Ransom, P. L. [480]
Rastogi, P. [221, 696]
Rastogi, P. K. [309, 320, 650, 670, 680, 681, 684]
—, see Kaufmann, G. H. [683]
—, see Spajer, M [699]
Ratnam, M. M. [595]
Rautenberg, M., see Haupt, U. [713]
Reeves, M., see Parker, R. J. [347]
Reichenbach, S. E. [742]
Reid, G. T. [372]
Reynolds, G. O. [194, 474]
Richter, J., see Marquardt, E. [771]
Rightley, G. S. [295]
Ringer, K., see Jüptner, W. [296]
Ritter, R., see Winter, D. [522]
Rizzi, M. L., see Vo, P. Del [356]
Robinson, D. W. [507, 509, 753]
—, see Ennos, A. E. [411]
—, see Green, R. J. [530]
—, see Kujawinska, M. [204, 205]
—, see Spik, A. [528]
Roddier, C. [469]
Roddier, F., see Roddier, C. [469]
Rogers, G. L. [4]
Romero, L. A., see Ghiglia, D. G. [513, 527]
Roosen, G., see Georges, M. [188]

Rose, H. W. [553]
—, see Neumann, D. B. [551]
Rosenfeld, A. [789]
Rosenfeld, D. P., see Bruning, J. H. [40]
Rottenkolber, H. [155]
—, see Osten, W. [392]
—, see Schörner, J. [517]
Ru, Q.-S. [454, 455]
Rubayi, N. [354]
Russev, I., see Shulev, A. [773]
Ryhon, S., see Georges, M. [188]
Rytz, H. [489]

Sabatino, C., see Cavaccini, G. [326]
Sadewasser, H., see Winter, D. [522]
Sadjadi, F. A. [397]
Saedler, J., see Osten, W. [38, 392]
Sainov, V., see Shulev, A. [773]
—, see Simova, E. [682]
Saito, H., see Nakadate, S. [620]
—, see Yatagai, T. [406]
Sajenko, P. [619]
Salathe, R. P., see Beghuin, D. [125]
Salazar, D., see Tentori, D. [477]
Salazar, R., see Tribillon, G. [662]
Saldner, H. O. [779]
Saldner, H., see Huntley, J. M. [533, 534]
Saleh, B. E. A. [162]
Sandeman, R. J., see Bone, D. J. [461]
Sander III, W. A., see Ramanath, R. [217]
Sandoz, J.-L., see Rastogi, P. [696]
Sandoz, P., see Jacquot, M. [120, 121]
Santoyo, F. Mendoza [777]
—, see Fröning, P. [776]
—, see Lopez, C. Perez [714, 715]
—, see Schedin, S. [306, 719]
—, see Schedin, St. [62]
Sato, T., see Ueda, M. [629]
Scauflaire, V., see Georges, M. [188]
Schaefer, L. F., see Macovski, A. [47]
Schedin, S. [306, 718, 719, 768, 770]
—, see Lopez, C. Perez [714]
—, see Pedrini, G. [222, 278, 279]
—, see Santoyo, F. Mendoza [777]
Schedin, St. [62]
Schemm, J. B. [419]
Schilling, B. W., see Poon, T.-C. [139]
Schmitz, E., see Hüttmann, G. [183]
Schnack, E. [321]

Schnars, U. [56–60]
—, see Kebbel, V. [552]
—, see Kreis, Th. [66]
—, see Pomarico, J. [668]
Schönebeck, G. [494, 495]
Schörner, J. [517]
Schultz, M. E., see Gilbert, J. A. [199]
Schulz, M., see Borner, H. [598]
Schulze, M. A. [750]
Schumann, W. [169, 570, 583]
—, see Cuche, D. [126]
—, see Lutz, B. [569]
Schwider, J. [443]
—, see Kinnstaetter, K. [445]
Schwomma, O. [48]
Sciammarella, C. A. [334, 374, 579, 580, 582, 585, 606, 695, 697, 698, 700]
—, see Chawla, S. K. [701]
Sciammarella, F. M., see Sciammarella, C. A. [582]
Scott, P. D., see Onural, L. [130]
Seebacher, S. [664]
—, see Osten, W. [687]
—, see Wagner, C. [229, 666]
Sekanina, H. [82]
Sepold, G., see Geldmacher, J. [707]
—, see Kreis, Th. [42]
Servaes, D. A., see Reynolds, G. O. [474]
Shang, H. M., see Hung, M. Y. Y. [225, 584]
Sharma, A., see Rastogi, P. [221]
Sharma, D. K., see Ganesan, A. R. [781]
Sharon, B., see Katzir, Y. [401]
Sheffer, D. B., see Hong, H. [748]
—, see Woisetschläger, J [749]
Shegelski, M. R. A. [74]
Shinoda, K., see Poon, T.-C. [139]
Shiotake, N., see Tsuruta, T. [27, 317]
Shough, D. M., see Kwon, O. Y. [432]
Shulev, A. [773]
Sikora, J. P. [704]
—, see Dhir, S. K. [577]
Simova, E. [682]
Simova, E. S. [685]
Singh, H., see Sirkis, J. S. [375]
Singh, K., see Gupta, P. C. [615–617]
Sirkis, J. S. [375]
Skande, P., see Abramson, N. [564]
Skarman, B. [119]
Skotheim, Ø. [232]

Skydan, O. A. [390]
Sladek, J., see Balas, J. [605]
Slaney, M., see Kak, A. C. [785]
Smigielski, P. [568]
—, see Stimpfling, A. [567]
Smith, G. E., see Boyle, W. S. [209]
Smith, R. G., see Schulze, M. A. [750]
Smorenburg, C., see van Wingerden, J. [444]
Snyder, R. [736]
Snyder, W. E., see Ramanath, R. [217]
Solitro, F. [198]
Sollid, J. E. [286, 287, 336]
Solomos, G. P., see Zanetta, P. [335]
Somasundaram, K., see Woisetschläger, J [749]
Sommargren, G. E. [203]
Sona, A., see Fossati Bellani, V. [486]
Soontaranon, S. [249]
Sovic, I., see Demoli, N. [230]
Spagnolo, G. Schirripa [516, 766, 767]
—, see Carelli, P. [651]
—, see Paoletti, D. [467]
—, see Zanetta, P. [459]
Spajer, M [699]
Spaulding, K., see Adams, J. [214]
Spears, K. G., see Abramson, N. [660]
Spik, A. [528]
Spolaczyk, R., see Schwider, J. [443]
Spooren, R. [211]
Springer, G. S., see Farrell, P. V. [422]
Srinivasan, V. [416]
Stadelmeier, A. [236]
Stearns, S. D. [791]
Steckenrider, J. S. [351]
Steffens, H.-D., see Meyer, L. W. [608]
Steinbichler, H. [322]
—, see Borner, H. [598]
Steinlein, P., see Geldmacher, J. [707]
Stephenson, P. [523]
Stetson, K. A. [17, 31, 35, 50, 308, 312–314, 542, 586, 590, 612, 613, 622, 639, 709, 762]
—, see Levitt, J. A. [634]
—, see Molin, N. E. [32–34]
—, see Powell, R. L. [16]
—, see Pryputniewicz, R. J. [574, 588]
—, see Saldner, H. O. [779]
—, see Sollid, J. E. [336]
Stimpfling, A. [567]
Stoev, K. N., see Simova, E. S. [685]
Streibl, N., see Kinnstaetter, K. [445]

Strojnik, M., see Munoz, J. [501, 502]
Stroke, G. W. [179]
Sturm, V., see Pflüger, S. [201]
Su, X.-Y. [439]
Subbaraman, B., see Sciammarella, C. A. [334]
Sun, H. [270]
—, see Cha, S. S. [740]
Sun, J., see Steinbichler, H. [322]
Süsstrunk, S., see Alleysson, D. [216]
Suzuki, K. [285]
Suzuki, Y., see Poon, T.-C. [139]
Swain, R., see Preater, R. [468]
Sweeney, D. W. [30, 728]
Swinson, W. F., see Weathers, J. M. [596]
Swint, J. B., see Sollid, J. E. [286]

Tajahuerce, E. [134, 135, 260]
—, see Frauel, Y. [756]
Takai, N. [99, 282]
Takaki, Y. [100, 101, 271]
Takayama, K., see Watanabe, M. [348]
Takeda, M. [43, 102, 207, 470–472, 508, 532]
Takezaki, J., see Toyooka, S. [473]
Tan, H. [800]
Taniguchi, K., see Takeda, M. [102]
Tatam, R. P. [646]
Taylor, L. H. [587]
Teich, M. C., see Saleh, B. E. A. [162]
Tentori, D. [477]
Thalmann, R. [424, 431]
—, see Dändliker, R. [305, 428, 576, 690]
Thieme, W., see Breuckmann, B. [427]
Thizy, C., see Georges, M. [188]
Thompson, B. J., see Reynolds, G. O. [194]
Tichenor, D. A. [752]
Tiziani, H. J. [156, 190, 710]
—, see Fröning, P. [776]
—, see Pedrini, G. [63–65, 222, 235, 277–279, 498, 535, 640, 641, 644, 720, 769]
—, see Santoyo, F. Mendoza [777]
—, see Schedin, S. [306, 718, 719, 768, 770]
—, see Schedin, St. [62]
Tiziani, H., see Lopez, C. Perez [714]
—, see Zou, Y. [61]
Tonin, R. [625, 626]
Torres, E. [226]
Torzynski, M., see Demoli, N. [745]
Totzauer, W., see Aswendt, P. [772]
—, see Will, P. [611]

Towers, C. E., see Towers, D. P. [435]
Towers, D. P. [435, 524, 525]
Toyooka, S. [302, 473, 571]
Tozer, B. A. [747]
Tribillon, G. [662]
—, see Jacquot, M. [120, 121]
Tribolet, J. M. [506]
Trillo, C. [774]
—, see Doval, A. F. [223]
Trolinger, J. D. [289, 342, 343, 476, 549, 746]
Tsujiuchi, J., see Nakadate, S. [405]
—, see Ohyama, N. [460]
—, see Ru, Q.-S. [454, 455]
—, see Tsuruta, T. [27]
Tsuruta, T. [27, 317, 705]
Tung, Z., see Takeda, M. [472]
Turner, J. L., see Weathers, J. M. [596]
Tyrer, J. R., see Moore, A. J. [609]

Ueda, M. [629]
Underwood, K. L., see Womack, K. H. [417]
Unger, S., see Winter, H. [408]
Unser, M., see Liebling, M. [129, 241–243, 245]
Upatnieks, J., see Leith, E. N. [7–9, 171]
Usry, W., see Schulze, M. A. [750]

van Haeringen, W., see Junginger, H. G. [734]
van Ingen, J. L., see Lanen, T. A. W. M. [341]
van Wingerden, J. [444]
Varade, A., see Fraile, D. [360]
Varner, J. R. [647]
—, see Zelenka, J. S. [28, 648]
Vasarhelyi, G., see Kornis, J. [754]
Verbiest, R., see Boone, P. M. [315]
Verhoeven, D. D. [732, 801]
Verrall, S. C., see Tajahuerce, E. [135]
Vest, C. M. [170, 725, 730, 735]
—, see Birnbaum, G. [323]
—, see Cha, S. S. [722]
—, see Farrell, P. V. [422]
—, see Prikryl, I. [741]
—, see Schemm, J. B. [419]
—, see Sweeney, D. W. [30]
—, see Watt, D. W. [349, 350]
—, see Xu, Y. [112]
Vest, Ch. M., see Lira, I. H. [731]
Vikram, C. S. [628, 635]
Villain, J., see Borner, H. [598]
Vo, P. Del [356]

Voelkl, E., see Schulze, M. A. [750]
Vogel, D. [575]
Vogel, J., see Vogel, D. [575]
Vogt, E. [708]
von Bally, G., see Kasprzak, H. [329]
—, see Lai, S. [269]
—, see Su, X.-Y. [439]
von Kopylow, Chr., see Kolenovic, E. [717]
Voorhis, C. C. Van, see Ladenburg, R. [726]
Vrooman, H. A. [510]
Vucic, Z., see Lovric, D. [456]
Vukicevic, D. [724]
—, see Demoli, N. [745]

Waddell, P. [711]
Wagner, C. [229, 666]
Wagner, J. W., see Ehrlich, M. J. [345]
—, see Steckenrider, J. S. [351]
Walker, J. G., see Green, R. J. [529, 530]
Walles, S. [307]
Walz, Th., see Krupka, R. [775]
Wang, B., see Long, P. [478]
Wang, H.-T., see Guo, C.-S. [80]
Wang, X. [191]
Watanabe, M. [348]
Watanabe, Y., see Takeda, M. [532]
Waters, J. P. [556]
Watson, A. B. [793]
Watson, J., see Sun, H. [270]
Watt, D. W. [349, 350, 361]
Weathers, J. M. [596]
Weber, D. C., see Trolinger, J. D. [549]
Webster, J. M., see Tozer, B. A. [747]
Wegner, H., see Becker, F. [402]
Welford, W. T. [318, 319]
Welling, H., see Jüptner, W. [296]
Wesolowski, P. J., see Woisetschläger, J [749]
Wesolowski, P., see Füzessy, Z. [545]
White, D. A., see Bruning, J. H. [40]
Widjaja, J., see Soontaranon, S. [249]
Wiesenfeldt, M., see Lauterborn, W. [159]
Will, P. [611]
Willemin, J.-F., see Dändliker, R. [305]
Williams, D. C., see Ennos, A. E. [411]
—, see Robinson, D. W. [509]
Winkler, J., see Ladenburg, R. [726]
Winter, C., see Schörner, J. [517]
Winter, D. [522]
Winter, H. [408]

Wise, C. M., see Landry, M. J. [483]
Wizinowich, P. L. [434]
Woisetschläger, J [749]
Woisetschläger, J., see Vukicevic, D. [724]
Womack, K. H. [417, 481, 482]
Wozniak, K., see Skarman, B. [119]
Wåhlin, A., see Mattsson, R. [763]
Wright, D. G., see Tozer, B. A. [747]
Wright, M. A., see Rubayi, N. [354]
Wu, M. H., see Poon, T.-C. [139]
Wuerker, R. F., see Brooks, R. E. [19]
—, see Heflinger, L. O. [20, 24]
Wyant, J. C. [688]
—, see Chang, M. [440]
—, see Cheng, Y. Y. [442]
—, see Lam, P. [691]
Wykes, C., see Jones, R. [310]

Xiao, X. [151]
Xu, L. [113, 114, 258, 276]
Xu, W. [77–79]
—, see Kreuzer, H. J. [76]
Xu, Y. [112]

Yamada, M., see Takai, N. [99]
Yamaguchi, I. [103, 106–110, 665]
—, see Zhang, T. [104, 105]
Yang, L., see Hung, M. Y. Y. [225]
Yang, X., see Lu, B. [352]
Yaroslavskii, L. P. [55]
—, see Kronrod, M. A. [53]

Yaroslavsky, L. [152, 153, 787, 788]
Yaroslavsky, L. P. [786]
Yasuda, N., see Murata, S. [96]
Yatagai, T. [393, 406]
—, see Lai, G. [143, 566]
Yeakle, J., see Rubayi, N. [354]
Yeo, S.-T., see Srinivasan, V. [416]
Yonemura, M. [653, 716]
Young, M. [220]
Yourassowsky, C., see Dubois, F. [69, 70]
Yu, L. [90, 91, 272]
Yu, Q., see Andresen, K. [511]
Yu, Y. H., see Becker, F. [403]

Zanetta, P. [335, 459]
Zarubin, A. M., see Su, X.-Y. [439]
Zelenka, J. S. [28, 648]
—, see Larson, R. W. [284]
Zhang, L., see Guo, C.-S. [80]
Zhang, T. [104, 105]
—, see Yamaguchi, I. [103]
Zhang, X., see Longere, P. [218]
Zhu, J., see Liu, Ch. [81]
Zhu, Y. Y., see Guo, C.-S. [80]
Ziolkowski, E., see Lu, B. [352]
Zirn, R., see Jüptner, W. [610]
Zou, Y. [61]
Zou, Y. L., see Pedrini, G. [63, 64]
Zozulya, A. A., see Owen, R. B. [146–148]
Zürcher, J.-P., see Schumann, W. [583]
Zürn, M., see Zanetta, P. [335]

Subject Index

2π ambiguity 224
1/f noise 72
2+1-technique 246

Abel transform 376
aberrations 47
abnormal dispersion 386
absolute phase 224
absorption coefficient 385
acoustooptic modulator 304
acoustooptical modulator 65, 236
addition theorem 415
adhesive bonding 321
aerodynamics 370
algebraic reconstruction technique 454
algorithm, FFT 234, 256, 435
algorithm, two-dimensional FFT 430
aliasing 424
ambiguity, 2π 224
ambiguity, sign 222, 224, 238, 255, 257, 261
amplitude division 17, 37
amplitude hologram 39, 58
amplitude modulation holography 330
amplitude spectrum 256, 413
amplitude spectrum, Fourier 390
amplitude, complex 421
amplitude, real 10
amplitude, vibration 217
analog-to-digital converter 228
analysis, correlation 419
analysis, particle 160
anamorphism 114
angular frequency 10
angular magnification 46
anti-phase 14
antinodes 324
approach, convolution 115

approximation, Fresnel 24
approximation, paraxial 27
arctan-function 222
arctan-function, principal value of 222
argon-ion laser 57
arrangement, in-line 47, 118
arrangement, off-axis 47
array, Bayer 77
artificial neural network 234, 393
associativity 419
astigmatism 114, 115
asymmetric refractive index field 377
autocorrelation function 420
autocorrelation theorem 420
automata, cellular 292
automatic qualitative evaluation 390
average component 413

B-spline 133
background variations 258
backprojection, filtered 164
backprojection, reconstruction by filtered 452
bacteriorhodopsin 60
band-limited function 424
bandlimit demodulation 294
bandpass filter 256
bandpass filtering 257
bandwidth, effective 381
Bayer array 77
beam 314
beam, cantilever 314
beam, laser 54
beam, nondiffracting 12
beat frequency 15, 235
bending moment 216, 218
bending waves 217
Bessel function 190, 326, 431, 459
Bessel function of the first kind 459

Bessel function of the second kind 459
Bessel wave 12
bilinear interpolation 235, 308, 309
birefringent 218
blooming 69
bonding, adhesive 321
boundary element method 322
boundary of object 231
Bragg cell 65
Bragg reflection 52
BSO 60
butterfly 435

camera, consumer still 139
camera, pyroelectric 60
camera-tubes, TV 61
cantilever beam 314
capacity, well 66
Carre formula 248
carrier frequency 263
cascaded reconstruction 150
cavity 55
CCD-target 61
cell, Bragg 65
cell, Pockels 56
cellular automata 292
central reconstruction formula of digital holography 96
centrifugal force 360
chamber, pressure 218
chamber, vacuum 218
characteristic function 329
charge coupled devices 66
check, consistency 294
chirp function 145, 441
chrominance 79
circular symmetric phase object 376
CMOS 73
coefficient, absorption 385
coefficient, thermal expansion 216
coefficients, stress-optical 373
coherence function, spatial 21
coherence function, spatio-temporal 21
coherence length 18
coherence time 18
coherence, degree of 18
coherence, mutual degree of 21
coherence, self 18
coherence, spatial 19

coherence, temporal 17
coherent superposition 13
color filter detector, lateral 77
color filter detector, vertical 77
color matching function 77
color sensor 76
commutativity 419
comparative holographic interferometry 350
comparative holographic moiré interferometry 352
comparative holography, digital 354
complementary metal oxide semiconductor (CMOS) 73
complex amplitude 421
complex exponential function 414
component, average 413
component, d.c. 413
computer aided tomography 377, 447
computerized tomography 164
concentration, mass 217
concentration, species 386
condition number, Hadamard 310
condition of a system of linear equations 310
conjugated image 41
conservation of energy 431
consistency check 294
constant function 411
constant, Gladstone-Dale 370
consumer still camera 139
continuation 224, 287
contour measurement 306
contouring 339
contouring by refractive index variation 336
contouring, digital holographic 340
contouring, holographic 333
contouring, projected fringe 338
contours 335
contours, three-dimensional 218
contrast 16
contrast, speckle pattern 33
control, feedback 304
control, integration time 72
converging lens 27
converter, analog-to-digital 228
convolution 230, 418, 431
convolution approach 115
convolution kernel 230
convolution theorem 419
convolution, two-dimensional 430

Subject Index

coordinate, spatial 429
coordinates, polar 447
correction, shading 229, 230
correlation analysis 419
cosine transform 439
cosine transform, separable 439
crack 322
crack propagation 217, 323
cross-correlation function 419
cross-reconstructions 201
crystals, photorefractive 60
current, dark 71
cutoff frequency 256, 258
CW laser 304
CW-laser 57

d.c.-term 102
damped vibration 217
dark current 71
data encryption 180
data truncation 425
data, truncated 424
decorrelation, speckle 227
defect detection 321
defect validation 322
deformation 191
deformation gradient matrix 319
deformation, homogeneous 318
deformations 216
degree of coherence 18
degree of coherence, mutual 21
degree of transmission 39
delta, Dirac 411
demodulation 224, 287
demodulation, bandlimit 294
demodulation, path dependent 288
demodulation, path independent 288, 289
demosaicking 79
density, electron 217
density, energy 60
dependent object motions 325
derivative, second 317
derotator, image 362
desensitized holographic interferometer 365
detection, defect 321
detection, spatial synchronous 265
detector, lateral color filter 77
detector, vertical color filter 77
devices, charge coupled 66

diagnostics, flow 217
diagnostics, plasma 217, 374
diagram, holo- 200
dichromated gelatin 58
difference holographic interferometry 350
difference, frequency 235
difference, interference phase 188
differences, finite 317
differentiation 416, 431
differentiation, numerical 317
diffraction efficiency 60
diffraction formula, Fresnel-Kirchhoff 22
diffraction grating 65, 243
diffractive optical element 365
diffuse illumination holographic interferometry 195
digital comparative holography 354
digital holographic contouring 340
digital holographic microscopy 170
digital holography 61, 81
digital holography without wavefront reconstruction 400
digital holography, central reconstruction formula of 96
digital holography, phase-shifting 127
digital image plane holography 400
digital shearography 405
digital speckle pattern interferometry 400
digitization 227
diode laser 57
Dirac delta 411
Dirac delta impulse 48
direct inversion 379
discrete Fourier transform 421
dispersion 217
dispersion, abnormal 386
displacement vector 192
displacement vector field 297
displacement, dynamic 216
displacement, known reference 298
displacement, one-dimensional 216
displacement, rigid body 307
displacement, static 216
displacement, three-dimensional 216
displacement, unknown reference 299
distortion, perspective 298, 308
distortions of holographic interferogram 225
distributed loads 219
distribution, refractive index 194

distributivity 419
diverging lens 27
division, amplitude 17, 37
division, wavefront 19, 37
domain, spatial frequency 257
Doppler shift 15
double aperture interferometer, Young's 399
double exposure hologram 266
double exposure method 186, 235, 242
doubly refractive prism 92
DSPI 400
dye laser 57
dynamic displacement 216
dynamic evaluation 301
dynamic evaluation method 266
dynamic load 219
dynamic range 72
dynamics, fluid 217

effect, stress-optical 217, 373
effective bandwidth 381
effective refractive index 374
efficiency, diffraction 60
efficiency, quantum 71
elasticity, modulus of 314, 373
elasticity, shear modulus of 315
electric field strength 9
electro-optic holography 400, 404
electron density 217
electron gas 217
electronic holography 400, 404
electronic noise 226
electronic speckle pattern interferometry 400
emission, spontaneous 53
emission, stimulated 53
emulsions, photographic 58
encryption, data 180
endoscopic holographic interferometry 364
energy 416
energy density 60
energy, conservation of 431
enhancement, image 258
equality, Parseval 443
equality, Plancherel 443
equation, Gladstone-Dale 217, 370
equation, ideal gas 217, 371
equation, Lorentz-Lorenz 217, 372
equation, phase sampling 242
equation, ray 367

equations, holographic imaging 45
equations, system of 298
equivalent wavelength 358
ESPI 400
ESPI methods 61
etalon 55
Euler formula 11, 412
evaluation method, dynamic 266
evaluation method, static 266
evaluation, automatic qualitative 390
evaluation, dynamic 301
evaluation, Fourier transform 256
evaluation, phase sampling 242
evaluation, qualitative 215
evanescent wave 29
event, transient 242
events, transient 217
exposure, noise equivalent 72
exposure, saturation equivalent 72
exposure, time of 38

Fabry-Perot interferometer 55
factor, stress intensity 322
far-field region 50
fast Fourier transform 434
feature map, self organizing 234
feedback control 304
FFT 434
FFT algorithm 234, 256, 435
FFT algorithm, two-dimensional 430
FFT-subroutine 436
fibers, monomode 62
fibers, multimode 62
fibers, optical 61
field strength, electric 9
field, speckle 30
field, strongly refracting refractive index 382
fill-factor 74
filter, bandpass 256
filter, finite impulse response 230
filter, linear low-pass 230
filter, median 230, 274
filter, nonlinear 230
filter, nonrecursive 230
filter, recursive 230
filter, spatial 61
filtered backprojection 164
filtered backprojection, reconstruction by 452
filtered projection 452

Subject Index

filtering, bandpass 257
finite differences 317
finite discrete Fourier transform 258
finite element method 321
finite Fourier transform 422
finite fringe interferogram 224
finite fringes 224
finite impulse response filter 230
fixed pattern noise 72
flexural rigidity 317
flow diagnostics 217
flow visualization 217, 370
fluid dynamics 217
focal length 27
force 312
force, centrifugal 360
formula of Euler 412
formula, Carre 248
formula, of Euler 11
Fourier amplitude spectrum 390
Fourier integral theorem 410
Fourier series 421
Fourier slice theorem 379, 450
Fourier spectrum 234
Fourier synthesis 379
Fourier transform 410
Fourier transform evaluation 256
Fourier transform hologram 48
Fourier transform holography 47
Fourier transform holography, lensless 49, 91, 100
Fourier transform pair 410
Fourier transform, discrete 421
Fourier transform, fast 434
Fourier transform, finite 422
Fourier transform, finite discrete 258
Fourier transform, inverse 257, 410
Fourier transform, inverse finite 422
Fourier transform, short time 132
fracture mechanics 322
fracture toughness 322
frame transfer 68
Fraunhofer diffraction region 50
Fraunhofer hologram 50
free space propagation, transfer function of 29
free-space propagation, impulse response of 29
frequency 10
frequency difference 235

frequency domain, spatial 257
frequency modulation 304
frequency plane restoration 380
frequency shift 62, 65
frequency translated holography 329
frequency, angular 10
frequency, beat 15, 235
frequency, carrier 263
frequency, cutoff 256, 258
frequency, fundamental 420
frequency, local 443
frequency, Nyquist 423
frequency, one-dimensional spatial 410
frequency, spatial 234, 260
frequency, temporal 410
Fresnel approximation 24
Fresnel diffraction region 50
Fresnel hologram 50
Fresnel-Kirchhoff diffraction formula 22
Fresnelets 133
fringe locus function 222, 318
fringe numbering 234
fringe order 234
fringe tensor 319
fringe tracking 229
fringe-vector 307, 318
fringes, finite 224
fringes, infinite 224
fringes, interference 14
function, arctan- 222
function, autocorrelation 420
function, band-limited 424
function, Bessel 190, 326, 431, 459
function, Bessel of the first kind 459
function, Bessel of the second kind 459
function, characteristic 329
function, chirp 145, 441
function, color matching 77
function, complex exponential 414
function, constant 411
function, cross-correlation 419
function, fringe locus 222, 318
function, Gaussian 414
function, harmonic spatial 256
function, modulation transfer 74
function, periodic 411
function, point spread 141, 227, 417, 430
function, rectangular 411
function, rectangular pulse 414

function, rectangular window 426
function, sensitivity 200
function, sigmoid 393
function, spatial coherence 21
function, spatio-temporal coherence 21
function, transfer 29, 419
function, transient 411
function, triangular pulse 414
function, two-dimensional rectangular 429
function, unit-step 414
fundamental frequency 420

Gabor transform 445
gas laser 57
gas, density of 217
gas, electron 217
Gaussian function 414
Gaussian least squares 301
gelatin, dichromated 58
generalized phase shifting interferometry 263
generalized projection theorem 449
Gladstone-Dale constant 370
Gladstone-Dale equation 217, 370
global iteration 293
gradient matrix, deformation 319
gradient operator 367
grating, diffraction 65, 243
grating, radial 65
gravity 218
gray-values 227, 429
grid method 380
grog method 337
ground state 54

Hadamard condition number 310
Hankel transform 431
Hanning window 426
harmonic spatial function 256
harmonic vibration 190, 217, 323
harmonic wave 10
Hartley transform 439
Hartley transform, inseparable 440
Hartley transform, separable 440
heat transfer 217
helium-cadmium laser 57
helium-neon laser 57
Hermitean 256
heterodyne holographic interferometry 235, 326

heterodyne holography, numerical 127
heterodyne method 62, 201
heterodyne method, quasi 242
heterodyning, spatial 263
heterodyning, temporal 235
HNDT 215
holo-diagram 200
hologram 38, 40
hologram interferometry, sandwich 305, 338
hologram, amplitude 39, 58
hologram, double exposure 266
hologram, Fourier transform 48
hologram, Fraunhofer 50
hologram, Fresnel 50
hologram, image 50
hologram, in-line- 42
hologram, multiplexed 280
hologram, off-axis 42
hologram, phase 39, 58
hologram, rainbow 53
hologram, three-dimensional 52
hologram, volume 52
hologram, white light 51
holographic contouring 333
holographic contouring, digital 340
holographic imaging equations 45
holographic interference pattern 185
holographic interferogram 185
holographic interferogram, distortions of 225
holographic interferometer, desensitized 365
holographic interferometric metrology 303
holographic interferometry 185
holographic interferometry, comparative 350
holographic interferometry, difference 350
holographic interferometry, diffuse illumination 195
holographic interferometry, endoscopic 364
holographic interferometry, heterodyne 235, 326
holographic interferometry, real-time 188, 236, 260, 326
holographic interferometry, stroboscopic 325
holographic interferometry, time average 326
holographic interferometry, two-wavelength 357
holographic microscope, optoelectronic 405
holographic moiré 359
holographic moiré interferometry,comparative 352

holographic nondestructive testing 215, 303, 387
holographic vibration analysis 190
holography, amplitude modulation 330
holography, digital 61, 81
holography, digital comparative 354
holography, digital image plane 400
holography, digital without wavefront reconstruction 400
holography, electro-optic 400, 404
holography, electronic 400, 404
holography, Fourier transform 47
holography, frequency translated 329
holography, lensless Fourier transform 49, 91, 100
holography, numerical heterodyne 127
holography, optoelectronic 405
holography, phase modulation 330
holography, phase-shifting digital 127
holography, rainbow 53
holography, sandwich 276
holography, single beam 47
holography, spectroscopic 331
holography, split beam 47
holography, stroboscopic 304
holography, TV 400, 404
holography, two beam 47
holography, two reference beam 201, 235, 242, 255
homogeneous deformation 318
homogeneous medium 367
homologous points 204
homologous rays 204
Hopfield network 294
Huygens principle 12
hybrid method 321

ideal gas equation 217, 371
ill-conditioned matrix 310
illumination point 192, 306
image 429
image derotator 362
image enhancement 258
image hologram 50
image plane holography, digital 400
image, conjugated 41
image, orthoscopic 42
image, pseudoscopic 42
image, real 42, 97

image, virtual 42, 98
images, twin 98
imaging equations, holographic 45
imaging, polarization 314
immersion method 336
impulse 411
impulse load 219
impulse response 141, 230, 417
impulse response filter, finite 230
impulse response of free-space propagation 29
impulse, Dirac delta 48
in-line arrangement 47, 118
in-line-hologram 42
in-phase 14
in-plane strains 216
incoherent part 16
incoherent superposition 16
independent object motions 325
index, refractive 366, 385, 447
infinite fringe interferogram 224
infinite fringes 224
inseparable Hartley transform 440
integral theorem, Fourier 410
integral, superposition 417
integration time 67
integration time control 72
intensity 12
intensity modulation 251
intensity, short time 13
intensity, speckle pattern 33
interference 13
interference fringes 14
interference pattern 14
interference pattern, holographic 185
interference pattern, macroscopic 185
interference pattern, microscopic 185
interference phase 188
interference phase difference 188
interferogram, distortions of holographic 225
interferogram, finite fringe 224
interferogram, holographic 185
interferogram, infinite fringe 224
interferogram, space-time 265
interferometer, desensitized holographic 365
interferometer, Fabry-Perot 55
interferometer, Michelson 17
interferometer, reflective grating 114
interferometer, rotating 361

interferometer, Twyman-Green 17
interferometer, Young's double aperture 19, 399
interferometric metrology, holographic 303
interferometry digital speckle pattern 400
interferometry, comparative holographic 350
interferometry, comparative holographic moiré 352
interferometry, difference holographic 350
interferometry, diffuse illumination holographic 195
interferometry, electronic speckle pattern 400
interferometry, endoscopic holographic 364
interferometry, generalized phase shifting 263
interferometry, heterodyne holographic 235, 326
interferometry, holographic 185
interferometry, Mach-Zehnder 217
interferometry, real-time holographic 188, 236, 260, 326
interferometry, sandwich hologram 305, 338
interferometry, stroboscopic holographic 325
interferometry, time average holographic 326
interferometry, two-wavelength holographic 357
interline transfer 68
interpolation 235
interpolation by triangulation 235
interpolation, bilinear 235, 308, 309
interpolation, one-dimensional spline 235
invariant, shift 429
inverse finite Fourier transform 422
inverse Fourier transform 257, 410
inversion, direct 379
inversion, population 54
isoplanatic system 417
isotropic material, optically 218
iteration, global 293
iteration, local 292

J-integral 323
jitter 73
Jones matrices 62

kernel, convolution 230
kernel, Ramachandran-Lakshminarayanan 454
kernel, Shepp-Logan 454
krypton laser 57

Laplace operator 9
Laplacian 431
laser 53
laser beam 54
laser, argon-ion 57
laser, CW 57, 304
laser, diode 57
laser, dye 57
laser, gas 57
laser, helium-cadmium 57
laser, helium-neon 57
laser, krypton 57
laser, neodymium:YAG 57
laser, pulsed 56, 304
laser, ruby 56
lateral color filter detector 77
lateral magnification 46
law, Maxwell-Neumann stress-optical 218, 373
law, Snell's 367
leakage 425
least squares method 299
least squares phase unwrapping 295
least squares, Gaussian 301
length, coherence 18
length, focal 27
lens formula 86
lens, converging 27
lens, diverging 27
lens, negative 27
lens, positive 27
lens, thin 26
lensless Fourier transform holography 49, 91, 100
level, trigger 237
light, speed of 9, 366
light-in-flight recording 338, 377
limit, refractionless 377
line address transfer 67
line integral transform 370
linear 429
linear low-pass filter 230
linear system 416
linearity 431
linearity theorem 415
liquids 372
load, dynamic 219
load, impulse 219

load, periodic 219
load, point 218, 219
load, static 219
load, tensile 323
load, thermal 219, 321
load, transient 219
load, vibrational 219
loading, pressure 218
loads, distributed 219
local frequency 443
local iteration 292
localization 205
longitudinal magnification 46
longitudinal modes 55
Lorentz-Lorenz equation 217, 372
low-pass filter, linear 230
luminance 79

Mach-Zehnder interferometry 217
macroscopic interference pattern 185
magnification formula 86
magnification, angular 46
magnification, lateral 46
magnification, longitudinal 46
mass concentration 217
mass transfer 217
material, optically isotropic 218
materials, recording 58
matrix, deformation gradient 319
matrix, ill-conditioned 310
matrix, sensitivity 299, 306
matrix, singular 310
Maxwell-Neumann stress-optical law 218, 373
measurement, contour 306
measurement, temperature 371
measuring time 13
mechanical stressing 218
mechanics, fracture 322
media, transparent 194
median filter 230, 274
medium, homogeneous 367
medium, nonhomogeneous 367
MEMS 405
method, boundary element 322
method, double exposure 235, 242
method, dynamic evaluation 266
method, finite element 321
method, grid 380

method, grog 337
method, heterodyne 201
method, hybrid 321
method, immersion 336
method, least squares 299
method, phase lock 233
method, phase shift 201, 242
method, phase step 201, 242
method, quasi heterodyne 242
method, real-time 255
method, static evaluation 266
method, time average 191, 333, 460
method, two-wavelength 333
metrology, holographic interferometric 303
Michelson interferometer 17
microelectromechanical systems 311, 405
microscope, optoelectronic holographic 405
microscopic interference pattern 185
microscopy, digital holographic 170
mode, vibration 217, 360
mode-shape 324
mode-shapes 324
modes, longitudinal 55
modes, transverse 55
modulation transfer function 74
modulation, frequency 304
modulation, intensity 251
modulator, acoustooptic 304
modulator, acoustooptical 65, 236
modulo 2π 224
modulus of elasticity 314, 373
moiré pattern 424
moiré, holographic 359
moment, bending 216, 218
monomode fibers 62
mother wavelet 133
multimode fibers 62
multiplexed hologram 280
mutual degree of coherence 21

NDT 215
near-field region 50
negative lens 27
neodymium:YAG laser 57
network, artificial neural 234, 393
network, Hopfield 294
network, neural 294
neural network 294
neural network, artificial 234, 393

neuron 234, 393
nodes 324
noise equivalent exposure 72
noise, 1/f 72
noise, electronic 226
noise, fixed pattern 72
noise, quantization 72
noise, reset 72
noise, shot 72
noise, speckle 258
noise, white 72
non-vibration isolated object 304
nondestructive testing 215
nondestructive testing, holographic 215, 303, 387
nondiffracting beam 12
nonhomogeneous medium 367
nonlinear filter 230
nonlinear vibration 217
nonrecursive filter 230
nonuniformity, photoresponse 72
normal strain 311
normal stress 312
number, Hadamard condition 310
numbering, fringe 234
numerical differentiation 317
numerical heterodyne holography 127
numerical reconstruction 93
Nyquist frequency 423
Nyquist rate 423

object motions, dependent 325
object motions, independent 325
object motions, separable 324
object related triggering 304, 361
object wave 37
object, boundary of 231
object, circular symmetric phase 376
object, non-vibration isolated 304
object, phase 194
object, rotating 360
object, transparent 194
objective speckles 35
objects, transparent 217
observation point 192, 306
observer projection theorem 214, 268
off-axis arrangement 47
off-axis-hologram 42
one-dimensional displacement 216

one-dimensional spatial frequency 410
one-dimensional spline interpolation 235
operator, gradient 367
operator, Laplace 9
optical element, diffractive 365
optical fibers 61
optical path difference 191
optically isotropic material 218
optoelectronic holographic microscope 405
optoelectronic holography 405
order, fringe 234
orthogonally polarized wave 16
orthoscopic image 42

pair, Fourier transform 410
parallelly polarized wave 16
parameter, scale 132
parameter, translation 132
paraxial approximation 27
Parseval equality 443
Parseval's theorem 416, 431
particle analysis 118, 160
path dependent demodulation 288
path difference, optical 191
path independent demodulation 288
path length transform 370
pattern, holographic interference 185
pattern, interference 14
pattern, moiré 424
pattern, speckle 30
period 10
periodic function 411
periodic load 219
periodic replications 421
perspective distortion 298, 308
phase 10, 413
phase conjugated wave 42
phase hologram 39, 58
phase lock method 233
phase modulation holography 330
phase object 194, 217
phase object, circular symmetric 376
phase relations 217
phase sampling 62
phase sampling equation 242
phase sampling evaluation 242
phase shift 62
phase shift method 201, 242
phase shifting 62, 244

Subject Index 537

phase shifting interferometry, generalized 263
phase spectrum 413
phase step method 201, 242
phase stepping 244, 404
phase unwrapping 224, 287
phase unwrapping, least squares 295
phase unwrapping, temporal 295
phase velocity 11
phase, absolute 224
phase, relative 10
phase, speckle pattern 33
phase-shifting digital holography 127
photochromics 59
photographic emulsions 58
photography, speckle 399
photopolymers 58
photorefractive crystals 60
photoresists 58
photoresponse nonuniformity 72
photothermoplastics 59
picture 429
piezoelectric transducer 65
pinhole 61
pixel number 74
pixel size 74
pixels 227
Plancherel equality 443
plane polarized wave 10
plane stress 216, 314
plane wave 11
plasma diagnostics 217, 374
plastic zone 323
plate 316
Pockels cell 56
point load 218, 219
point source 429
point spread function 141, 227, 417, 430
point, illumination 306
point, observation 306
points, homologous 204
Poisson ratio 216, 315, 373
polar coordinates 447
polarization 9, 61
polarization imaging 314
pollution, wrap-around 258
polynomial, Zernike 303, 460
population inversion 54
positive lens 27
power spectrum 420

pressure 217
pressure chamber 218
pressure loading 218
pressure vessel 305
principal value of arctan-function 222
principle of superposition 417
principle, Huygens 12
prism, doubly refractive 92
prism, wavelength selector 55
prism, Wollaston 92
problem, random walk 30
processing, regional 292
processing, tile 292
product, space-bandwidth 73, 427
projected fringe contouring 338
projection theorem for Fourier transforms 450
projection theorem, generalized 449
projection, filtered 452
propagation, crack 217, 323
propagation, impulse response of free-space 29
propagation, transfer function of free space 29
property, sampling 411
property, symmetry 415
pseudoscopic image 42
pulsed laser 56, 304
pumping 54
pyroelectric camera 60

Q-switch 56
qualitative evaluation 215
qualitative evaluation, automatic 390
quantization 227, 258
quantization noise 72
quantum efficiency 71
quasi heterodyne method 242

radial grating 65
Radon transform 370, 447
rainbow hologram 53
rainbow holography 53
Ramachandran-Lakshminarayanan kernel 454
random walk problem 30
range, dynamic 72
rate, Nyquist 423
ratio, Poisson 216, 315, 373
ratio, signal-to-noise 226
ray equation 367
Rayleigh's theorem 416

rays, homologous 204
real amplitude 10
real image 42, 97
real-time holographic interferometry 188, 236, 260, 326
real-time method 255
reconstruction 40
reconstruction by filtered backprojection 452
reconstruction technique, algebraic 454
reconstruction, cascaded 150
reconstruction, numerical 93
reconstruction, series expansion 454
recording materials 58
recording, light-in-flight 338, 377
rectangular function 411
rectangular function, two-dimensional 429
rectangular pulse function 414
rectangular window function 426
recursive filter 230
reference displacement, known 298
reference displacement, unknown 299
reference wave 37, 89
reflection, Bragg 52
reflective grating interferometer 114
refractionless limit 377
refractive index 366, 385, 447
refractive index distribution 194
refractive index field, asymmetric 377
refractive index field, strongly refracting 382
refractive index variation 217
refractive index variation, contouring 336
refractive index, effective 374
refractivity, specific 372
region, far-field 50
region, Fraunhofer diffraction 50
region, Fresnel diffraction 50
region, near-field 50
regional processing 292
relation, uncertainty 444
relative phase 10
replications, periodic 421
reset noise 72
resolution, spatial 40, 74, 255
resonance 385
response, impulse 141, 230, 417
responsivity 72
responsivity, spectral 71
restoration, frequency plane 380
rigid body displacement 307

rigid body rotation 208
rigid body rotations 216
rigid body translation 208, 300
rigid body translations 216
rigidity, flexural 317
rotating interferometer 361
rotating object 360
rotation 312, 431
rotation, rigid body 208
rotational symmetry 431
rotations, rigid body 216
ruby laser 56

sampling property 411
sampling theorem 82, 228, 379, 423
sampling theorem, two-dimensional 431
sampling theorem, Whittaker-Shannon 427
sampling, phase 62
sandwich hologram interferometry 305, 338
sandwich holography 276
saturation 258
saturation equivalent exposure 72
scalar wave equation 10
scale parameter 132
scaling 431
scaling theorem 415
second derivative 317
segmentation 229, 231
self coherence 18
self organizing feature map 234
self-reference 405
sensitivity function 200
sensitivity matrix 299, 306
sensitivity vector 193, 299, 306
sensitivity vector, varying 300
sensor, color 76
separability 430, 431
separable cosine transform 439
separable Hartley transform 440
separable object motions 324
series expansion reconstruction 454
series, Fourier 421
shading correction 229, 230
shear modulus of elasticity 315
shear strain 311
shear stress 312
shearography, digital 405
Shepp-Logan kernel 454
shift invariant 429

Subject Index

shift theorem 415
shift, Doppler 15
shift, frequency 62, 65
shift, phase 62
shift-invariant system 417
shifting 431
shifting, phase 62, 244
shock waves 217
short time Fourier transform 132
short time intensity 13
shot noise 72
sigmoid function 393
sign ambiguity 222, 224, 238, 255, 257, 261
signal-to-noise ratio 226
silver halide 58
similarity theorem 415
single beam holography 47
singular matrix 310
sinusoidal vibration 323
sinusoids 420
size, pixel 74
size, speckle 34
skeletonizing 229
smoothing 229
Snell's law 367
solids, transparent 373
source, point 429
space-bandwidth product 73, 427
space-invariant system 417
space-time interferogram 265
spatial coherence 19
spatial coherence function 21
spatial coordinate 429
spatial filter 61
spatial frequency 234, 260
spatial frequency domain 257
spatial frequency, one-dimensional 410
spatial heterodyning 263
spatial resolution 40, 74, 255
spatial synchronous detection 265
spatial transform 308
spatio-temporal coherence function 21
species concentration 386
specific refractivity 372
speckle decorrelation 227
speckle field 30
speckle noise 258
speckle pattern 30
speckle pattern contrast 33

speckle pattern intensity 33
speckle pattern interferometry, digital 400
speckle pattern interferometry, electronic 400
speckle photography 399
speckle size 34
speckle-shearing methods 405
specklegram 399
speckles 30
speckles, objective 35
speckles, subjective 35
spectral responsivity 71
spectroscopic holography 331
spectrum manipulating method 183
spectrum, amplitude 256, 413
spectrum, Fourier 234
spectrum, Fourier amplitude 390
spectrum, phase 413
spectrum, power 420
speed of light 9, 366
spherical wave 12
spline interpolation, one-dimensional 235
split beam holography 47
spontaneous emission 53
state, ground 54
static displacement 216
static evaluation method 266
static load 219
stepping, phase 244
STFT 132
still camera, consumer 139
stimulated emission 53
strain 216, 311
strain, in-plane 216
strain, normal 311
strain, shear 311
streamlines 217
stress 217, 312
stress intensity factor 322
stress tensor 312
stress, normal 312
stress, plane 216, 314
stress, shear 312
stress, tangential 312
stress, tensile 218
stress, torsional 218
stress-optical coefficients 373
stress-optical effect 217, 373
stress-optical law, Maxwell-Neumann 218, 373

stresses 216
stressing, mechanical 218
stroboscopic holographic interferometry 325
stroboscopic holography 304
strongly refracting refractive index field 382
subjective speckles 35
subroutine, FFT 436
superposition integral 417
superposition principle 10
superposition, coherent 13
superposition, incoherent 16
superposition, principle of 417
surface tension 216
symmetry property 415
symmetry, rotational 431
synaptic weight 393
synthesis, Fourier 379
synthetic wavelength 358
system 416
system of equations 298
system of linear equations, condition of 310
system, isoplanatic 417
system, linear 416
system, shift-invariant 417
system, space-invariant 417
systems, microelectromechanical 311, 405

tangential stress 312
target, CCD- 61
technique, 2+1- 246
temperature 217
temperature measurement 371
temporal coherence 17
temporal frequency 410
temporal heterodyning 235
temporal phase unwrapping 295
tensile load 323
tensile stress 218
tension, surface 216
tensor, fringe 319
tensor, stress 312
testing, holographic nondestructive 303, 387
testing, nondestructive 215
theorem, addition 415
theorem, autocorrelation 420
theorem, convolution 419
theorem, Fourier integral 410
theorem, Fourier slice 379, 450
theorem, generalized projection 449

theorem, linearity 415
theorem, observer projection 214, 268
theorem, Parseval's 416, 431
theorem, projection, for Fourier transforms 450
theorem, Rayleigh's 416
theorem, sampling 228, 379, 423
theorem, scaling 415
theorem, shift 415
theorem, similarity 415
theorem, two-dimensional sampling 431
theorem, Whittaker-Shannon sampling 427
theorem, Wiener-Khinchine 420
thermal expansion coefficient 216
thermal load 219, 321
thin lens 26
three-dimensional contours 218
three-dimensional displacement 216
three-dimensional hologram 52
tile processing 292
time average holographic interferometry 326
time average method 191, 333, 460
time of exposure 38
tomography, computer aided 377, 447
tomography, computerized 164
torsional stress 218
toughness, fracture 322
tracking, fringe 229
transducer, piezoelectric 65
transfer function 29, 419
transfer function of free space propagation 29
transfer, frame 68
transfer, heat 217
transfer, interline 68
transfer, line address 67
transfer, mass 217
transform pair, Fourier 410
transform, Abel 376
transform, cosine 439
transform, discrete Fourier 421
transform, fast Fourier 434
transform, finite discrete Fourier 258
transform, finite Fourier 422
transform, Fourier 410
transform, Gabor 445
transform, Hankel 431
transform, Hartley 439
transform, inseparable Hartley 440
transform, inverse finite Fourier 422

Subject Index 541

transform, inverse Fourier 257, 410
transform, line integral 370
transform, path length 370
transform, Radon 370, 447
transform, separable cosine 439
transform, separable Hartley 440
transform, spatial 308
transform, wavelet 132
transient event 242
transient events 217
transient function 411
transient load 219
translation parameter 132
translation, rigid body 208, 216, 300
transmission, degree of 39
transparent media 194
transparent object 194, 217
transparent solids 373
transverse modes 55
transverse wave 9
triangular pulse function 414
triangulation, interpolation by 235
trigger level 237
triggering, object related 304, 361
tristimulus value 77
truncated data 424
truncation, data 425
TV camera-tubes 61
TV holography 400, 404
twin images 98
two beam holography 47
two reference beam holography 201, 235, 242, 255
two-dimensional convolution 430
two-dimensional FFT algorithm 430
two-dimensional rectangular function 429
two-dimensional sampling theorem 431
two-wavelength holographic interferometry 357
two-wavelength method 333
Twyman-Green interferometer 17

uncertainty relation 444
unit-step function 414
unwrapping, least squares phase 295
unwrapping, phase 224, 287
unwrapping, temporal phase 295

vacuum chamber 218

validation, defect 322
value, tristimulus 77
variation, refractive index 217
variations, background 258
vector field, displacement 297
vector, displacement 192
vector, sensitivity 193, 299, 306
vector, varying sensitivity 300
velocity, phase 11
vertical color filter detector 77
vessel, pressure 305
vibration amplitude 217
vibration analysis, holographic 190
vibration mode 217, 360
vibration, damped 217
vibration, harmonic 190, 217, 323
vibration, nonlinear 217
vibration, sinusoidal 323
vibrational load 219
vibrations 217
virtual image 42, 98
visibility 16
visualization, flow 217, 370
volume hologram 52

watermarking 183
wave equation 9
wave equation, scalar 10
wave number 10
wave vector 11
wave, Bessel 12
wave, evanescent 29
wave, harmonic 10
wave, object 37
wave, orthogonally polarized 16
wave, parallelly polarized 16
wave, phase conjugated 42
wave, plane 11
wave, plane polarized 10
wave, reference 37, 89
wave, spherical 12
wave, transverse 9
wavefront 11
wavefront division 19, 37
wavefront reconstruction, digital holography without 400
wavelength 10
wavelength selector prism 55
wavelength, equivalent 358

wavelength, synthetic 358
wavelet transform 132
waves, bending 217
waves, shock 217
weight, synaptic 393
well capacity 66
white light hologram 51
white noise 72
Whittaker-Shannon sampling theorem 427
Wiener-Khinchine theorem 420

window function, rectangular 426
window, Hanning 426
Wollaston prism 92
wrap-around pollution 258

Young's double aperture interferometer 19, 399

Zernike polynomial 303, 460
zone, plastic 323

New Textbook on Optics:
From Classical Electrodynamics to Laser Cooling

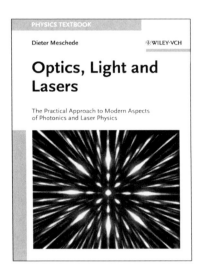

DIETER MESCHEDE, University of Bonn, Germany
Optics, Light and Lasers
The Practical Approach to Modern Aspects of Photonics and Laser Physics

2003. IX, 410 pages, 281 figures, 18 tables. Softcover.
ISBN 3-527-40364-7

This textbook attempts to link the central topics of optics that were established 200 years ago to the most recent research topics such as laser cooling or holography. From the concepts of classical optics, the author summarises the properties of modern laser sources in detail. Several examples from the scope of current research are provided to emphasize the relevance of optics in current developments within science and technology. In scientific education, this textbook may serve as a reference for the foundations of modern optics: classical optics, laser physics, laser spectroscopy, nonlinear optics as well as applied optics may profit. The text has been written for newcomers to the topic and benefits from the author's ability to explain difficult sequences and effects in a straightforward and readily comprehensible way.

Wiley-VCH
P.O. Box 10 11 61 • D-69451 Weinheim, Germany
Fax: +49 (0)6201 606 184
e-mail: service@wiley-vch.de • www.wiley-vch.de

Integrated Optics and Modern Microoptics

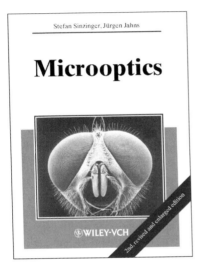

STEFAN SINZINGER, Technical University of Ilmenau, Germany, and JÜRGEN JAHNS, University of Hagen, Germany
Microoptics
2nd revised and enlarged edition

2003. XIV, 433 pages, 209 figures, 12 tables. Hardcover.
ISBN 3-527-40355-8

The updated second edition of this modern text and reference book by Stefan Sinzinger and Jürgen Jahns expertly and comprehensively presents the basics and applications in microoptics. The authors have taken into consideration and incorporated the most important developments in past years as well as the continuously improving manufacturing technology for microoptical components and the rapid progress being made especially in the field of materials research. An additional chapter covers the characterization of microoptical components, in particular lenses and lens arrays, while new sections include photonic crystals and materials for microoptics. A must-have for physicists and electrical engineers, from advanced students right up to designers working in the field.

Wiley-VCH
P.O. Box 10 11 61 • D-69451 Weinheim, Germany
Fax: +49 (0)6201 606 184
e-mail: service@wiley-vch.de • www.wiley-vch.de

www.optics-encyclopedia.com

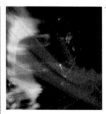

The Optics Encyclopedia
Basic Foundations and Practical Applications

Edited by Thomas G. Brown, Univ. of Rochester, USA; Katherine Creath, Creath Optineering Services, USA; Herwig Kogelnik, Lucent Technologies, USA; Michael E. Kriss, Sharp Laboratories of America, Inc., USA; Joanna Schmit, Veeco Instruments, Inc., USA; Marvin J. Weber, LBNL, USA

3527-40320-5   2004   3530 pp with 1798 figs, 39 in color   Hbk
€ 1099.- / £ 590.- / US$ 985.-

THE Reference Work in Optics!

5 Volume Set

With 94 expert articles in 5 volumes, this is both a comprehensive review as well as an introduction to the entire field. The contributions range from classical optics right up to the latest applications, including:

- IT and telecommunications
- Optical sensing and metrology
- Material processing
- Biomedicine
- Optical components and systems
- Laser design and technology.

The international editor team paid great attention to ensuring fast access to the information, and each carefully reviewed article features:

- an abstract
- a detailed table of contents
- continuous cross-referencing
- references to the most relevant publications in the field, and
- suggestions for further reading, both introductory as well as highly specialized.

In addition, a comprehensive index provides easy access to the enormous number of key words beyond the 94 headlines.

The result is a rapid reference for skilled professionals on all topics of modern photonics, while newcomers from physics and engineering will appreciate the readily comprehensible style and structure.

Read more in a free sample chapter at: www.pro-physik.de

Highlights

Laser Cooling and Trapping of Neutral Atoms, Harold Metcalf, Stony Brook, State Univ. NY, USA, and Peter van der Straten, Debye Inst., Utrecht Univ., The Netherlands

Electrodynamics, J. David Jackson, Univ. of California at Berkeley, California, USA

Optical Design, Pantazis Mouroulis, California Inst. of Technology, Pasadena, USA

Geometric Optics, Roland Shack, Univ. of Arizona, Tucson, USA

Photography, Digital, Michael E. Kriss, Sharp Lab. of America, Inc., Camas, WA, USA

Solid State Lasers, Tso Yee Fan, Lincoln Lab., Massachusetts Inst. of Technology, Lexington, USA

Optical Metrology, Peter de Groot, Zygo Corporation, Connecticut, USA

Optoelectronics, Safa O. Kasap, Univ. of Saskatchewan, Saskatoon, CAN

Physiological Optics, Martin Jüttner, Neuroscience Research Inst., Aston Univ., Birmingham, UK

X-ray Optics, Alan Michette, Dep. of Physics, King's College London, UK

John Wiley & Sons, Ltd. • Customer Services Department • 1 Oldlands Way • Bognor Regis • West Sussex • PO22 9SA England
Tel.: +44 (0) 1243-843-294
Fax: +44 (0) 1243-843-296
www.wileyeurope.com

Wiley-VCH • Customer Service Department
P.O. Box 101161 • D-69451 Weinheim,
Germany • Tel.: +49 (0) 6201 606-400
Fax: +49 (0) 6201 606-184
e-Mail: service@wiley-vch.de
www.wiley-vch.de

WILEY   **WILEY-VCH**